Design in Canada Since 1945

Fifty Years from Teakettles to Task Chairs

Rachel Gotlieb and Cora Golden

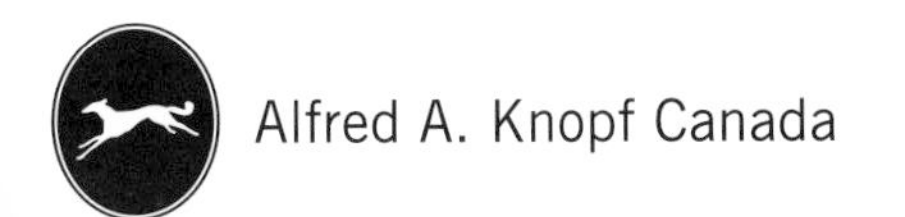

Alfred A. Knopf Canada

PUBLISHED BY ALFRED A. KNOPF CANADA

National Library of Canada Cataloguing in Publication Data

Gotlieb, Rachel
Design in Canada since 1945:
fifty years from teakettles to task chairs

ISBN 0-676-97138-5

1. Design—Canada—History—20th century.
2. Design, Industrial—Canada—History—20th century.
I. Golden, Cora. II. Title.

NK1413.A1G67 2001   745.4'4971'09045   C2001-930570-2

First Edition

National Library of Canada Cataloguing in Publication Data

Gotlieb, Rachel
Design in Canada since 1945 :
fifty years from teakettles to task chairs

ISBN 0-676-97452-X

1. Design, Canada.
2. Design, Industrial—Canada.
I. Golden, Cora. II. Title.

NK1413.A1G67 2002   745.4'4971   C2001-930735-7

Grateful acknowledgment is made to the following sponsors for their generous financial support: The McLean Foundation, Michael Stewart, The Canada Council for the Arts, The Macdonald Stewart Foundation, The Peter Munk Charitable Foundation and Yabu Pushelberg.

Pages 267 to 269 constitute a continuation of the copyright page.

www.randomhouse.ca

Jacket front image: Reed lamp. Design Exchange Collection.
Photo by Pete Paterson. Jacket back image: Deacon, Tom chair.
Design Exchange Collection. Photo courtesy of Keilhauer.
Photo by Karen Levy.
Endpapers: Gallop Meadow textile. Design Exchange Collection
(001.7.1). Gift of Joanne Brook. Photo by Pete Paterson.

Printed and bound in Canada

10 9 8 7 6 5 4 3 2 1

CONTENTS

Preface vii

Essays

1 Modernism 2

2 New Materials and Processes 14

3 Craft, Design and Industry 30

4 Canadian Design in the Pop Era 36

5 From Postmodernism to Pluralism 44

Surveys

6 Furniture 58

7 Lighting 118

8 Textiles 134

9 Consumer Electronics 146

10 Ceramics 170

11 Glass and Miscellany 184

12 Small Appliances 202

13 Metal Arts 212

Biographies & Corporate Histories 226

Notes 256

Sources 262

Photography Credits 267

Objects from the Design Exchange Collection 270

Acknowledgments 272

Index 273

PREFACE

A vibrant new design consciousness emerged in Canada after the Second World War. At that time, this country was one of the world's largest manufacturing nations. Post-war consumer demand and federal government reconstruction policies combined to create opportunities for industrial designers. Good design became recognized by both the media and the public as a necessary and positive development, and the profession was legitimized via professional associations and educational initiatives.

New materials like moulded plywood, plastic and aluminum profoundly changed the manufacturing landscape by permitting a single design to be mass-produced. Designers took on a myriad of roles that crossed disciplines like design, engineering, production and even marketing, engaging in unprecedented collaborations. The designers' exploration of the materials—while balancing the diverse conditions set by artistic, technical and economic concerns —resulted in innovative designs, a process that continues to this day.

In the post-war world, a *Zeitgeist* called modernism dominated the cultural agenda. Industrial designers embraced the movement's tenets—functionalism and truth to materials—with gusto. Modernist pioneers like Jan Kuypers and Stefan Siwinski brought their European training to Canada, while others, such as James Donahue and Lawrie McIntosh, studied with transplanted Europeans at institutions like Harvard University and Chicago's Institute of Design. By the sixties, design training programs were established in Canada, allowing a second generation of designers like Thomas Lamb and Michel Dallaire to study at home.

Although a cold climate and expansive spaces have contributed to design achievements in telecommunication, agricultural and transportation equipment, geography has rarely influenced the style of Canadian consumer goods. The best Canadian product designers approach their task from a global perspective.

The international marketplace was protected by various local tariffs, but a few enterprising companies like Sunar Industries (Douglas Ball), Clairtone Sound Corporation (Hugh Spencer, Al Faux and others) and C. P. Petersen & Sons (Carl Poul Petersen) exported product design—a tactic that with hindsight seems remarkable. As trade barriers around the world are now being lowered or eliminated, more Canadian companies are able to cultivate new worldwide markets.

Some designers have chosen artistic expression over mass production. Canada has a long legacy in studio manufacturing and a reputation for creating limited production designs with panache. By the eighties and nineties, advances in technology and processes made market-specific manufacturing economically feasible. This allowed some Canadian companies to exploit niches and others to lead entire industry sectors. Designers could create a variety of products in greater quantities with increased design content.

Canadian design is now in its third wave, with each successive movement building on the achievements of its predecessors. New designers continue to grapple with international issues in

LEFT: Fred Moffatt's enduring chrome dome kettle of 1940.

this age of pluralism. Although much manufacturing has moved offshore, designers like Tom Deacon, Helen Kerr and Karim Rashid, supported by locally founded manufacturers such as Teknion, Umbra and Keilhauer, can freely participate in the global marketplace.

The Book's Scope and Structure

Design in Canada recognizes and honours the work of design pioneers of the second half of the twentieth century. Both the quality of the work and its enduring appeal should inspire contemporary designers and connect them to the tradition of design in Canada.

Principally, the focus is on objects destined for the home, with important exceptions. Some interesting designs, for example, were made for Canada's booming institutional market, and these occasionally crossed over into residential. Additionally, successful systems furniture designers are included here, as their designs have been broadly influential.

Although the book concentrates on excellence in design, it also features design distinguished as representative of its time. Popular goods like housewares, consumer electronics and other everyday objects present design challenges and thus often form the bulk of designers' portfolios. Collaborations between designers and industry frequently translate into increased sales, market share and even industry dominance.

For ease of use, the text is divided into sections. The first five essays provide an overview (roughly in chronological order) of the way Canadian designers have handled new materials, interpreted stylistic movements and responded to opportunities in manufacturing and the marketplace. Surveys of specific sectors (lighting, ceramics and so on) follow, accompanied by photographs and descriptions of objects that are noteworthy and/or reflect an important aspect of Canadian design. Completing the text are biographies of designers and significant companies.

In addition to visual appeal, the selection criteria for the mass-produced objects were functionality, imaginative use of material and design innovation. How a design solved the problem presented in its design brief and its reception in the marketplace—one of many indicators of its success—were also considered. Some craft work, particularly where artisans collaborated with industry, is included because the objects were made in multiples and were of exceptional design quality.

Featured designers, many with diverse careers spanning thirty or forty years, have each produced a significant portfolio of work over time. On the issue of Canadian provenance, the decision was simply that if a substantial portion of a designer's body of work was created while he or she was living in Canada, it was accepted as Canadian.

Most of the designs have won national and international awards, and most have appeared in design publications around the globe. Many are also held in the collections of the Design Exchange, the Musée du Québec, the Musée des Beaux-Arts in Montreal, the Royal Ontario Museum in Toronto and other institutions.

Research and Methodology

The task of providing a thorough survey of a half century of Canadian product design was both surprising and humbling, as little original research or writing exists on topics such as consumer electronics, lighting and textiles. Research came from a wide array of primary and secondary sources, including corporate and public archives across the country and trade periodicals. Countless interviews with designers, and their oral histories and personal archives, have greatly enhanced the text.

The process of identifying the designer of an object, its manufacturer and other details tends to favour objects that have been reasonably well documented. In a few situations, it was impossible to uncover anything more substantial than a faded newspaper clipping noting an extraordinary object. In other instances, there will likely be regrettable—though unintentional—oversights. This book presents a diversity of designs and attempts to balance region, designer and medium, but no single text can do justice to the richness of Canadian accomplishments in the field.

Design in Canada also celebrates a new spirit in collecting, as objects from the mid-century and the recent past are being rediscovered and newly appreciated. We hope this book encourages more preservation of Canada's extensive design heritage and inspires indigenous new design for the future.

Essays

CANADA

1 Modernism

Post-war Optimism Spurs Design

By most accounts, modernism arrived late in Canada and, in some respects, never really left.[1] Although the movement was debated in architectural circles in the late 1920s, the initial Canadian response to modernism was reserved and moderate. With the exception of some art deco structures in Montreal, there was little built architecture or design that was modernist prior to the Second World War. The appropriate materials and manufacturing technology, and attendant skilled workers, simply weren't available. Not until the mid-sixties did the style noticeably affect city skylines. Unadorned, restrained and egalitarian, modernism by then communicated an appropriate image for many businesses.

While early modernism was clinical and scientific in its rigorous enforcement of utility over aesthetics, this didactic approach gave way to a more inclusive form after the Second World War. The objective of proponents of post-war modernism in North America was to make inexpensive, well-designed products available to the masses.[2] Modernism's "style without a style" attempted to be devoid of any national or cultural references.

The furniture industry capitalized on this universal appeal, initially modifying existing designs. Traditional Colonial-style maple furniture was pared down to simple lightweight forms that seemed contemporary yet familiar. Canadian firms like Imperial Furniture Manufacturing Company combined handcrafting with machine work to create furniture from solid wood. Buyers bought the compromise, and manufacturers were spared expensive retooling.

Promoting Modernism to Canadians

Modernism, the International Style, began in Canada as a fringe movement, embraced by only the vanguard in urban centres. Donald Buchanan, with support from Alan Jarvis, set about changing this with a missionary zeal. Buchanan was an editor of *Canadian Art* magazine and played a central role in founding the National Film Board. He joined the staff of the National Gallery of Canada in 1947 and retired as its associate director in 1960. Jarvis was an artist who became director of the National Gallery in 1955. Over the post-war decade, Buchanan and like-minded Canadians would shape the evolution of "good design," which in their definition mainly meant modernism.

On the West Coast, a design revolution was also under way, although it was decidedly more participatory. It drew its influences from the California modernism popularized by Charles Eames. The formation of the Community Arts Council (CAC) in Vancouver in 1946 provided an active and powerful voice for cultural practitioners to work toward improving design standards.

PREVIOUS SPREAD: 1.1 Children play on Vello Hubel's plastic Playsphere of the sixties.

LEFT: 1.2 Canada's booth at the Milan Triennale, 1957, featured designs by Robin Bush, Robert Kaiser and others.

It engaged a broad spectrum of disciplines in the process—architecture, design, landscape architecture and urban planning—and was collaborative in its approach. In short order, the CAC became a vibrant modernist environment, where a muralist like B. C. Binning might influence the lobby design of a building and city planners could acquire workable ideas for streetscapes. The cross-pollination (later extended to music, dance and more) resulted in a cohesive forum with a singular interdependent goal: a thriving cultural community. The barrier of the Rocky Mountains, the newness of Vancouver itself and its surging population growth contributed to local vitality.[3]

Similarly, Winnipeg became an important training ground for budding modernists, although its influences often emanated from the architecturally savvy Chicago area. At the University of Manitoba, the Department of Interior Design was in 1948 the first accredited program in Canada. One of the earliest graduates, Alison Hymas, became a noted interior designer, working on Montreal's Place Ville-Marie, Habitat and other signature buildings. The U. of M.'s School of Architecture also launched the careers of noted West Coast modern architects like Douglas Simpson and Harold Semmens (Semmens Simpson), Winnipeg's James Donahue and Canada's most successful modernist, John C. Parkin of Toronto.

Art or Industry or Mission?

In Eastern Canada there were more traditions to uphold, so proselytizing about modernism was a more bureaucratic affair. In 1947 Buchanan formed the Industrial Design Information Division, which, orphaned by more appropriate government agencies, fell under the auspices of the National Gallery, an unfortunate coupling that aligned design with art rather than commerce, where it sensibly belongs. A year later, the fledgling organization changed its name to the National Industrial Design Committee (NIDC) (later the National Industrial Design Council, the National Design Council and finally Design Canada) and launched a *Design Index* to showcase "suitable" products. For the next ten years, the NIDC would publish frequently and mount numerous exhibitions that crossed the country, all purporting to tell Canadians what constituted good design. The first was *Canadian Design for Everyday Use* in 1948–49. Viewers could be forgiven if they found the tone of the literature somewhat hectoring.

The NIDC buttressed its stance by eliciting support from the converted. The president of the American Society of Industrial Designers (ASID), Egmont Arens, warned the committee against exposing potential designers to an education filled with "pedagogical conditioning...that would turn their eyes backwards instead of forward." If designers couldn't be "fired up with a passion for modernity, their education might better have been dispensed with."[4]

On the positive side, the NIDC exhibition provided many people with their first look at Canadian-designed products such as a coffee maker by Jack Luck (FIG. 11.12) and a kettle by Sid Bersudsky, engendering some national pride. The public discourse spawned by the publications and exhibitions increased awareness of the role of design for both government and industry. About the same time, the NIDC published two booklets aimed at industry: "Good Design Will Sell Canadian Products" and "How the Industrial Designer Can Help You in Your Business."

In 1949 Vancouver's CAC launched an exhibition, *Design for Living*. More humanist than the NIDC in its approach, it had a catalogue that featured four modernist homes fully outfitted with appropriate art, furniture and decorative arts. The exhibition illustrated how the principles of modern design could be incorporated into every room from a child's bedroom to a kitchen.

Included were local designs such as custom furniture by the architect Catherine Wisnicki and an extraordinary copper table lamp by Murray Dunne (which never went into production). Canadian modernism had arrived, but with it came the critics. Herbert Irvine, an influential Toronto designer for the T. Eaton Company, sniffed, "Modernism is all right for some rooms, like a bedroom for a young girl or boy where they get a kick out of a chest of drawers flowing into bookcases, desks, and cupboards. But all-out modernism is stark."[5]

The resistance didn't dampen curatorial interest in modernism. In 1946 the Art Gallery of Toronto (AGT) (now the Art Gallery of Ontario) mounted an exhibition, under the sponsorship of the Canadian Manufacturers' Association and the Toronto Board of Trade, called *Design in the Household* that glorified mass-produced Canadian-designed products. Using panels borrowed from the Museum of Modern Art (MOMA) in New York, it also attempted to educate viewers about modern design practices.[6] Five years later, Toronto's Royal Ontario Museum, in conjunction with the Association of Canadian Industrial Designers (ACID), hosted an exhibition, *Industrial Design: 1951 B.C.—A.D. 1951*. Along with modern products, it highlighted the careers of emerging professional designers.

While these efforts may have piqued homeowner interest in modernism, a national study on the residential furniture industry (conducted in 1950 by John Low-Beer and James Ferguson on behalf of the NIDC,[7] with support from the Furniture Manufacturers' Association) found that Canadian manufacturers were largely indifferent to original design of any stripe. Most continued the ritual of attending the annual Chicago Furniture Mart and copying the popular models at the show.

Competitions Increase Awareness

To prod manufacturers into innovation, the NIDC launched a national design competition in 1951, with funding supplied by the National Gallery, the Aluminum Company of Canada (Alcan) and the Canadian Lumbermen's Association. The program's design criteria represented a manifesto for modernism. The jury's selections included an aluminum sling chair by Julien Hébert, which caught the attention of the ski-pole manufacturer Siegmund-Werner, Montreal, beginning a thirty-year collaboration.

As rigid as the process was (there were no winners, only honourable mentions), the competition was well received in most quarters and garnered considerable media attention. The second annual competition (1952) attracted a jury of blue-chip modernists like the architect John C. Parkin, who, in 1960, designed Terminal 1 at the Toronto International Airport. Another jury member was George Nelson, design director of the Herman Miller Furniture Company of Zeeland, Michigan, the firm most associated with introducing modernism to America.

Capitalizing on this interest, the NIDC opened the first Design Centre in Ottawa in 1953. (It would showcase design and host exhibitions until superseded by Design Centres in Toronto in 1964 and Montreal in 1966.) The following year, Canadian modernist design appeared for the first time at the Milan Triennale, a prestigious international showcase for design and architecture.

In 1954 the Royal Ontario Museum hosted *Design in Scandinavia*, a twenty-two-city travelling exhibition of more than seven hundred objects, many of which have become modernist icons. It was well received and perhaps pointed up the distance Canadian designers and manufacturers would have to travel to compete effectively on the world stage.[8] The following year, the Winnipeg Art Gallery also looked internationally for modern design leadership, showing two exhibits gleaned from the Milan Triennale, *The Modern Movement in Italy* and *Designs from France*.

The Trend House Program Succeeds

Members of the British Columbia wood industry[9] initiated the Trend House program in 1952. Ostensibly a three-dimensional advertisement for wood products, the first show home attracted 200,000 visitors. Following the American Case Study House model (which launched the careers of avowed American modernists), ten houses were built across Canada, and between 1952 and 1954 over one million people viewed them. In retrospect, it's clear that a largely self-aggrandizing effort to gain market share became the principal vehicle by which ordinary Canadians discovered modernism.

The design brief for the houses was non-interventionist: architects were to be locally based, the design was to be appropriate to its community and each house was to emphasize one species of wood. The ten houses were, by necessity, modest. All the designs, to varying degrees, were modernist in their approach, and some are now considered good examples of the era, such as the John di Castri–designed Victoria, B.C., residence.[10]

These innovative smaller settings demanded an uncluttered, minimalist aesthetic. Furniture became more lightweight and movable so it could be rearranged to suit each successive use; sofas were replaced with groupings of chairs; and accessories (such as magazine and coat racks) helped to maximize space. Decorative arts performed double duty: a rug could also function as a wall hanging; a vase was displayed as fine art.

The Trend Houses, all temporarily fitted up by Eaton's design department, introduced Canadians to the new-style interior decor. Many of the objects were selected from the NIDC's *Design Index*, resulting in a veritable Who's Who of Canadian modern designers and their specialties: Russell Spanner's webbed chairs (FIG. 6.14), J & J Brook fabrics (FIG. 8.9) and Jan Kuypers's bedroom furniture (FIG. 6.26). There were designs from Montreal (Jean Desbarats) to Vancouver (Robin Bush and Earle Morrison). The show homes, without furnishings, were later sold.

The Trend House program earned favourable media attention. Modernism had graduated from industry journals to mainstream newspapers with nary a peep of dissent. The movement was translated into real-world solutions for Canadian homebuyers. It would never again appear to be as glamorous, or as sensible.

Manufacturers Are Reluctant

Despite the intense promotion, there were signs that interest in modernism for the home was waning. The NIDC sponsored a conference called "How Can We Sell More Modern Furniture?" Participants included the designers Robin Bush, Court Noxon and Russell Spanner; major manufacturers such as the Imperial Furniture Manufacturing Company, Electrohome/Deilcraft and Snyder Bros. Upholstery Company, and influential retailers like the T. Eaton Company, Simpsons-Sears and the Henry Morgan Company, Montreal.

At the conference, George Soulis, designer for Snyder's, identified seven potential buying groups for modern furniture, including young marrieds, empty nesters, European immigrants and corporate clients (because they "often deal with architects and decorators whose tastes are definitely modern"). W. A.D. Murray of Morgan's offered some sales statistics from the upscale department store. Eighty-two per cent of buyers of dining room suites preferred traditional designs, while only 8 per cent bought modern. The split in taste was less pronounced in the living room, largely because upholstered furniture is more difficult to classify. According to Murray, the bedroom was the only room where buyers preferred modernism by 66 to 34 per cent.[11]

TOP LEFT: 1.3 The 1957 *Contemporary Furnishings for the Home* exhibition at the Art Gallery of Toronto featured a dining set by John Stene.

BOTTOM LEFT: 1.4 Court Noxon played with atomic imagery in his wall-mounted coatrack, 1957, for Metalsmiths.

The NIDC, now led by Buchanan's former assistant, Norman Hay, responded the following year by launching *Designs for Canadian Living*, a portfolio of design images and text aimed at youngsters. It attempted to sensitize schoolchildren to good design—or, more particularly, modern design.

For the latter part of the 1950s, the Canadian design community took an internationalist view of modernism. The NIDC sponsored exhibitions like *Good Design in Switzerland* and *New Furniture USA*, while the AGT hosted *Contemporary Furnishings for the Home*, featuring the works of American and European designers. Such enthusiasm may have been puzzling to the majority of Canadian consumers, who continued to reject modern design.

Furniture designers looked for encouragement (and potential sales) in foreign markets. Canadian modern designs appeared at the 1957 Milan Triennale (Robin Bush's furniture for Alcan's company town of Kitimat, B.C.), and at the 1958 World's Fair in Brussels, which featured furniture by the designers Sigrun Bülow-Hübe, Jan Kuypers and Lawrie McIntosh. For the most part, the designs showed an accessible brand of modernism: more conservative and made from familiar woods rather than metal or plastic.

Inexpensive, replaceable objects like radios and cookware in the modern style would meet less resistance. Modernism was also well represented in important craft exhibitions throughout the fifties. The metal artistry of Harold Stacey, Douglas Boyd and Andrew Fussell (FIGS. 13.1, 13.5, 13.6) appeared frequently in publications—with few apologies or explanations about its style. Montreal's C. P. Petersen & Sons silversmithy also achieved national and some international success with its version of Georg Jensen–styled modern gifts and hollowware (FIGS. 13.7, 13.8). Even the plastic radios that dominated the decade reflected the modern style.

In the early sixties, the federal government, under the auspices of the Department of Trade and Commerce, became more actively involved in design, signalling a change from the dogmatic NIDC to a more businesslike approach. The National Design Council, descendant of the NIDC, gave out awards, hosted exhibitions and trade missions and published, but by 1963 reported to the newly formed Department of Industry. By the end of the decade, its mandate shifted from design education to "achieving immediate and measurable improvement in the quantity and quality of product design in Canadian industry."

Architecture Embraces Modernism

While designers like John Stene and Robert Kaiser were struggling to wrench Queen Anne chairs out of the dining rooms of the nation, modern design and construction were meeting far less resistance in the corporate and institutional worlds. In bank towers or museums, the precision, austerity and low maintenance of modernist design requirements found a more natural home.

The seemingly simple, repetitive nature of modernism's "grids" also helped to fast-track the construction of sorely needed buildings. The post-war baby boom made unheard-of demands for new schools and hospitals. In Vancouver, a single architectural firm, Thompson, Berwick, Pratt, designed and built more than two hundred schools during the fifties.[12] Prefabricated panels, precast concrete and standardized laminated roof trusses made it possible to erect these structures in record time. High demand also pushed modernism out of the urban core and into small-town Canada.

In 1951 the Massey Foundation sponsored the first Massey Medal Awards for architecture, under the guidance of the Royal Architectural Institute of Canada (RAIC). Over the next two

RIGHT: 1.5 Jan Kuypers used wood in the Skogan side chair for Imperial, 1955, to sell modernism to Canadians.

decades, eight competitions were held.[13] The architect receiving the most awards in the period was John C. Parkin, who had studied at Harvard under the influential Bauhaus modernists Marcel Breuer and Walter Gropius. With completed projects like the aforementioned Terminal 1 (now demolished), Ortho Pharmaceutical and the Ontario Association of Architects headquarters (all in Toronto), he was considered the institutional modernist. In addition to numerous schools, hospitals and industrial buildings, the Parkin firm also collaborated on Toronto's revolutionary City Hall (Viljo Revell) and the Toronto-Dominion Centre (Ludwig Mies van der Rohe). The recognition for Parkin's work helped to legitimize modern architecture in Canada.

Advocates for Modernism

Toronto's other significant modernist was a brash young English émigré, Peter Dickinson, who designed 1 Benvenuto Place apartments, the public housing high rises at Regent Park (both in Toronto) and the Canadian Imperial Bank of Commerce tower, Montreal. Dickinson's and Parkin's passionate commitment to modernism supported a host of like-minded firms producing the requisite furniture, lighting and textiles to complete their visions. Robin Bush, Court Noxon and Stefan Siwinski (FIGS. 6.37, 6.42, 6.35) gained credibility and prospered after being selected to create furniture in a Parkin or Dickinson project. Likewise, the textiles, lighting and decorative arts of designers such as J & J Brook, Karen Bulow and Gunnar and Lotte Bostlund (FIGS. 8.9, 8.4, 7.9) benefited from their association with these modern architects.

By the sixties, the manufacturing industry in Canada was maturing and was able to more fully exploit the unique properties of materials like flat-rolled steel. Mass-production runs like those in the United States were still not possible, but many firms such as Metalsmiths and Leif Jacobsen in Toronto had carved out niches as suppliers that could produce two hundred high-quality chairs in a hurry.

On the West Coast, architectural commissions were largely dominated by the two firms Semmens Simpson, and Thompson, Berwick, Pratt, whose major projects included Alcan's at Kitimat and the B.C. Electric building respectively. Many of the modern architects who later made their names as individual practitioners passed through these firms, including Barry Downs, Fred Hollingsworth and Duncan McNab. Two young modernists, Geoffrey Massey and Arthur Erickson, followed the same path en route to creating signature buildings such as the MacMillan Bloedel building and Simon Fraser University respectively.

Unlike in Toronto, where Canadian designers at least had a toehold, these buildings in B.C. were primarily outfitted by Herman Miller and Knoll International, which can be explained by the lack of an industry in Vancouver capable of producing the quantity of furniture required. Robin Bush, responsible for much of the furniture in Kitimat, was a notable exception. Where possible, the Vancouver community did support its own designers, but for practical reasons, all the goodwill rarely translated into significant contracts.

In Montreal, modernism's move into the corporate world was similarly driven by major architectural projects like Place Ville-Marie (I. M. Pei) and Dorval International Airport. Companies like Jacques Guillon & Associates outfitted the interiors and included a few of their own designs as well as specifying Herman Miller and Knoll.

The Designers Fight Back

The 1961 to 1966 competition for and subsequent awarding of the furniture contract for Toronto's New City Hall was both a blow to and a rallying cry for Canadian modernist designers.[14] After considerable wrangling and many resignations, the contract was awarded to Knoll International Canada over three Canadian bids: Mitchell-Houghton, Eaton's and the Robert Simpson Company.

While the losing companies proposed chair designs by the Canadians Stefan Siwinski, Robin Bush, Walter Nugent and others, Knoll offered chairs by modernist stalwarts like Mies van der Rohe and Warren Platner, an emerging talent from Eero Saarinen's office. The Canadian-born head of the Knoll International Canada office, John Quigg, designed the precast-concrete and wooden case goods (desks, cabinets and tables).

Ironically, the Canadian firms Deilcraft (Electrohome), Leif Jacobsen and Canadian Rogers Eastern manufactured Knoll's designs in Canada. These companies, plus C. B. Wrought Iron Manufacturing Company, Toronto (which produced the metal bases for Platner's designs), would make most of the chairs for City Hall, while Leif Jacobsen, Pre-Con Murray and J. F. Gillanders, all in Toronto, made the case goods. The implication was that Canadian companies were good enough to manufacture the furniture but not good enough to design it.

The City Hall design slight reinforced the long-standing need for designers to have ersatz classic modernism in their repertoire. Al Faux created his own version of Marcel Breuer's Wassily chair; Court Noxon, Jack Dixon (FIG. 6.38) and a host of others had Mies-style chairs that could be pulled out of the hat to meet a last-minute deadline or a tight budget.

The situation was moderately better in consumer electronics, where Clairtone Sound Corporation, Toronto, and Electrohome, Kitchener, were achieving stylistic successes with modern design. By the mid-sixties, Clairtone's co-owner, David Gilmour, designed a series of Scandinavian-style stereo cabinets. At Electrohome, Gordon Duern, Keith McQuarrie and other designers produced a series of forward-looking stereos (FIGS. 9.17, 9.20, 9.18)—some of which broke sales records and were marketed in the U.S. In interiors, Janis Kravis's modern design for the restaurant Three Small Rooms in Toronto's Windsor Arms hotel (since rebuilt) was influential (FIG. 1.6).

Expo's Impact on Design

For many designers, the "Buy Canadian" advocacy of Expo 67 was decisive. Simply put, the World's Fair, and the "grands travaux" connected with it, including the Montreal Métro and Place Bonaventure, dramatically altered the situation by creating full employment for the Canadian design community.[15] Despite appearing in the era of Pop, Expo was all about modernism, albeit a more youthful incarnation. The impressive cohesive urban planning was on a scale that had never been seen before in Canada.

Most Canadian designers working in the sixties contributed in some way to Expo—signage, pavilions, parks and visitor centres. The size and diversity of the fair enabled many industrial designers to expand their expertise beyond product design and move into areas like graphic and exhibition design. For instance, the Montreal firm Jacques Guillon & Associates produced the interiors for the new Métro cars. Dudas Kuypers Rowan (DKR), Guillon's counterpart in Toronto, designed the exhibition *Man the Producer*. Vancouver's Robin Bush Associates (with the help of the designers Douglas Ball and Thomas Lamb) created a pavilion for the Atlantic provinces, and Hugh Spencer, designer of landmark stereos for Clairtone, was responsible for the Western Canada Pavilion.

A federal program in anticipation of the country's centenary event invited Canadians to submit products to be either displayed or sold at Expo. Al Faux, for example, designed the restaurant furniture for the Ontario Pavilion, and the textile company Karen Bulow Ltd. draped the DuPont

EXIT

Canada auditorium. The National Design Council's efforts to cultivate new products resulted in *Canadian '67*, a two-volume illustrated catalogue that included six hundred products, from furniture to giftware, by the period's most notable designers.

Most important, Expo's optimism inspired Canada's designers to experiment. The architect Moshe Safdie created Habitat, based on his thesis design while at McGill University, and it revolutionized the concept of high-density living. The progressive furnishings and interior design of the Habitat suites also received considerable exposure. Ironically, Canadians were almost denied the opportunity to design suites. *Chatelaine*, the arbiter of Canadian lifestyles, was charged with co-ordinating the suites' decor and all but ignored contemporary Canadian design. The outspoken Jacques Guillon, then president of the Association of Canadian Industrial Designers, successfully negotiated for and won control of thirteen of the twenty-six suites. Sigrun Bülow-Hübe of Montreal's AKA Works designed and outfitted a suite, as did Christen Sorensen, Montreal, who created an interesting wood shelf system. Jacques Guillon & Associates showcased its new furniture lines in plywood and aluminum manufactured by Paul Arno, Montreal,[16] and Jerry Adamson of DKR displayed his experimental plastic Habitat furniture (FIG. 6.47).

The modernism torch was thrown to a new generation of designers such as Michael Stewart, Keith Muller and Thomas Lamb, who often humanized the style by employing natural materials.

Modernism's presence also continued to grow in office systems furniture. Second-generation designers like Jonathan Crinion and Manfred Petri (Geiger Brickel) built on the pioneering work of Douglas Ball and Robin Bush.

By the 1990s, Canadian designers ranging from Helen Kerr to Karim Rashid were resurrecting elements of the seventy-five-year-old movement. In retrospect, the "form follows function" dictum of modernism is regarded as the defining aesthetic of the twentieth century.

TOP LEFT: 1.6 Janis Kravis's 1964 design for the Three Small Rooms restaurant, Windsor Arms, dragged Toronto society into the modern world.

BOTTOM LEFT: 1.7 The sculptural Katimavik pavilion (Ashworth, Robbie, Vaughan and Williams, Ottawa) at Expo 67 reflected modernism's continuing grip.

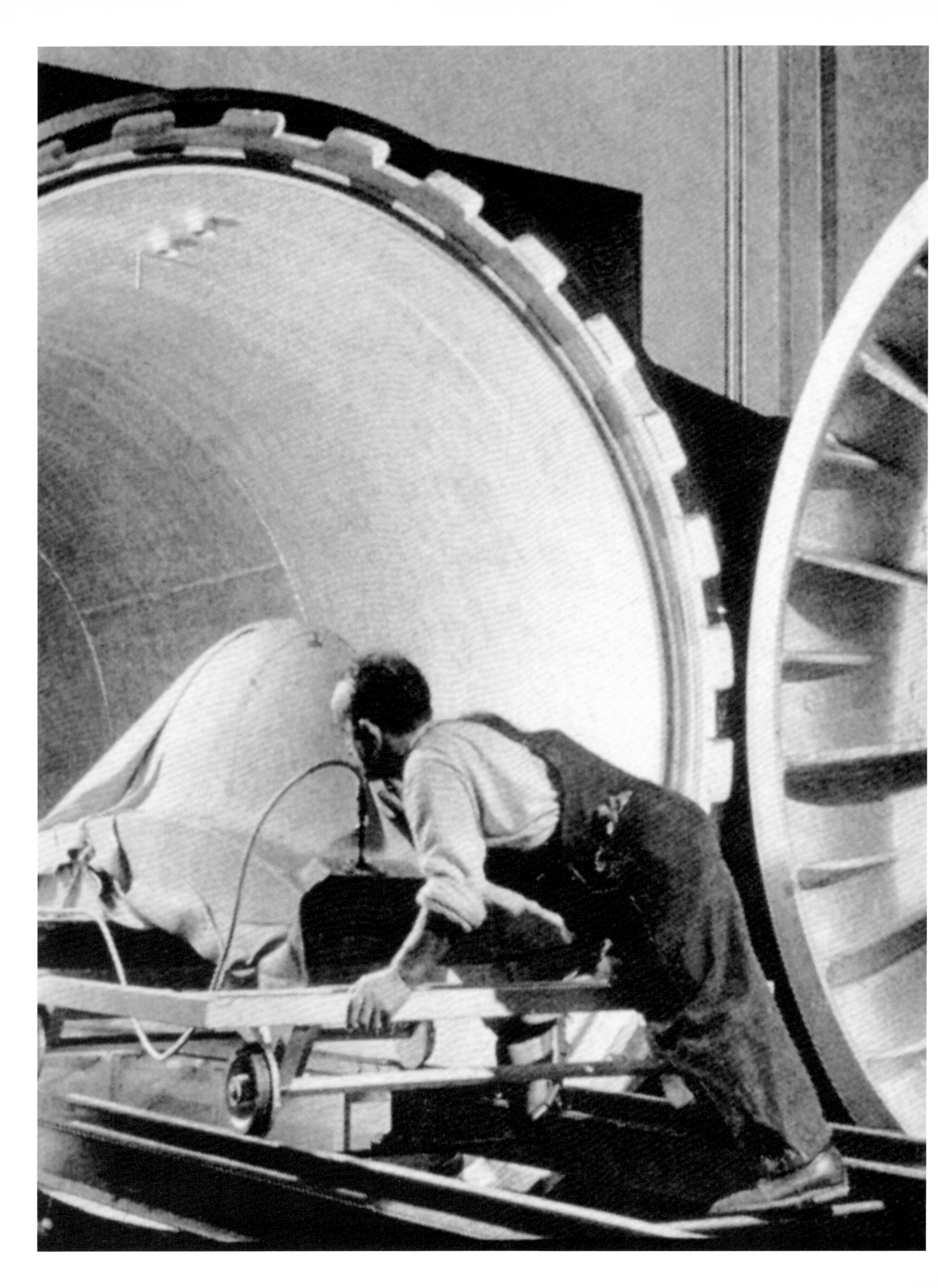

2 New Materials and Processes

Modern Materials Invigorate Design

Plastic, aluminum and moulded plywood transformed design in the last half of the twentieth century. Both commercial enterprise and the public sector were instrumental in elevating the status of these materials, while design reformers embraced them as the key to modernism. Immediately after the war, the media heralded all three materials, especially plastic, as the new "miracle materials" that won the war. Pioneering industrial designers took up surplus war materials in their latest products, and they came "marching home" to transform the household.[1] Although all three had been invented and used before the war, significant technical advances developed for a variety of military uses made them seem new and revolutionary.

Plastic, aluminum and moulded plywood were light, strong and malleable—ideal for mass production. In *Design for Use in Canadian Products* of 1948, Donald Buchanan argued that by responding to technical problems associated with modern materials, designers would avoid "mere styling."

George Englesmith, a professor at the University of Toronto School of Architecture and a friend of Buchanan's, was the first president of the newly founded Affiliation of Canadian Industrial Designers (known as the Association of Canadian Industrial Designers after 1948). He believed that the study of these materials would hasten the advance of modernism. In 1948 Englesmith had his students visit the moulded plywood facility at Aero Marine Industries and the Aluminum Goods Factory, both in Toronto.[2] Waclaw Czerwinski, a designer for Canadian Wooden Aircraft, Stratford, Ontario, lectured on moulded plywood.

For new materials to seem modern, the thinking of the day was that they had to avoid imitation. The influential 1950 NIDC "Survey of Design Requirements and Conditions in the Canadian Furniture Industry" discussed the appropriate use of plastic, wood and metal. While it maintained that new materials had to be treated with respect and imitation avoided, it acknowledged that those few manufacturers using new materials favoured traditional copying.[3]

New materials were so much in the public eye that they became the topic of major cultural institutions' design exhibitions, including *Design in Industry* held at the Royal Ontario Museum in 1945, *Design in Industry* at the National Gallery of Canada in 1946 as well as the AGT's *Design for the Household*.[4] The National Gallery show, organized by Buchanan, was by far the most supportive of aluminum, plastic and plywood. Photographs and samples of unprocessed materials along with illustrations of their wartime uses, such as plywood airplane wing tips, created the framework for the display of new peacetime products. Even the exhibition signage panel was made of plywood and extruded aluminum poles.

LEFT: 2.1 In the forties, the National Research Council in Ottawa used the autoclave oven to mould plastic and wood components.

Commercial retailers endorsed the new materials. In 1948 Simpson's in Toronto presented a documentary on new plastics and devoted its display windows to plastic household products, as did Morgan's in Montreal in the fifties (FIG. 2.3).[5]

Plastic Goes to War

When celluloid, the first man-made partially synthetic plastic, came on the market in the late nineteenth century, society perceived it as a magical, less expensive alternative to scarcer materials such as ivory and tortoiseshell. Moulded into "fancy goods" like stylish combs and letter openers, celluloid plastic made these frills affordable to the middle class.[6] Decades later, radio manufacturers catered to the craze for plastic luxuries by providing Bakelite and Catalin cabinetry for desirable home radios. These materials seemed well adapted to the imitative styles of the forties, when radios could look like pyramids or toasters.

During the Second World War, plastic's reputation was highly favourable, and new forms of it were used for army provisions, including just-invented vinyl raincoats, melamine dishes and acrylic canopy tops for fighter planes. Transparent plastic bubble domes became the signature feature of the futuristic dream cars of the early fifties. Acrylic (invented in 1933), often advertised as "the aristocrat of plastics," provided the first decorative plastic accents in home interiors, notably light fixtures, side tables and vanity sets for "the peacetime boudoir." In 1945 the Simpson's "Apartment of Tomorrow" featured an acrylic cylinder "slumber wing."[7]

Plastic's ever-growing popularity had transformed it from novelty to pedestrian material, but low quality limited its acceptance. Suppliers, like the multinational company American Cynamid, were justifiably concerned that moulders didn't employ proper production methods and often could not recognize the various kinds of plastic suitable for specific applications.[8] The moulders in turn believed the blame lay with consumers because they used plastic products incorrectly. Ultimately, buyers lost out.

Canadian Homes and Gardens devoted its entire November 1948 issue to promoting the new "Plastics Age." Despite the magazine's enthusiasm, it still recommended that household plastics come with warnings: "becomes warped if dunked in hot water...shapeless if boiled...will fade or darken if left too long in bright sunlight; keep clear of red-hot stoves." Even so, the problems with plastic did not stop an industry boom after the war.

Industrial Designers Embrace Plastics

Industry suppliers introduced branding to help consumers identify superior-quality plastic. More important, the suppliers hired industrial designers and offered their design and rendering services to mould makers free of charge.[9] As the consumer goods industry converted to plastic, the fortunes of the fledgling industrial design profession rose.[10]

Sid Bersudsky, Toronto, was one of Canada's first industrial designers, and he chose plastic as his specialty. In the fifties, he worked as an industrial design consultant to Dow Chemical Canada and managed his own design firm. He even considered setting up a plastic moulding plant, but ultimately decided against it since machine time was available practically at cost from existing fabricators.[11] Henry Finkel of Montreal parlayed his wartime experience in munitions into the founding of the Die-Plast Company. It offered "under-one-roof...complete plastic service," including product design evaluation and mould making.

TOP RIGHT: 2.2 Plastics pioneer Sid Bersudsky shows off his 1955 NIDC award.

BOTTOM RIGHT: 2.3 Morgan's in Montreal promoted nylon products in a fifties window display that featured Jacques Guillon's "Parachute" nylon cord chair.

rose during the 1974 embargo by the Organization of Petroleum Exporting Countries, designers moved away from the material.

Transparency Revitalizes Plastic

By the nineties, technology developed for the auto industry, combined with efforts to protect the environment, breathed new life into the material. New thermoplastics can be re-formed for greater strength and versatility, use less energy for processing and are cleaner to produce. Recycling has made plastic acceptable, if not fashionable. Improved moulding techniques initiated by computer-aided design and lower tooling costs introduced a stunning new look for polypropylene.[25] Plastic can now be blown into curved shapes so thin that they're almost see-through. Translucence appealed to fin-de-siècle designers' fascination with lightness and transparency.

Konstantin Grcic, working for the German manufacturer Authentics, introduced the new plastic look to Europe. Here in North America, Umbra in Scarborough, Ontario, hired Kerr Keller Design and Karim Rashid (FIGS. 5.9, 11.22). On the West Coast, Martha Sturdy also turned to plastic, but preferred cast resin for small-scale production (FIG. 11.21).

The Canadian contract furniture industry finally converted to plastic in the latter half of the nineties. Wider markets provided an incentive for industry to invest in the expensive machinery required for large production runs. Two Toronto contract seating specialists, Keilhauer and Allseating Corporation, acquired equipment to produce plastic components for furniture designed by Tom Deacon (FIG. 6.64) and Miles Keller (FIG. 2.10). On the residential side, Umbra equipped its Buffalo plant with expensive specialty moulds to launch its first mass-produced plastic furniture collection by Canadian designers, including Paul Epp and Jonathan Crinion.

Aluminum Furniture Becomes Popular

Since the Second World War, aluminum has competed with plastic, which shares similar properties of lightness and strength, as well as the ability to be shaped into a variety of forms. Aluminum-clad products from flying Zeppelins to fuel-efficient electric cars have come to symbolize the modern imagery of the twentieth century.[26] A brochure from Alcan in 1946 boasted that "aluminum has contributed...to the victories of the United Nations and the tremendous output of aluminum is recognised as one of Canada's great contributions to the war effort."[27] Aluminum smelting as well as fabrication had expanded dramatically to meet wartime demands. After the war, aluminum, more than plastic and plywood, needed to secure new markets. The potential influx of inexpensive imports of raw aluminum, combined with competition from plastic, threatened the industry here. Former war machinists took up the cause of reapplying aluminum components to the domestic market.

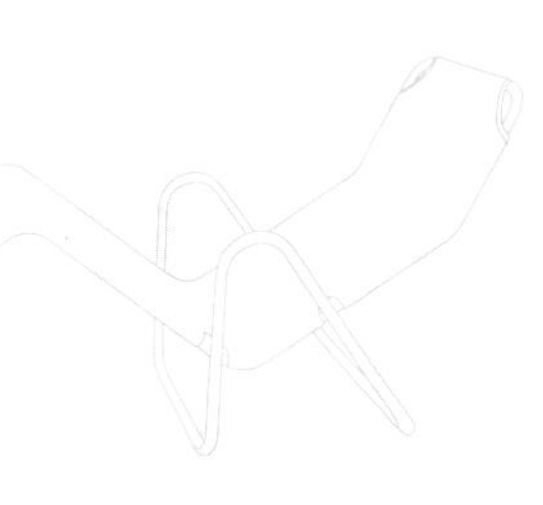

Taking advantage of the army-surplus tubing (used for structural aircraft framing) supplied by Alcan and other mills, entrepreneurs equipped their garages with simple but efficient bending machines. In Winnipeg, an electrical engineer named William Trott produced a line of floor lamps using aluminum tubing (FIG. 7.1), which he sold by mail order.

Tubular garden, office and cafeteria furniture provided the necessary boom for the industry. William Holtzman, a former toolmaker for an aircraft factory, founded H&K Metal Products, Toronto, in 1946. Over the next few years the company grew to become one of the largest producers of aluminum garden furniture in Canada, securing a much-coveted contract from Eaton's.[28] Holtzman soon renamed the company Featherweight Aluminum Products to emphasize the material's special asset, and his marketing was aimed directly at women. The May 1952 issue of *Canadian Homes and Gardens* trumpeted, "Women like it [aluminum] because it's 'light' and ideal 'for casual living.'"

Siegmund-Werner, Featherweight's competitor, offered the greatest design content in its line of aluminum furniture (FIG. 6.23). B. F. Harber of Fort Erie, Ontario, who had a background in airplane mechanics, developed a mechanism for folding garden furniture. He also introduced coloured and anodized coating for protection as well as style. Simpson's distributed the line, which included an award-winning utility chair. In the sixties, Harber sold the licence to an American manufacturer.[29]

Cookware Emerges as a Market

The family firm of Orr Associates in Toronto secured a contract with Supreme Aluminum Industries, Toronto, to design kitchen products for the next decade,[30] a duty they shared with Sid Bersudsky. Among the post-war designers, however, it was Jack Luck who earned a venerable reputation in the aluminum industry. Luck spent most of his career in Aluminum Laboratories, the engineering department of Alcan, based in Kingston. In the forties, he persuaded the company to open an industrial design division "to promote the use of aluminum by improvements and inventions in the field of design."[31] Luck became a spokesman for the industrial design profession and appeared in *Design for Living*, a documentary produced by the National Film Board in 1956. The naively propagandistic film presents Luck as a brave new modernist: in his crisp white shirt and tie, he sketches at his drafting table, designing aluminum cookware.

Luck also assisted associated companies and their customers to create competitive products "in any line in which aluminum can be used."[32] The U.S. giants Reynolds Metal Company and the Aluminum Company of America (Alcoa) had also formed in-house design studios to promote the use of aluminum to the design profession.[33] Luck contracted his services to the newly opened Jean Raymond Manufacturing in Lachine, Quebec, and designed a series of products, including extrusion-pulled door handles. The company, specializing in custom products for the architectural trade, still exists today.[34]

Notably, after the war Luck designed cookware for Aluminum Goods, an Alcan sub-industry, under the name Mayfair. His discreetly modern metal coffee pots and teakettles equipped North American kitchens for more than two decades (FIG. 14.1). Luck's premature death in 1963 appears to have put an end to Alcan's effort to stimulate the use of aluminum in industrial design within its own departments. Instead, the corporation focused its attention on its new flagship headquarters in Montreal.

Place Ville-Marie—Alcan's Salute to Design

After the war, architects and builders embraced aluminum as a construction material for skyscrapers and prefab housing. Standardized aluminum components became the new material for door and window frames, while aluminum siding promised that you need "never paint your house again." Eaton's and the federal government's Central Mortgage and Housing Corporation (CMHC—later the Canada Mortgage and Home Corporation) toured an Aluminum House across Canada in 1949, offering free plans to anyone interested in building the inexpensive if traditional design.[35] Meanwhile, architects began to use aluminum to finish the exterior facades of steel-and-glass office towers.[36] In 1958 the noted American architect I. M. Pei (in association with the Montreal firm Affleck, Desbarats, Dimakopoulos, Lebensold, Michaud and Sise) designed

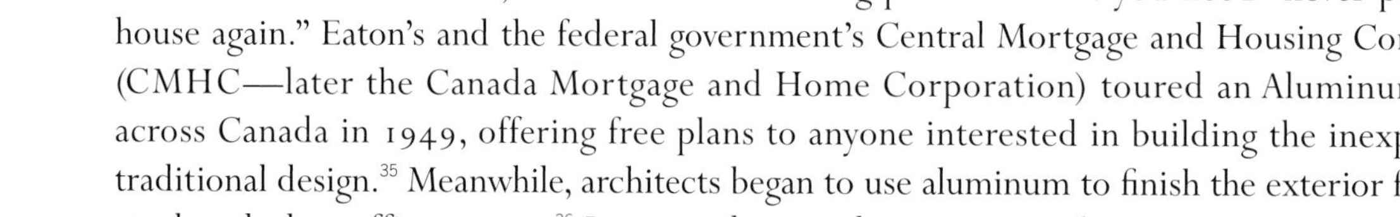

RIGHT: 2.7 The dramatic aluminum staircase in Alcan's showpiece interiors, opened in 1962.

Place Ville-Marie. The cruciform-shaped office tower demonstrated the use of aluminum cladding for a facade on a monumental scale.

Alcan's eight-floor penthouse office suite opened in 1962. Jacques Guillon & Associates co-ordinated the showpiece interiors. A floating aluminum staircase (FIG. 2.7), designed with the help of Alcan fabricators, featured machine-extruded stair treads. The architect Norman Slater created a series of tactile and sculptural reception screens that defined each floor. Jacques Guillon persuaded Alcan to let his firm develop some furniture, including the Alumna office desk group (FIG. 6.41). Unfortunately when Alcan moved in the eighties, the Alumna desks were relocated to the basement and Slater's screens have been placed in storage.

Techniques Shape Aluminum Design

Aluminum can be processed in a multitude of ways. Not surprisingly, the techniques influence the look of the design: spinning is round, extrusions are hard edged and casting is fluid. Although the industry is not as sophisticated as in Europe, Canadian industrial designers have worked with a large network of fabrication mills to develop new products.

The cookware industry has used the spinning technique for years: the process gave Luck's coffee pot its simple cylindrical shape, and William Wiggins's seventies Ball-B-Q (FIG. 4.3) for the company Shepherd Products in Toronto its distinctive shiny round dome top. In the fifties, the electrical engineer Frank Reed took advantage of the technology, creating the Ring Master (FIG. 7.2) hanging lamp for the institutional market. A typical example of post-war industrial design ingenuity is Reed's adaptation of the aluminum air-duct grille to create a handsome lamp defined by three concentric rings. In the nineties, Scot Laughton and James Bruer of Toronto chose the spinning process to distinguish the seat design of its Jim stool series for Pure Design, Edmonton. The resourceful designers went to a local frying pan factory to produce the initial run of spun aluminum seats (FIG. 2.9).

Canadian manufacturers and designers mostly favour the extruded-bar technique, chiefly because it requires little tooling. In this popular method, molten aluminum is mechanically squeezed, then forced through a cylinder into the desired shape. Toronto's subway cars, launched in the late fifties, offer a famous example of the use of aluminum extrusion. Produced by Hawker Siddeley Canada, the cars were made of the longest extruded aluminum shapes yet manufactured in Canada. Alcan made the extrusions for the cars at its mill, Kingston Works.[37] Aluminum extrusion defined the door pulls designed by Jack Luck for Jean Raymond Manufacturing and the cube shapes of the Tukilik salt and pepper shakers by Marcel Girard and Ian Bruce, in Montreal (FIG. 11.18). Koen de Winter of Montreal used an extruded aluminum profile to create the horizontal ceiling wing light for Axis Lighting (FIG. 2.8). Michael Santella, also in Montreal, admires the material for what he calls its "noble duality in strength and lightness" (FIG. 13.9). He reconfigures and customizes extruded tubes and bars into household products.

Cast aluminum is relatively inexpensive and allows for creative shapes, although results can be inconsistent. It's poured by hand into the mould, dries almost instantly and must be buffed and polished by hand. The freedom of the medium permitted Miles Keller to create the futurist Os[5] chair for Allseating Corporation, featuring sand-cast legs and arms (FIG. 2.10). Hoselton Studio in Colborne, Ontario, has been sand-casting tableware and souvenir ware since 1970, albeit with more pedestrian results. The artisan foundry produces twenty-five hundred products a year, including a line of organically shaped dishes, bowls and figurines.[38]

TOP RIGHT: 2.8 Koen de Winter employed extruded aluminum in an institutional light, in 1998.
BOTTOM RIGHT: 2.9 Spun aluminum reappeared in Bruer and Laughton's Jim stool of 1997.

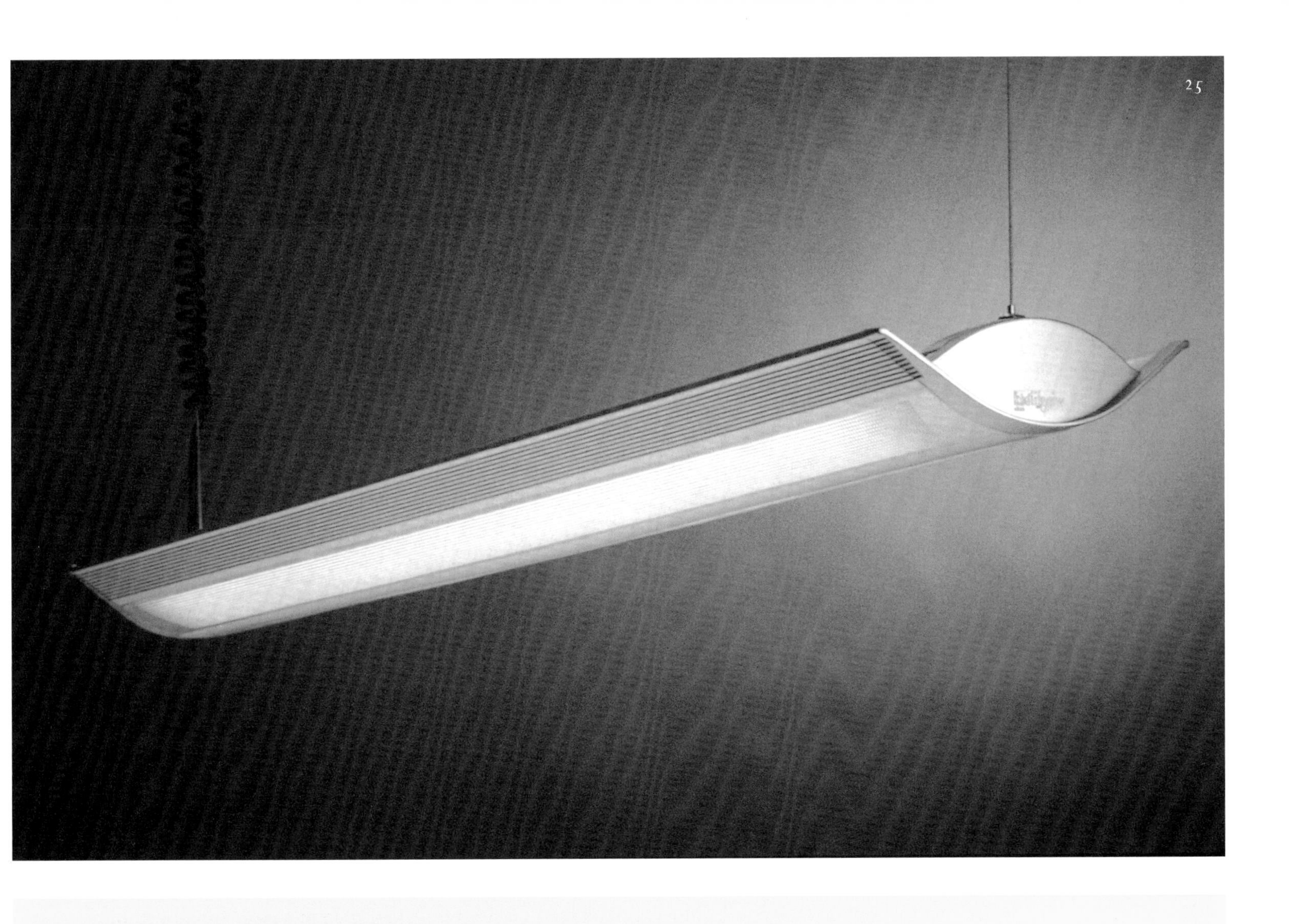

Die-casting offers a reliable alternative to the alchemy of sand-casting. While considerably more expensive, it permits greater uniformity and longer runs by using steel moulds. Todd Wood of Ottawa, with Allen Simpson Marketing and Design, developed a sophisticated casting procedure for his sinuous Plus Four garden tools (FIG. 2.11). Wood originally intended the tools to be ceramic, but the design suited the cast-aluminum process.[39] Nienkämper, Toronto, introduced aluminum hardware to accent its new collection of wooden furniture. With the success of such products as the Vox Conference table and Tangent shelving (FIG. 6.63), the designer Mark Müller has persuaded the company to change from sand-casting to permanent metal moulds.[40]

Moulded Plywood's Newfound Markets

Light, strong and free of labour-intensive finishings, moulded plywood is an inexpensive process that involves bending several layers of wood veneer, with grains running at right angles to each other, under pressure.

Moulded plywood reached new potential in the 1940s through the fabrication of aircraft components. As by-products of the plastic industry, new glues allowed for larger sheets of plywood with greater strength and versatility. Before the Second World War, the Canadian furniture industry used laminated plywood in limited capacity. Because it was perceived as a cheap material, manufacturers hid it, using it to line the inside of drawers or the backs of furniture.[41]

In 1942 southwestern Ontario's furniture industry began to use moulded plywood in the wings and fuselages of glider planes, as well as parts for bomber planes. With shortages in metal hardware like springs and coils, the companies needed to find a new market and formed an alliance to win a lucrative government contract to build aircraft components. At the same time, in British Columbia, the wood industry produced components for bombers near Vancouver.[42] After the war, the British Columbia industry focused on construction.

In Ontario, several aircraft manufacturers, including Canadian Wooden Aircraft, switched to furniture production (FIGS. 6.18, 6.17, 6.16). Hugh Dodds, operating as Aero Marine Industries, made a variety of moulded plywood products and even a prototype for a plywood automobile. Dodds also created a popular institutional chair (FIG. 6.21).

Beyond these few companies, the furniture industry initially avoided plywood, considering it a poor substitute for solid wood. Slow acceptance of the material sidelined the much-anticipated "technical revolution" in the furniture industry.[43] Plywood appeared in bench seating for churches, restaurants and bowling alleys, but it remained a novelty inside the home. Plywood's blatantly modern aesthetic meant that even the progressives at the Ottawa Design Centre often confined moulded plywood furnishings to kitchen displays, rather than including them in living or dining room settings.[44]

Specialty manufacturers like Curvply Wood Products of Peterborough, Mac Craft Industries of Sarnia (later known as Marlin Superior Products in Gravenhurst) and Crown Zellerbach Building Materials of Quebec became long-term producers of moulded plywood components. After the war, they acquired high-frequency wood-laminating machines and established factories to supply institutions with stacking furniture.[45] Because Americans preferred folding metal stacking chairs, the fabrication of wooden seats and backs for metal furniture became an all-Canadian industry.

These companies began to work closely with Canadian designers. In the fifties, Mac Craft Industries produced plywood seats and backs for Jan Kuypers (FIG. 6.25) and made parts for the

TOP RIGHT: 2.10 Miles Keller's Os[5] chair drew on a Darth Vader aesthetic to create a sci-fi chair with punch, 1998.

BOTTOM RIGHT: 2.11 Todd Wood's Plus Four garden tools demonstrated cast aluminum's appeal, 1992.

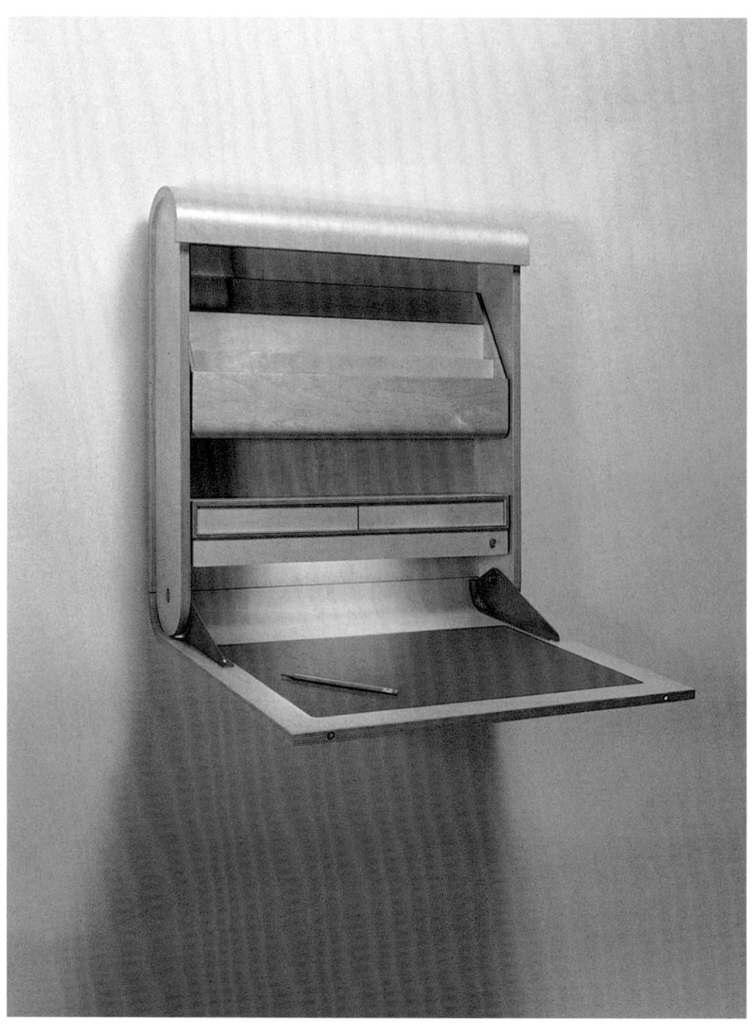

popular furniture designs of Russell Spanner. (FIGS. 6.14, 6.15)[46] In the late sixties, Curvply, promoting itself in *Canadian Interiors*, invited designers to send drawings and samples for competitive quotes.[47]

Canada's most adventurous designers prefer moulded plywood. The material's comparatively inexpensive tooling has allowed them to unite modernist concerns with efficient production and original design. Peter Cotton in Vancouver combined moulded plywood with jet-black steel in his Perpetua furniture line of the early fifties. He promoted the material's resilience by naming his most recognized product the Spring-back chair (FIG. 6.17). In the late sixties, Michael Stewart and Keith Muller constructed their MS stacker (FIG. 6.48) from just three moulded plywood parts. Working with Curvply, they formed the company Ambiant Systems, Toronto, which produced a line of bent plywood furniture, including Al Faux's swing-back chairs for the Ryerson Institute of Technology. The furniture designer Thomas Lamb cut his teeth creating a series for Plydesigns, the furniture division of Curvply. Lamb produced a collection of children's furniture in bent plywood, including a rocking horse and a high chair, and the Roo school chair (FIG. 2.13). The designs expressed his love for the material and its ease of assembly. Each sculptural component could be inserted into another to either fold or stack, eliminating most metal hardware.

Veneer Supersedes Bent Plywood

Designers in the eighties and nineties eschewed moulded plywood in favour of richer veneers like cherry and mahogany, in more traditionally carved shapes. Moreover, new reconstituted woods entered Canadian manufacturing and caught the eye of emerging and veteran designers alike. For instance, MDF (medium-density fibreboard), a recomposed particle board invented in the United States for the construction industry, created a new sensation because its strength was ideal for machine carving. The designer Vello Hubel chose it for his figurative Clover-Leaf nesting tables (FIG. 6.60), and new designers like Michael Santella combined it with his aluminum furniture. Typically, designers lightly stained the material to allow the wood pattern to come through or painted or laminated the surface with colourful patterns.

The modern revival in the late nineties brought blond machine-moulded plywood furniture back into fashion. Karim Rashid designed a stool for Pure Design featuring a wave-shaped plywood seat with a steel base. It pays homage to the sculptor Jan Arp, one of the founders of Biomorphic art, which inspired the plywood furniture tradition. Pure Design subcontracts the manufacturing to Sylvaplex, a descendent of Marlin Superior Products. Andrew Jones, in Toronto, attracted considerable attention using moulded plywood for his poetic designs. He chose a blond birch plywood veneer for his minimal Wall-bureau desk (FIG. 2.14). The prototype now belongs to the collection of the Royal Ontario Museum.

Undoubtedly, the flexibility, low cost and sheer volume of materials like plastic, aluminum and moulded plywood will ensure their continued pre-eminence. The difference will increasingly be the way these twentieth-century materials are combined with other substances. New combinations (known as composites) will give additional strength and texture and ultimately new functionality to products.[48]

TOP LEFT: 2.12 In the sixties Jack Dixon used laminated plywood to form a curvaceous chair and table.
BOTTOM LEFT: 2.13 Thomas Lamb moulded the well-regarded Roo stacking chair, 1970, from plywood.
BOTTOM RIGHT: 2.14 Andrew Jones's wall-hung desk reintroduced moulded plywood in the late nineties.

MADE
PATENT NO

3 Craft, Design and Industry

The Craft Revival

Since the 1930s craft revival, many artisans have contributed to Canadian design. Choosing to standardize their process of production, either by hand or with the assistance of tools, craftmakers made hundreds, sometimes thousands, of functional pieces. Craftmakers, however, have rarely collaborated with the manufacturing sector, owing in part to the anti-industry perspective that shapes the craft movement.[1] Within Canadian industry, a separate trend arose—the "studio manufacturer," an industrial designer or architect who establishes a fabrication studio to offer small-scale production. Studio manufacturers prefer to have their designs mass-produced but often have no alternative to limited production to have their designs enter the marketplace.

Craftmakers are influenced by two traditions: the nineteenth-century Arts and Crafts movement, led by William Morris, and the post-war Studio Craft movement. The first encouraged handmade pieces produced in multiples, while the latter elevated craft to the status of fine art and called for total control of the entire production process.

The Arts and Crafts tradition came to Canada via enthusiastic hobbyists in 1900, when many well-to-do Canadians began to weave and make pottery for relaxation. The Women's Art Association of Canada, based in Montreal, founded the Canadian Handicrafts Guild in 1906. Over the next thirty years, the guild developed into a national body, with chapters forming in cities across the country.

At the same time, provincial governments saw craft as a means to support ailing farming communities. Weaving and pottery, followed by metal arts, became the favoured media, lauded by governments as a means for household members to supplement income as well as create souvenirs for the budding tourist industry.

Quebec institutions were the first to effectively stimulate craft revival. In the past, the Catholic Church had served as an important patron of artisans, commissioning ecclesiastical wood carvings and metalwork. Worried about the effects of industrialization, the provincial government sponsored new schools like the École des Arts Domestiques, formed to provide instruction for home weaving across the province, while the École du Meuble (opened in Montreal in 1930) united cabinetry and other decorative arts. The school's last and most significant commercial commission was the Queen Elizabeth Hotel in the fifties (since refurbished).[2] In 1950 Jean-Marie Gauvreau, the former director of the École du Meuble, founded the Centrale d'Artisanat in Montreal, which operated as a strong exhibition and distribution agency promoting Quebec craft.

The rest of Canada looked to Quebec as a model. For example, in the forties, the Nova Scotia government formed the Handcraft Centre in Halifax, run by Mary Black, to cultivate a local craft culture.[3] Provincial governments also saw craft as a means for the sick and the wounded to get well.

LEFT: 3.1 Andrew Fussell demonstrates the craft of silversmithing in the fifties.

Black herself was trained as an occupational therapist. After 1945 schools like the Ontario College of Art received new provincial grants to expand their craft courses and make them available to returning soldiers.

Craft Pioneers Focus on Function

As crafts grew in popularity, there was increasing concern about maintaining professional standards. Committed professional artisans soon emerged, such as Erica and Kjeld Deichmann (FIG. 10.9) in New Brunswick, Karen Bulow (FIG. 8.4) in Montreal and Harold Stacey in Toronto (FIGS. 13.1, 13.4). They managed to eke out a living in pottery, weaving and metalwork respectively but still relied on private money, commissions and teaching to survive. The Deichmanns and Bulow (who were friends) represented the tradition of Scandinavians coming to Canada and forming their own companies to produce standard patterns as well as custom designs. Canada's most successful production-ware silversmith was Carl Poul Petersen (FIGS. 13.7, 13.8) in Montreal, and he came from a similar Nordic background. These early artisans were comfortable with the position that craft is both functional and artistic. Stacey, Petersen and Bulow all collaborated with industry at some point in their careers.

Associations and retail stores across the country contributed to the professionalism of handicrafts. Craftmakers' associations catered to their particular medium, facilitating the sharing of materials and equipment and improving standards. Early ones include the Canadian Guild of Potters, formed in 1936 (later known as the Pottery Guild), and the Metal Arts Guild of Ontario, established in 1946.

The Canadian Handicrafts Guild ran a shop in Eaton's College Street store from 1932 to 1937 and then moved to Bloor Street. In 1945 the Handicraft Guild's Ontario branch made its first serious attempt to link craft and industry by organizing *Design in Industry* at the ROM. Co-ordinated by the arts patron Nora Vaughan, who led the selection committee, the exhibition was intended "to show the place of the designer-craftsman in successful industrial production."[4] In fact, the three-week exhibit ignored industry, focusing on craft-based materials such as leather, silver and glass by the likes of Petersen, Bulow and the Deichmanns and unintentionally contributed to a rift between craft and industry. Donald Buchanan demanded to know "where were the aluminum sheets and the laminated wood?"[5] Even museum staff noted the conspicuous omission of Canadian ceramics manufacturers like Medalta Potteries and Sovereign Potters.[6] After this effort, the guild rarely united craft and industry in its programs. Its descendant, the Ontario Crafts Council, founded in 1974, continued to promote this separation. For the most part, the NIDC followed suit. Though it occasionally included craft in programs and exhibitions, linking craft and industry was never part of its mandate.

Craft Takes on Fine Art

Craft experienced its greatest growth in the late sixties and seventies, simultaneously forging a populist direction and an artistically elite route. New community colleges opened for Canada's centenary fuelled this surge. Ontario alone built twenty-one new schools, all offering craft instruction. Responding to the counterculture hippie movement, amateurs took up craft. Souvenir mugs and macramé plant holders became fodder for regional markets and souvenir stores.

The community college programs brought an influx of American and British craft teachers well versed in the new Studio Craft movement. Their impact was enormous and formed the foundation for a studio glass and furniture revival. They also repositioned the teaching of craft from material- and function-based concerns to artistic ones. By the eighties, potters explored issues of "containment," designing teapots not intended for pouring. Weavers became "fibre artists," creating installations for commissions and galleries.

The desire to elevate craft into the realm of fine art did not come without struggles. Sheridan College's School of Design (known as School of Crafts and Design after 1978) was a battleground for such concerns from its founding in 1967. It was initially financed by a grant from the Ontario Craft Foundation (the new provincial agency), but when the federal government provided additional funding, it insisted that Applied Arts and Technology be added to the school's name and mandate. Administrators, primarily industrial designers, also tried to disassociate the design school from its craft legacy. Robin Bush and Giles Talbot Kelly, who served as directors of the school, led the drive to maintain a strong industrial focus. Their leadership sparked a rebellion among faculty members. The crafts movement won over Donald McKinley, the school's founding director and originator of the furniture program, who had up to that point maintained a neutral position. According to Michael Fortune, Sheridan's most accomplished furniture graduate, the American-Danish designer James Krenov moved the department away from manufacturing issues and inspired a romantic vision of furniture making that emphasized the spirituality of wood and handicraft.[7]

Contributing to craft's fine art status, the federal government launched a program in 1964 to devote 1 per cent of the capital budget for public buildings to the arts, stimulating the market for tapestries and ceramic murals. The Massey Foundation, the Samuel and Saidye Bronfman Family Foundation and the Chalmers Fund have all played prominent roles in improving the profile of craftmakers as fine artists by either creating collections or establishing awards programs. Both the Massey Collection and the Bronfman prize winners' works are in the Canadian Museum of Civilization in Hull.

In the mid-eighties economic boom, new support emerged for craft as fine art. This led to the creation of the specialized Canadian Craft Museum in Vancouver and the Canadian Clay and Glass Gallery in Waterloo, which opened in 1992 and 1993 respectively. The influential private Prime Gallery, founded in 1979 in Toronto by Suzann Greenaway, has achieved notable success. In the nineties, it sold contemporary conceptual ceramics by Canadians to the Victoria and Albert Museum in London, including works by Steven Heinemann, Matthias Ostermann and Walter Ostrom.

Collaborations with Industry in Current Practice

In the postmodernist eighties, craftmakers began to embrace synthetic materials and digital technology. Despite this liberal attitude, few craftmakers collaborated with industry. Michael Fortune and Paul Epp, however, have managed to cross boundaries separating craft from industry. In the past twenty years, Michael Fortune has spearheaded the fine-art cabinetmaking tradition in Canada. Fortune calls himself a "designer-maker," a term first coined in Britain in the late seventies. A protégé of Thomas Lamb's, Fortune worked under his direction at Plydesigns in 1976, learning about moulded plywood. He adopted a craft approach when he apprenticed with Alan Peters in England (FIG. 6.58). He also acts as a "manufacturing consultant" and designs furniture prototypes to help underdeveloped communities establish new businesses using sustainable resources.

Paul Epp, by contrast, works entirely for industry, but not without misgivings. A student at the Sheridan College School of Design in the early seventies, Epp hoped to specialize in plastic furniture but switched to wood to be in step with the Canadian furniture industry, which had yet to convert to plastics. When Epp apprenticed with James Krenov, he discovered one-of-a-kind furniture. However, after he returned to Toronto, he worked for industry, including Ambiant Systems (FIG. 6.59). While Epp still dreams of making art furniture, Umbra in 1998 began mass-producing his celebrated Luna nesting tables in candy-coloured plastic, fulfilling his college goal.

In the early nineties recession, Canadian craft suffered some setbacks. Notably, the Ontario College of Art closed its hot-glass and enamelling programs and the Ontario Crafts Council was forced to sell its new building, which had been a showcase for craft installations. The positive outcome of these events has been that more craftmakers, like the Toronto glass maker Jeff Goodman (FIG. 11.5), are returning to production work (but still not commercial industry) as the means to support their artistic endeavours.

Studio Manufacturing Bridges Craft and Industry

The "studio manufacturer" has become an adventurous counterpoint to Canada's manufacturing industry. The term describes an individual or a collective that establishes a modest company which balances craft and industrial practices. Typically, the designer-manufacturer acquires finished components, or else subcontracts these to small specialized industries, then assembles and fine-tunes the product in his studio. Studio manufacturers are prevalent in the furniture and housewares industries. In the fifties, the architect James Donahue produced his modern upholstered lounge chairs (FIG. 6.20) from his house in Winnipeg. Jacques Guillon founded Modernart in Montreal to fabricate his nylon cord chair (FIG. 6.2).

Studio manufacturers continued to put down roots in industry over the next four decades. In the sixties, the Toronto designer Stefan Siwinski (FIGS. 6.35, 6.36, 6.46) founded Korina Designs, a small craft-based factory producing high-end furniture for the contract industry. Similarly, Gunnar and Lotte Bostlund (FIGS. 7.8, 7.9) from Denmark opened a factory on their farm in Oak Ridge, Ontario, to produce their designs for ceramic and spun-fibreglass lighting. In the seventies, Keith Muller and Michael Stewart ran Ambiant Systems, a marketing company, to co-ordinate the subcontracting of their award-winning furniture design (FIG. 6.48, 6.49), and the company eventually became a manufacturer in its own right. A decade later, Scot Laughton co-established Portico to manufacture small runs of coffee tables, lighting and seating. Today he designs for Keilhauer, Nienkämper and Umbra.

Since the nineties, Niels Bendtsen has run his Vancouver furniture company Bensen (formerly known as Inform) and produces modular upholstered couches, as well as tubular metal seating. Patty Johnson, a graduate of the furniture program at Sheridan College, eschews fancy custom pieces typically associated with craft commissions in favour of designs featuring modest materials like plywood and plastic laminate so her work can be industrially manufactured (FIG. 3.3).[8]

Virtu, Directions in Canadian Design, a private, not-for-profit annual national design competition established in 1984 by Esther Shipman and Allan Klusacek, has been instrumental in the development of studio manufacturing. Portico and Patty Johnson have participated in the forum to gain exposure, as have Michael Fortune and Jeff Goodman. Shipman, who runs the competition, includes the category "limited edition" to encourage new ideas and make entering attractive to craftmakers and studio manufacturers.

The benefits of operating a studio factory can be enormous, from exposure to actually securing new contracts from manufacturers. Since manufacturing encompasses many non-design issues,

BOTTOM LEFT: 3.2 The Mantis CD rack, 1992, propelled Pure Design from studio collective to manufacturer.

BOTTOM RIGHT: 3.3 Patty Johnson's stool/table, 1999, reflects a synthesis of craft and mass production.

including quality control, merchandising and distribution, many designers straddle the worlds of craft and industry.[9] Tom Deacon withdrew from his Toronto furniture company, AREA, to concentrate on design, but without AREA he would never have had the opportunity to meet Keilhauer, today his primary client. The notable systems furniture designer Jonathan Crinion, Toronto, also tried manufacturing his Bebop chair on his own but ultimately licensed the product to focus on what he does best, design. Alternatively, Thomas Lamb had a long career as both a freelancer and as an in-house furniture designer, first for Plydesigns in the seventies and then for Nienkämper in the eighties. By the end of his career, he was an entrepreneur co-producing his furniture designs in Malaysia. Geoffrey Lilge of Pure Design—a vibrant studio manufacturer that emerged in the nineties—no longer designs but serves as marketing director, commissioning products from outside the company and from designers such as Constantin Boym in New York City (FIG. 3.2). Pure is now a full-fledged manufacturer, perhaps more recognized for the designs of its freelancers.[10]

Some proclaim that designer-manufacturers have minimal impact on Canadian industry, but their contribution is not inconsequential. Studio manufacturing helps Canadian designers launch their careers and forge ties with industry. Self-employed designers create prototypes and by example set high design standards for manufacturing. By serving niche markets, they also provide custom products for architects and interior designers.

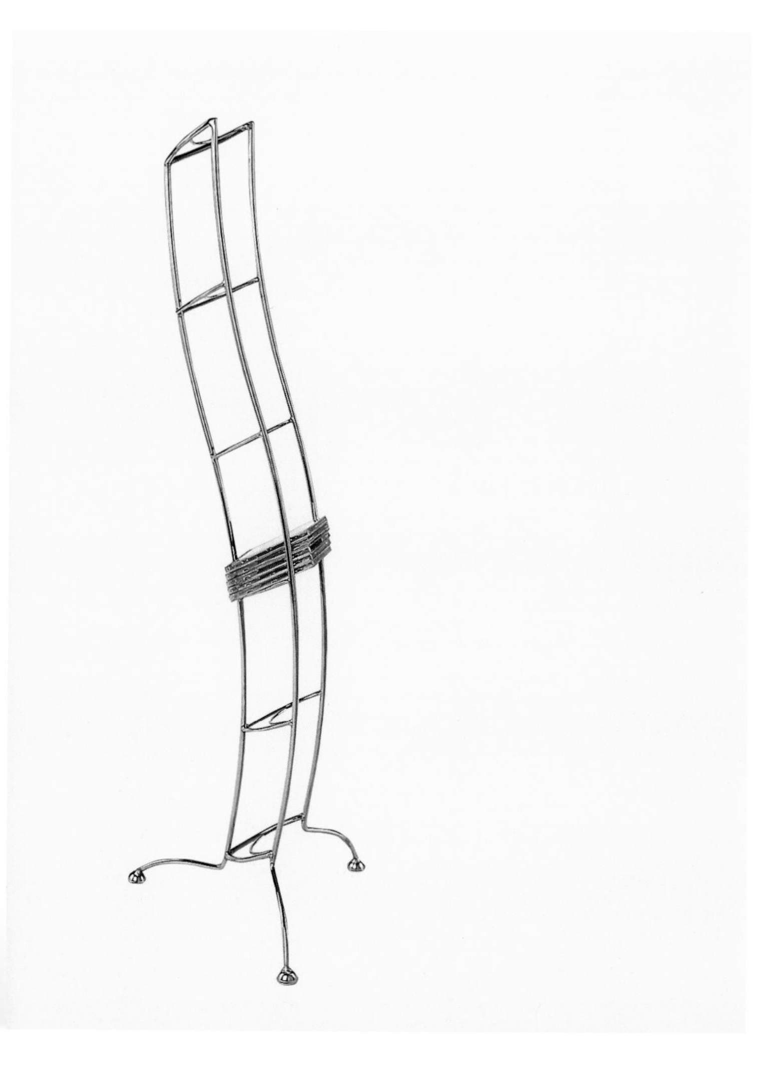

4 Canadian Design in the Pop Era

Youth Culture Adopts Playful Style

Modernism and Pop flourished at once in Canada. Short for "popular," the term *Pop* was coined in the early fifties. Often called an anti-design movement, Pop embraced society's consumerism, and it revered the ephemeral, especially in Day-Glo colours, supergraphics and disposable plastics. The swinging British led the wave of miniskirts and theoretical space-age architecture, while the Italians produced playful sculptural furniture. In many crucial respects Pop was anti-modernist. The movement rejected the "tired format of functionalism" of the International Style. Pop artists and designers scorned modernism's "good taste" and instead adopted the eclecticism of American consumer culture and the youth market.[1]

For Canadians, the Pop movement, nourished by economic prosperity, was really about the impact of the prevailing free-spirited approach to life. Pop styling emerged in Canada in the latter part of the sixties. Designers contributed to this dynamic scene, showing a lighter touch in their work and favouring flexible and sculptural forms.

The avant-garde Pop artist François Dallegret eliminated boundaries between sculpture, design and architecture (FIG. 4.2). A French citizen, Dallegret came to Montreal from New York City in 1964. He was friends with the fashion designer Paco Rabanne, inventor of the "micro-mini," and Reyner Banham, the British critic and supporter of Pop. With Banham, Dallegret collaborated on the hypothetical Unhouse, an inflatable plastic shell that celebrated mobility and the mechanical "guts" of architecture, including electricity and plumbing. First published in *Art in America*, Dallegret's drawing/collage of the Unhouse became an enduring image of Pop architecture.[2]

A preoccupying theme for Dallegret was "KiiK," which he identified as the energy created by opposites (FIGS. 7.13, 8.10). Like so many interventions of the anti-design movement, few of Dallegret's designs were built and most remained on paper or as prototypes.

The young swingers Peter Munk and David Gilmour established Clairtone Sound Corporation, Toronto, and by 1963, relatively early for the Pop movement, the company launched Hugh Spencer's Project G stereo (FIG. 9.15). Katia, a former go-go dancer from Montreal, became famous as the exotic model in the Clairtone catalogues—the later the catalogue, the shorter her skirt. She is often positioned on top, underneath or in front of the stereo while she fondles the ball-shaped speakers.

Space-Age Spheres Predominate

The Project G line was influential in the electronics industry, sparking interest in the motif of spheres. At Electrohome, Clairtone's Kitchener-based competitor, staff designers like Keith McQuarrie employed ball speakers and introduced a Plexiglas bubble top on its Apollo 861 stereo (FIG. 9.20).

LEFT: 4.1 The sphere appears most provocatively in Clairtone's G2 advertisements of the mid-sixties.

Space-age styling became a major component of Canadian Pop design, no doubt reacting to Canada's active role in the space race.[3] In 1962 the nation was the third (after the U.S.S.R. and the U.S.) to design and build a scientific satellite, the Alouette I. From the silver satellite to the astronaut helmet, the sphere was a prevalent icon in the technology of space exploration. It replaced previous forms of futuristic imagery, notably the starburst, nuclear rings and the Atomium, the molecular ball-and-stick monument of the 1958 Brussels World's Fair. Space-age styling was as definitive of the era as jet tailfins were of the fifties.

The sphere had become the utopian symbol for progress and change and spread from consumer electronics to lighting and interior design. In William Wiggins's burnt-orange Ball-B-Q (FIG. 4.3), the motif dominates the design. Toronto's Stefan Siwinski created plastic ball lamps. For Simpson's tony Arcadian Court restaurant in Toronto, John C. Parkin, the keeper of International Style modernism, specified three gigantic spherical chandeliers.

Although Dallegret's anti-art products and the upscale Clairtone G series of stereos sold in small quantities, boutiques and shops that espoused Pop styling sprouted up across Canada. In the seventies, Curved Space opened in Montreal, Vancouver and Toronto, selling beanbag chairs and other Italian inspired design. Terence Conran, the British arbiter of taste, located a Habitat shop in Toronto in 1968 and sold modular knock-down furniture. These cash-and-carry lifestyle stores transformed the purchase of furniture from custom-ordered future family heirlooms to youthful consumer products, ready immediately or assembled simply with a screwdriver.

National home shows and the Canadian media featured bachelor pads, bedrooms and recreation rooms in Pop styling. Deep pile carpeting, colourful chairs and transparent acrylic dome footstools represented just some of the elements of the new home decor. The Better Living Centre at Toronto's Canadian National Exhibition presented "The Pad" with vinyl coloured walls and Op art cube furniture. The media joined in. *Chatelaine* was "turned on" to rooms painted with big bold stripes and chevrons.[4] *Canadian Interiors*, the magazine for the trade, devoted editorials and features to designer apartments and family rooms with free-form furniture.

Pop Enters Homes and Commercial Interiors

Pop styling, largely an urban movement, entered the retail and hospitality industries by the late sixties. The ephemeral nature of the Pop movement suited the world of commercial retail, restaurants and nightclubs. Retailers like the Big Steel clothing stores used large circles for their window displays,[5] as did Toronto's Hazelton Lanes, by the architectural firm Diamond and Myers. The Electric Circus in Toronto, a Marshall McLuhan–inspired discothèque that opened in 1968, presented multi-screen computer-driven slide shows with live music. For the club, the architect Jerome Markson gutted two attached warehouses and linked them with continuous ramps and balconies.[6] Frank Davies, with his partner, Gloria Collison, formerly of Clairtone, supplied the supergraphics for the walls and co-ordinated the installation of a body-paint salon, adult playpens and padded "meditation" tubes. Even the developer George Minden, who gave Toronto the Scandinavian-inspired classic Three Small Rooms restaurant in the Windsor Arms hotel, shocked clientele with his new milestone restaurant Noodles (now demolished).[7] Designed in 1972 by Blakeway Millar, it introduced exposed shiny chrome ductwork, tangerine Op art ceramic walls and pink neon lights.

Not surprisingly, the illusory nature of Pop made it ill-suited for corporate architecture, nor was it typically found in residential architecture. One notable exception was the apartment

TOP RIGHT: 4.2 Le Drug Discothèque & Café, 1965, in Montreal evoked both cavernous womb and space-age labyrinth.

BOTTOM RIGHT: 4.3 The Ball-B-Q, 1970, funked up backyard cooking with its radical potbelly.

buildings by Uno Prii, who was known for his work in Toronto's downtown Annex district. The white concrete buildings are remarkable for their parabolic curved sculptural shapes and decorative balconies with Op-pattern metal strapwork.[8]

Fat Albert Lamps and Flower-Power Fabrics

As elements of Pop styling began to penetrate the broader Canadian market, mobility and figurative forms defined the movement. New design-focused companies like Danesco in Montreal and Interiors International Limited (IIL) in Toronto employed the second generation of designers trained in local design schools. The Image series furniture design by Keith Muller and Michael Stewart epitomized this new relaxed sensibility (FIG. 6.49).

Light, movable cube furniture and its variations were also popular. The Danish-born designer Christen Sorensen in Montreal was a master of upholstered furniture, an example being his experimental 1 + 1 series (FIG. 6.40). For the corporate market, Metalsmiths, Toronto, offered a modular cube furniture system; it was the company's best-seller for many years. Niels Bendtsen, the Vancouver designer, created the Ribbon chair when he returned to Denmark in the seventies. The tubular metal frame, wrapped with a quilted blanket cover, folded flat to facilitate shipping, and eventually more than thirty thousand were made. The Museum of Modern Art selected the design for its Study Collection. Mid-level furniture companies using less well known designers (or none at all) followed the trend. Henderson Furniture, working with the plastic moulding company IPL, both in Quebec, offered Go-go chairs in polyurethane foams.[9] Avanti Furniture Manufacturing, Montreal, also created commercial Pop furniture, shaping polyurethane cores into fantastical forms.

Youth culture brought a radical new look to the lighting industry. Electrohome produced an array of Pop lighting in polished plastics and painted metals. D. S. Griffin created "infinite mood" lamps for the company, and Michael Baldwin designed lollipop and mushroom lamps with large white Fat Albert bulbs (FIG. 7.3). Another popular shape was the cylinder pillar base designed to complement the fluorescent light tube, as in the case of Al Faux's and Gustavo Martinez's designs (FIG. 7.15). Even Douglas Ball, the champion of Eames modernism, dabbled in Pop with his candy-coloured acrylic Glo-Up lamps (FIG. 7.14) for Danesco.

Canada's small textiles industry offered Pop imagery. John Gallop, who ran Parkin's interior design division, created a bold flower-power silkscreened pattern for Toronto's J & J Brook (FIG. 8.8). Monique Beauregard and Robert Lamarre, two young designers based in Montreal, formed SÉRI + in the early seventies to cater to the taste for bold supergraphics (FIGS. 8.11, 8.12). For the office, textile manufacturers like Unifab, Montreal, created screens with optical-patterned fabrics by Maryanne Cain.[10]

Plastic Enters the Furniture Market

Plastic brought a dramatic new look to sixties furnishings. The rise of Italian design and its mastery of production technology profoundly influenced progressive Canadian designers. *Canadian Interiors* devoted an entire issue to plastic, noting that "Plexiglas furniture is leading the way," while *Canadian Home* identified this trend for "floating chairs and invisible tables" as "the big style news in furniture."[11] The Canadian furniture industry first used Plexiglas because its sheet form could more easily adapt to the industry's wood-panel machinery. Kaufman Furniture, Collingwood, Ontario, among the larger furniture manufacturers, capitalized on this trend. Known mostly for

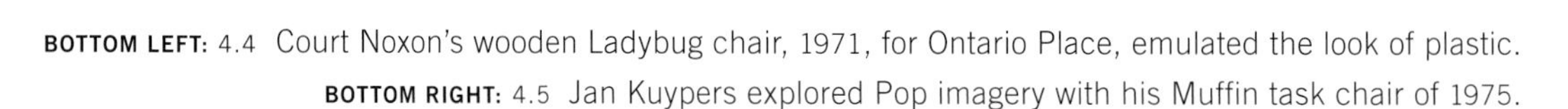

BOTTOM LEFT: 4.4 Court Noxon's wooden Ladybug chair, 1971, for Ontario Place, emulated the look of plastic.
BOTTOM RIGHT: 4.5 Jan Kuypers explored Pop imagery with his Muffin task chair of 1975.

case goods, the company worked with Sidney Gibson to create a line of transparent acrylic furniture.[12] Similarly, Stephen Harrison won a 1968 EEDEE furniture award (an Ontario provincial government award program for excellence in design) for his limited run of transparent acrylic tables made by Dunlea Plastics in Toronto. Acrylic was also embraced by boutique manufacturers like Toronto's L'image Design (FIG. 6.54)

Reyner Banham called plastic our "second skin," similar to fur. It was the "software" amid the space-tech "hardware."[13] Stefan Siwinski, the fastidious Polish émigré designer, captured Banham's sensual conceit with his remarkable design for an acrylic chair. An image of the plastic bubble chair (FIG. 4.6) by the Toronto photographer Wim Vanderkooy suggests the metaphor of plastic as bare skin: a nude woman sits demurely behind the chair, enhancing its transparency, its space-age quality of weightlessness and, of course, its eroticism. Siwinski's design showed the highest sculptural finesse, but he produced fewer than twenty.

Not equipped with the technology to make plastic furniture in volume, manufacturers began producing glossy lacquered wood as a substitute. Innovative examples include the previously mentioned Image series. Max Magder, who ran Du Barry Furniture in Toronto, eventually converted his Ovation line from solid wood to plastic components. Court Noxon of Metalsmiths produced the curved lacquered wood Ladybug (FIG. 4.4) and Minibug club chairs for Ontario Place in Toronto.

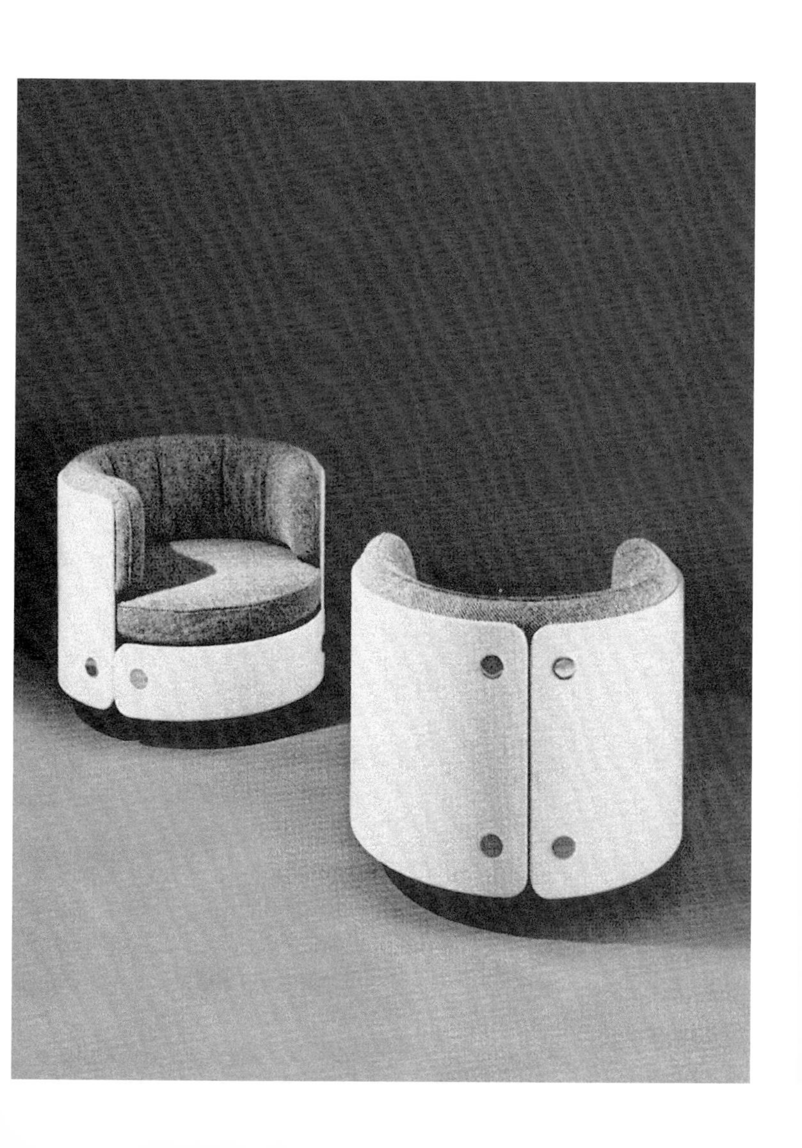

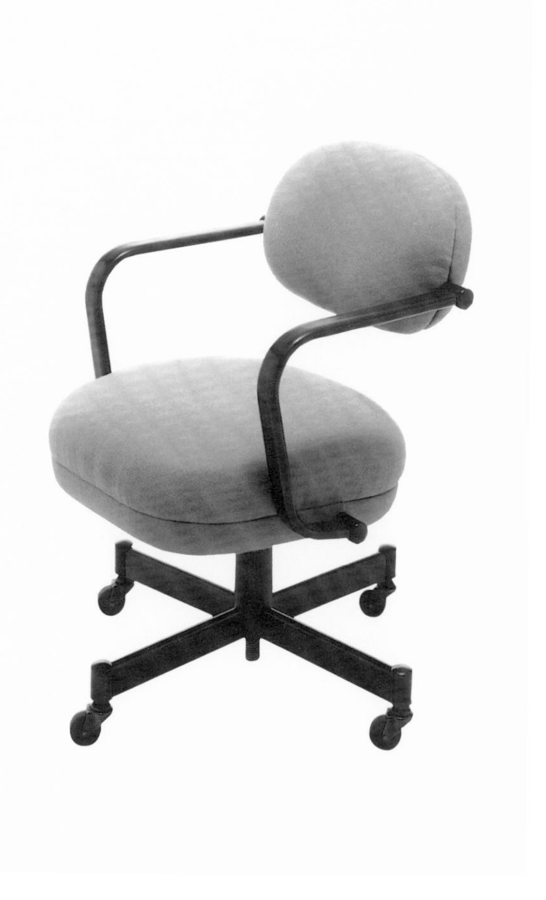

Notable Successes in the Seventies

Today Pop may appear naive and commercial, but its legacy is important. The celebration of spontaneous popular culture displaced modernism's elitist canon of "good design," preparing for pluralism. Although the wit and style of many Canadian Pop products achieved considerable international exposure, too few examples reached a wide audience. The seventies, in comparison with Canada's decade-long birthday party, seemed far less optimistic, affected by political changes in Quebec and a worldwide recession. Nonetheless, those Canadian designers who managed to overcome setbacks created iconic products that have endured.

The 1974 oil crisis had some positive impact on design. The American design ecologist Victor Papanek had launched a movement for socially responsible ecological design with his seminal 1971 publication *Design for the Real World*. Jerry Adamson, Toronto, of KAN Industrial Design, created an early example of ecological design with his electric kettle for Proctor-Silex (FIG. 12.10).

In the latter part of the seventies, other bold innovations emerged. Jan Kuypers, of the same firm, cultivated clients outside Canada, notably Harter, the American contract furniture company for which he designed the Pop-inspired Muffin task chair (FIG. 4.5) in 1975. Produced with advanced blow-moulding plastic technology, the chair, with its oversize stylish muffin-shaped seat, demonstrates Kuypers's best work since his designs for Imperial Furniture, Stratford, in the fifties. New talent developed, like Michel Dallaire, who designed the furniture for the Montreal Olympic Village.

In response to the weakened economy, Design Canada closed its public showcases, the design centres in Montreal and Toronto,[14] and transferred its funding to the Industrial Design Assistance Program (IDAP). Although this fund contributed to the creation of some exceptional products, it was controversial. For designers to become eligible for grants, they needed to be recognized by the government Record of Designers, a list controlled by bureaucrats.

Thomas Lamb's Steamer furniture reflected government design promotion's new strong-armed sponsorship of design in industry. In 1975 Design Canada (formerly the National Design Council) and the Nova Scotia Design Institute hired Lamb to create a renewal strategy for the hundred-year-old Dominion Chair Company in Bass River. He made his recommendations to upgrade its wood-bending technology and create an exportable Canadian furniture line. He then won the commission to develop it, which resulted in the now famous Steamer collection (FIG. 6.57).[15]

Douglas Ball identified the demoralizing and isolating effects of corporate office furniture. In 1976 he developed Race for Sunar Industries in Waterloo, a new office system that lowered and opened partition panels as well as hid cumbersome wires in "raceway" tracks (FIG. 6.5). The resulting royalty payments, an uncommon occurrence in Canadian design, were substantial.[16] The commercial and critical success of Race south of the border proved the importance of Canadian design to industry and prefigured what would become one of the most critical issues for design at the end of the century, free trade.

LEFT: 4.6 The swinging sixties arrived in a clear plastic chair, 1965, by Stefan Siwinski.

5 From Postmodernism to Pluralism

Design Captures the Public's Imagination

Toward the end of the twentieth century, most corporate presidents understood the value of design, and more sophisticated consumers enjoyed contemporary design. For better or worse, Canadian industry lost the protection of tariffs, forcing manufacturers to become tougher, leaner and more mechanized in order to export to the U.S. Meanwhile, the digital revolution put the personal computer—a powerful creative tool—into many hands and facilitated the new era of pluralism (anything goes) in design.

The eighties are often referred to as the Designer Decade. Big-name designers, such as the American architect Michael Graves, expanded their offerings to include everything from bedsheets to teakettles. Public awareness of prestige design increased, fuelled by conspicuous consumption and the further globalization of the design marketplace. After surviving an economic downturn, companies expanded, new ones formed and many hired local designers. Nienkämper ceased manufacturing Knoll furniture under licence and hired Thomas Lamb to design an original line. Danesco expanded its manufacturing arm by sponsoring Koen de Winter, a designer recruited from Holland. Ambiant Systems focused on international markets, securing a new manufacturing plant and opening two showrooms in the U.S. AREA, a small design-focused company in Toronto, got its feet wet producing Le Corbusier furniture, then moved into original Canadian design with the Gazelle chair by Jonathan Crinion (FIG. 6.61).[1] As a sign of the mergers and acquisitions to come, the American company Hauserman bought Sunar Industries in 1978.

Postmodernism Reinstates Decoration

During these flourishing times, postmodernism established itself as the dominant expression for design discourse. The American architects Robert Venturi and Denise Scott Brown argued that rational problem solving denied meaning and metaphor. Postmodernism spread from architecture to interiors and finally to product design, celebrating historicism, decoration and above all irony. In 1981 the renegade designer Ettore Sottsass founded the Milan-based Memphis group, which created irreverent, anti-functionalist product designs.

In Canada, the seventies architectural preservation movement represented an early expression of the postmodern thinking and reflected a new sensitivity to the past. A trend emerged to restore the exposed interior brick walls and wood plank floors of warehouses and skinny Victorian row houses. Gradually, postmodern idioms reached residential architecture.[2]

Postmodernism arrived officially in 1982, when the Toronto architectural partnership of Edward Jones and Michael Kirkland won an open competition, for Mississauga's City Hall,

LEFT: 5.1 Kerr Keller's Rocket pepper mills, 1992, for Umbra are part of the trend for novelty housewares.

a neo-classical monument to civic architecture (FIG. 5.2). It was soon followed by Paul Merrick's 1989 château-roofed Cathedral Place in Vancouver.

Commercial interiors picked up the trend. In the late eighties, the firm Yabu Pushelberg, Toronto, designed flamboyantly decorated boutiques and restaurants. Custom arts and crafts, art deco colours and patterned terrazzo floors were standard for such projects as Les Cours Mont-Royal shopping centre in Montreal in 1987 and Stilife nightclub in Toronto in 1990.

Exuberant Furniture Is a Mixed Blessing

For decorative, whimsical furniture, young Quebec designers led the way. Jean-François Jacques admired the amusing but garish designs of Memphis and created some overtly decorative pieces, like the Circus table of 1983, that spoofed the notion that form follows function by using playful shapes and colourful laminates. In Toronto, the newly formed studio Portico cultivated the practice of product semantics, symbolism and ritual in everyday objects. Founded by Scot Laughton, James Bruer and Scott Lyons, it advocated artistic expression rather than design as a commercial business tool. The Strala lamp (FIG. 7.17) remains the company's most recognized product and embodies many of these concerns.

To gain recognition, young designers typically submitted their work to Virtu, the annual competition. In its early days, selection committee members showed disdain for postmodern excess and complained about the overabundance of self-indulgent designs.[3] Nevertheless, notable designers did surface, including Helen Kerr, Karim Rashid and Martin de Blois. Although the award-winning designs were produced in limited quantities, the designers caught the attention of Canadian manufacturers, who would hire them a decade later.

Established furniture designers and manufacturers were slow to embrace postmodernism, but Thomas Lamb was an exception. Inspired by the traditionalist art furniture movement of the eighties, his 1982 handcrafted sculpted armchair for Nienkämper exhibits a strong art nouveau quality. Baronet Corporation, the high-volume residential furniture manufacturer based in Quebec, hired Vello Hubel to update its historical bedroom furniture and he acknowledged Memphis design (FIG. 5.3).

When a worldwide recession hit in the late 1980s and the Free Trade Agreement was ratified in 1989, all sectors of Canadian manufacturing suffered. Industry was forced to compete with a wave of less expensive imports from the Pacific Rim and the U.S. Anticipating the recession, Black & Decker shut its Barrie, Ontario, plant in 1985, ending the half-century-long production of Fred Moffatt–designed electric kettles. In the furniture industry, Canadian successes such as Sunar Hauserman and Ambiant Systems faltered and closed. Electrohome focused on the institutional market, closing its Deilcraft lighting and furniture division and its home stereo line. Not surprisingly, the smaller industries like ceramic tableware did not survive. Sial II, a relatively new Quebec pottery manufacturer, founded in the late seventies, soon closed, as did old-timers like Hycroft China in Alberta and Céramique de Beauce, also in Quebec.

Government cutbacks dealt another serious blow to design culture. The Conservative government of Brian Mulroney closed Design Canada. A promised new policy to have the professional design associations distribute public funding directly to designers never appeared.[4] Design awards and catalogue publishing decreased as grants to designers vanished.

Excessive ornamental design was seen as out of place in a recession. Even as postmodernist design made headway in Canada, critics dismissed "PoMo" as glib and nostalgic. In reaction,

TOP RIGHT: 5.2 The Mississauga City Hall, 1982, introduced postmodernism to Canada.

BOTTOM RIGHT: 5.3 Vello Hubel alluded to the Memphis movement with this eighties dresser for Baronet.

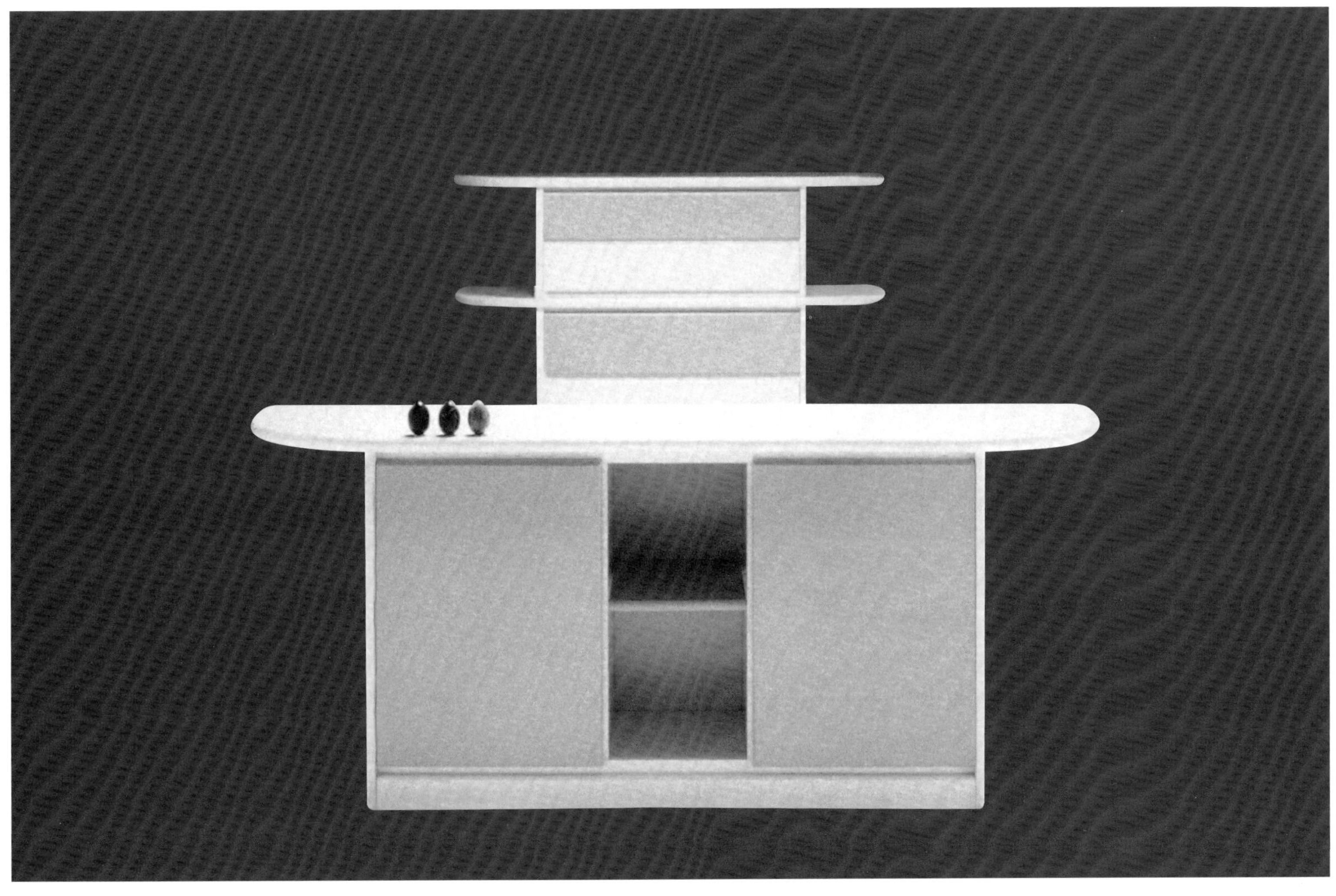

so-called deconstructivism made an appearance. Architecture was torn apart and then reassembled to find new forms. In a seminal exhibition in 1988, the MOMA established deconstructivism as a movement, presenting the architecture of Frank Gehry, Daniel Libeskind and other leaders. The media gobbled up the show, reducing the radical movement to a style celebrating semi-finished construction.

The impact of deconstructivism was minimal in Canada. The CBC Broadcast Centre in Toronto, by the Americans Philip Johnson and John Burgee, is a corporate example, while Raymond Moriyama's Bata Shoe Museum, also in Toronto, is the micro version.

Pluralism Increases Choice

The new era of pluralism and tolerance arrived. Local traditions and materials were to be embraced, but with a "critical" understanding of the "vernacular" or region, to stimulate new forms and imagery. Patkau Architects articulated these values in their Seabird Island Band School in Agassiz, B.C. The architects drew inspiration from the animal-lore tradition of the island community and used regional post-and-beam construction.[5]

With the advent of pluralism, designers object to being pigeonholed into a style or movement. Tom Deacon is a case in point. When he graduated from the University of Toronto School of Architecture in 1982, dissent raged between the modernists, led by Professor Peter Prangnell, and Professor George Baird's postmodernists. Deacon was a proponent of the former and, to this day, finds postmodernism "excessive, self-conscious pastiche,"[6] but his elegant designs owe something to pluralism as he favours traditional styles and reveals a sensitivity to historical precedent. One of his first products for AREA, the Spider side table, features witty, angular spider legs that terminate in copper ball feet. For Keilhauer's Danforth series, he combined the timeless look and comfort of the Boston rocker with ergonomic task chairs, employing a wood frame rather than the standard plastic or metal. With the Tom task chair (FIG. 6.64) of 1996, Deacon used moulded plastic but still managed to create a turn-of-the-century look in this otherwise high-tech chair.

Despite the criticism and confusion, postmodernism prevailed well into the nineties. For the Max chair of 1994 (FIG. 5.8), Nienkämper's design director, Mark Müller, updated the heavy, upholstered art deco chair by employing a simple wood-rail frame. Also for Nienkämper, Yabu Pushelberg designed the art deco Carlisle club chair and the whimsical Court Jester table (FIG. 5.5).

Postmodern romanticism and humour remained strongest among Quebec designers who catered to the restaurants and cafés of the province's lively urban nightlife. Jean-François Jacques created the capricious 1993 Bowling chair, named for its three-ringed bowling-ball holes bored into the seat back. Distributed by DISMO International, Montreal, the chair was one of many furniture designs commissioned by the Quebec government for the Casino de Montréal. Plouk Design, founded by Christian Bélanger and Jean-Guy Chabauty, dabbled in neo-medieval designs, like their velvet pillow chair of 1996, titled Fou du Roi. Designs by Plouk and other Montrealers weren't widely produced and rarely appear outside the province. Quebec's large residential furniture manufacturers produce in quantity but design with less panache.

David Burry, a transplanted Torontonian, continues to create and manufacture some of the most flamboyant designs through his firm, Design Emphasis, in Montreal, enjoying the city's relatively low overhead. Burry's interest in sculpted upholstery steered him away from Sheridan College's wood-focused furniture program to Toronto's Humber College. His oversize figurative furniture,

LEFT: 5.4 David Burry's Stiletto Shoe chair, 1998, appeared on *The Oprah Winfrey Show.*

including the 1998 Stiletto Shoe chair (FIG. 5.4), follows the Surrealist tradition of figurative furniture, like the Mae West Hot Lips Sofa by Salvador Dalí. Fabricated in shocking colours, Burry's designs suit the province's Continental flair.[7]

Design Resurgence in the Nineties

The early-nineties recession inspired baby boomers to stay at home and "cocoon." As the economy improved, Canadian manufacturers capitalized on this trend and produced designer housewares. Hothouse Design Studio in Edmonton created quirky curvilinear products in metal, including CD holders shaped like otter tails and clothes racks resembling English butlers. In the late nineties, Jeff Wilson of Ripple Design, Toronto, created Tong in Cheek salad servers, made into cartoon creatures.

Umbra emerged as one of the most important housewares firms by offering well-designed lifestyle products at affordable prices. In contrast to manufacturers like Danesco, the company understood that nesting meant design would no longer be confined just to the kitchen. In 1990 it hired Helen Kerr and her then partner/husband, Miles Keller. As principal designers for Umbra, they turned salt and pepper shakers into Rockets (FIG. 5.1), rubber doormats into Picket Fences (an ironic reference to suburban dwelling) and soap dishes into Fish and Pebbles. One of Umbra's most fruitful collaborations has been with Karim Rashid, and their fortunes rose together with the success of his Garbo collection (FIG. 11.22).

Rashid became Canada's most internationally recognized designer in the late nineties, but he needed to move to New York City to achieve this distinction. His relationship with Canadian companies like Umbra and Pure Design has permitted him to maintain strong ties to the country. Initially he made one-off pieces that gently parodied consumption and fashion trends. His 1989 chest of drawers called Viator (FIG. 5.7) launched his individual design vocabulary. He repeated Viator's voluptuous curves in his 1997 plastic purses for the Japanese fashion designer Issey Miyake, and the shape turns up again in his 1999 collection of crystal vases for the New Mexican company Nambé. Rashid admits that many of his themes derive from his personal obsession with "sensual minimalism."[8]

Computers Aid Design and Manufacturing

Designers' ability to manifest a strong style in mass-produced pieces owes much to the advance of computer technology. For example, through new versatile software programs, Rashid executes his organic style across a variety of media and products—from sheet glass and foam upholstery to aluminum alloy and plastic. As well as empowering designers with greater flexibility, computers offer three-dimensional modelling and reduce the number of cumbersome production steps. "Just in time" manufacturing processes allow for customized production and low inventories. Despite greater accuracy and consistency, the downside can be a certain sameness to designs.

The need for flexibility, versatility and mobility has transformed both the home and the office. Teknion's design department, led by the vice-president, John Hellwig, initially conceived its Ability furniture system of 1998 for home offices (FIG. 5.6). Hellwig realized that because of living-space constraints, large, expansive work surfaces ideally should disappear when not in use.[9] Ability includes a group of tables, desks and wall panels that glide on castors to allow some units to be stored under the table and to be instantly reconfigured. Ability's kidney-shaped table looks good from any angle and defies the notion that systems furniture must fit into right angles.

TOP LEFT: 5.5 The legs of Yabu Pushelberg's mid-nineties Court Jester table resemble the tips of a jester's hat.
BOTTOM LEFT: 5.6 Teknion's Ability furniture system, 1997, sells into the international market.

Competition Alters Global Markets

The global economy has profoundly reshaped design and manufacturing practice worldwide, increasing competition but also expanding markets. The result for Canadian design is new mobility: Umbra, for instance, manufactures in Buffalo, New York. The Trudeau Corporation, a producer of kitchen products in Quebec, uses local designers like Michel Morelli. All Trudeau's manufacturing takes place offshore, making the company a broker of design and manufacturers. Designers, too, work around the globe. Canadian-based ones like Douglas Ball and Jonathan Crinion design internationally, and others like Karim Rashid and Stephan Copeland work in the U.S.

Canada's undervalued dollar has given manufacturers a competitive edge, particularly with the American market. Wider distribution and greater industrialization have been crucial, since generally 80 per cent of the industry's market is the U.S. Keilhauer invested millions in machinery to produce high-volume runs for North America. Nienkämper (saved from bankruptcy) joined the American company International Contemporary Furniture (ICF) to broaden its distribution arm. Umbra has become a $100 million company and moved beyond housewares into residential plastic furniture, working with an even larger roster of Canadian designers.

The changing retail landscape may well encourage Canadian design. Department stores are being replaced by retail chains for home furnishings. IKEA opened its first store in North America in 1976, in Vancouver, and since then it has almost single-handedly expanded the market for contemporary design. Upscale copycats include Crate & Barrel and Pottery Barn in the U.S., which carry Canadian-designed products. In Canada, Roots and the now Ralph Lauren–owned Club Monaco have entered the world of branding home accessories.

For Canadian design and manufacturing, these high-volume stores have tapped into enormous and lucrative markets. Even discount chain stores want a piece of the action. High-end designers disappointed with the limited editions from the eighties have looked to such stores as a means of democratizing design. Often criticized for the lavish excess of his eighties products, Michael Graves now designs kettles and toasters for the American discount chain Target—quite a contrast to his $200 models for the Italian manufacturer Alessi. Companies like Umbra are attracted to this market, but fear losing brand identity and their fashion-forward reputation.[10]

Government Support for Design

Design promotion in Canada has similarly undergone changes, becoming more regionally based. Ontario and Quebec took over from the federal government's Design Canada in 1989. The Ontario Arts Council initiated a well-endowed Design Arts Program, but severe budget cuts reduced its potential impact. The Design Exchange, a not-for-profit institution that receives municipal funding, opened in 1993 in the former Toronto Stock Exchange. As well as promoting the commercial value of design, the institution documents and collects post-war Canadian design.

The province of Quebec vigorously supports its designers. The 1986 federal government Picard Report (on the development of the Montreal region) spawned the Institut de Design Montréal. It has become since 1989 the province's major funding source for innovative design projects. The provincial government also now offers a 40 per cent tax rebate to industry using local design. Since 1993 the Musée du Québec in Quebec City has become an active collector of industrial design from the province.

RIGHT: 5.7 The voluptuous shapes of Karim Rashid's 1989 Viator dresser reflect his sensuous obsessions.

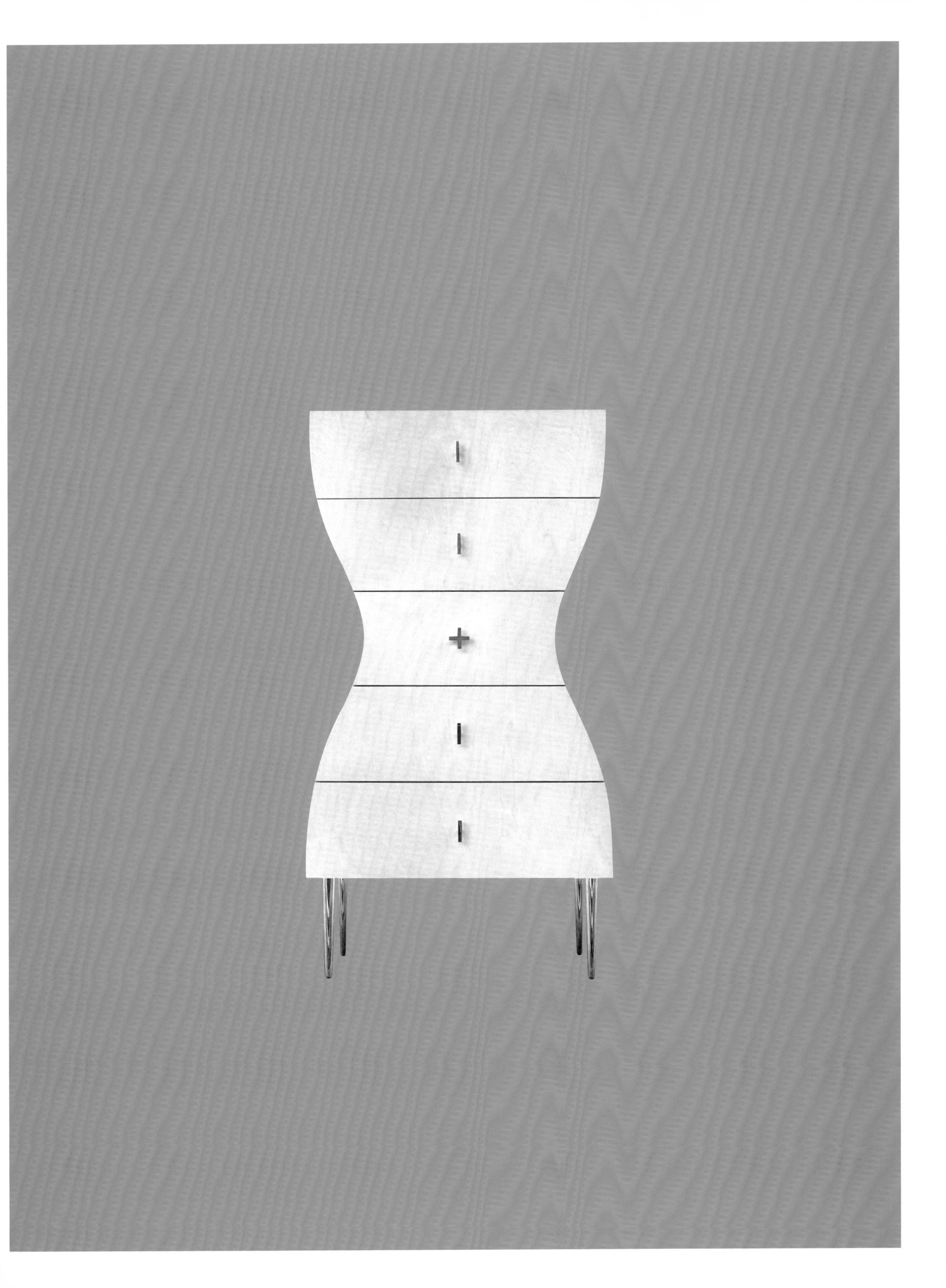

Design and the Environment

Some designers and manufacturers now exercise more restraint by using less material and recycling. Toronto played host to the prestigious Humane Village Congress for the twentieth International Council of Societies of Industrial Design (ICSID) in 1997. Its theme was designing for a sustainable world. Companies like Keilhauer (whose president, Mike Keilhauer, served as ICSID chairman) have adopted some responsible approaches to manufacturing—for example, it uses only wood from non-endangered, easily regenerated forests and avoids the pollutant process of chrome plating in favour of aluminum.[11] Bensen, the Vancouver-based studio manufacturer of residential furniture, has also shown some sensitivity to the environment, employing CFC-free foam in its upholstery.

Independent designers who produce their own work sometimes pursue measures of restraint by appropriating inexpensive found or surplus products. The results can represent an inspired handling of materials. The Toronto architect Andrew Jones has also designed with found objects, turning china teacups into pendant lamps and plates into wall sconces.

Still, environmental issues have had much less clout in the marketplace than might be expected. Buyers show enthusiasm for ecologically sensitive approaches, but often the drawback is price. Creating furniture using water-based (non-toxic) paints and organically tanned leather means higher costs for the consumer.[12] The current interest in translucent plastic in consumer goods also contradicts good environmental practice since only virgin plastic, rather than recycled, can be used to achieve this look.

Modernism Redux

Modernism has reasserted itself, as its original aim—good original design for the masses—seemed honourable and still true.[13] Today, the legacy of modernist thinking is an attraction to new abstract and simple solutions. Exploring lightness, transparency and form has become popular among current designers and architects. In 1999 Umbra commissioned the architects John Shnier and Martin Kohn to renovate the exteriors of their new headquarters. The architects framed Umbra's banal commercial box exterior with translucent plastic panels that glow when the lights are on at night, obscuring the boundaries between the interior and exterior. Similarly, with the 1999 translucent plastic Oh chair (FIG. 5.9), Rashid investigated the possibilities of new materials, production and design in the same spirit as the early modernists.

New interest in modernist theories have, of course, led to the phenomenon of a retro-neo-modernist style, most notoriously espoused by the magazine *Wallpaper**, whose editor, Tyler Brûlé, is Canadian. Re-released modern classics and new retro designs cater to the desire for minimalist, clean-surfaced furniture.

Nevertheless, no one style dominates a complex marketplace. Branding and global marketing versus design rooted in local cultures will certainly evolve as the two principal directions of design practice, and the debate between individual expression and ideology will persist. Manufacturing will also continue to move offshore, and more Canadian designers will leave the country to gain international recognition. This exodus is a reversal of the past: Canada now exports design talent rather than importing outside expertise. While no one can predict where design will lead us, the increased public response to contemporary design will continue to nurture the next generation of designers.

TOP RIGHT: 5.8 Mark Müller's 1994 Max chair for Nienkämper recalls the art deco era.

BOTTOM RIGHT: 5.9 Rashid's 1999 Oh chair plays with our perception of solid and void.

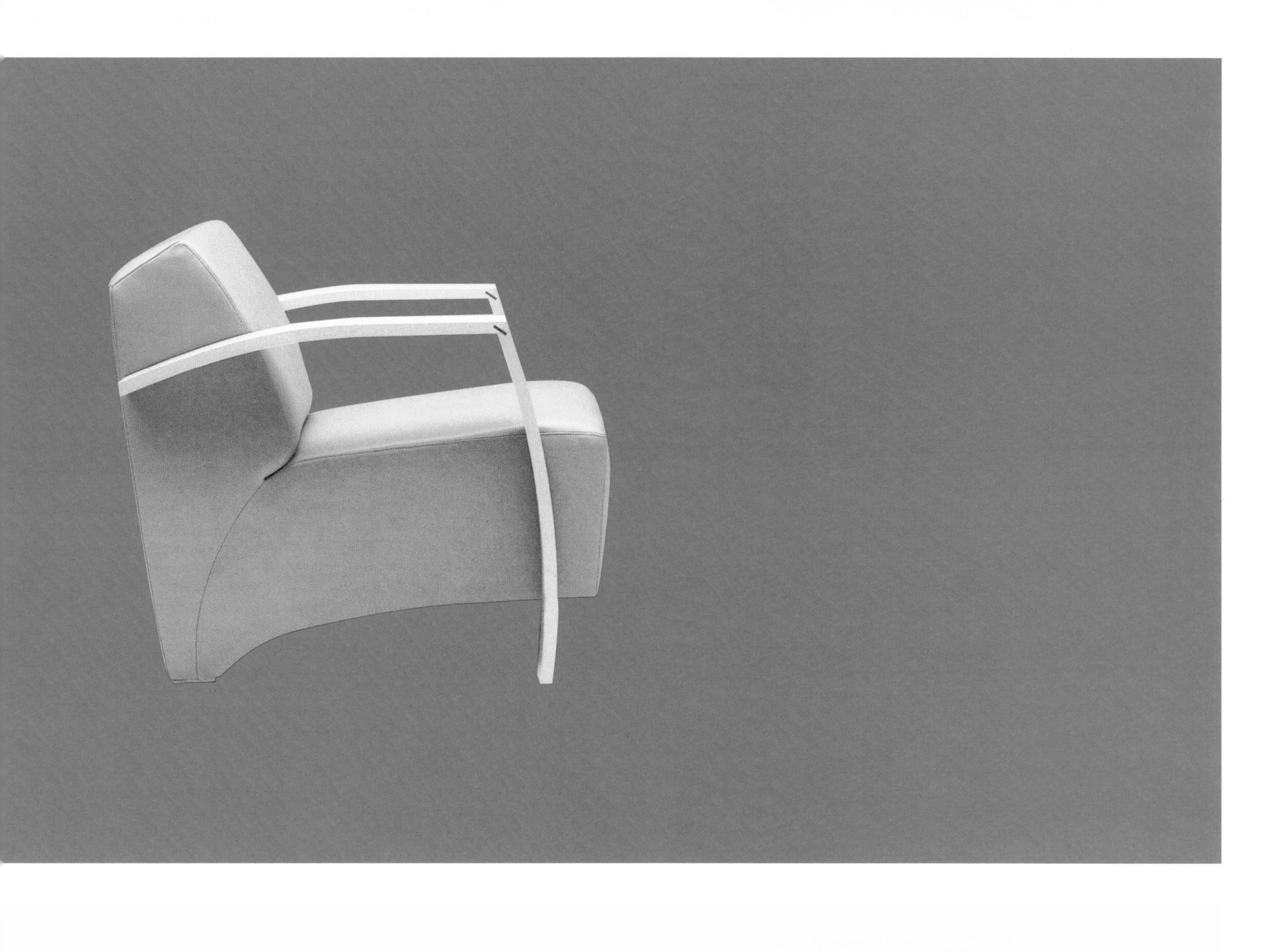

Surveys

6.2 **JACQUES GUILLON**

Cord chair

The architect Jacques Guillon's cord chair is a near-perfect balance of fragility and strength. Taut lines of parachute nylon cord, widely available as army surplus, hold its intersecting boomerang-shaped legs firmly in place in a rare feat of precise craftsmanship. The chair was professionally load tested and found to support 3,380 pounds (1,533 kg), an advantage the company then advertised.[1] Riding a wave of interest in "low-cost" furniture design, the chair was shown at the 1954 Milan Triennale, earned a U.S. patent (168,501 Washington) and was licensed to a company that, not coincidentally, manufactured wooden tennis rackets. It simplifies a 1942 patented design by Alexei Brodovitch, a New Yorker, by using its leg as the seat brace. The wood was available in natural or black finish and the cord seat in white or brown.

Tens of thousands of the chairs were constructed in Montreal and assembled in Greenwich, Connecticut, over a four-year period.[2] It sold at Morgan's and Ogilvy's in Montreal and at American department stores such as Macy's and Lord & Taylor, which advertised it in the Sunday *New York Times*. It appeared in the 1954–55 *Studio Year Book of Furnishing and Decoration* and was recently collected by the Art Museum of Dallas. At home, it won an NIDC award and is in the Liliane and David M. Stewart Collection at the Musée des Beaux-Arts, Montreal. Fuelled by Guillon's clarity of vision, this deceptively simple seven-pound (3kg) chair is a modern masterpiece.

1950

MANUFACTURER(S): Modernart of Canada, Montreal; A. A. Sporting Goods Corporation, Rouses Point, New York; andreef chairs, Greenwich, Connecticut
MATERIALS: natural laminated walnut and birch, brown nylon cord
DIMENSIONS: h83 x w46 x d38 cm (h32 ¾ x w18 x d15 in.)
MARKINGS: none

6 Furniture

Residential Furniture Design

The production of household furniture has been part of Canadian industry since Confederation. Until the mid-twentieth century, it was characterized by numerous small plants owned by local families, often for generations. Many of these artisan-style shops employed cabinetmakers in the guild tradition, with a single worker cutting, planing and finishing the lumber for each chair, largely by hand.

Initially, furniture plants set up shop in southern Ontario and southern Quebec to be near small hardwood forests containing black walnut, white oak and black cherry. (Contrary to expectation, most Canadian forests have few of the hardwoods suitable for fine furniture.)[3] As these woodlots diminished and other hardwoods such as walnut and mahogany came into favour, companies imported much of their lumber, primarily from the United States. The existence of manufacturing facilities and trained workers encouraged the industry to stay in these areas, resulting in a tradition that continues to this day.

Early manufacturers often specialized in one type of furniture: wooden case goods (tables, chairs and cabinets), upholstered furniture or bed frames. The case goods market was further segmented into the type of wood utilized on the shop floor. For example, the Krug Company, Kitchener, Ontario, principally used red oak; Vilas Industries, Cowansville, Quebec, used maple; and Imperial Furniture Manufacturing Company of Stratford, Ontario, used yellow birch.[4]

Of the few available local woods, maple was used for Colonial designs, birch was stained to resemble other woods, and elm (until it was wiped out by disease) was used for secondary or low-cost purposes. Designs emulated popular styles from other countries and periods such as Regency, Provincial and Chippendale. The only indigenous design produced in any volume was Quebec Habitant furniture, a local reinterpretation of Colonial style.

Furniture was handcrafted from solid wood, making it heavy and therefore expensive to ship. As a result, manufacturers catered almost exclusively to regional buyers. There was little incentive to look farther afield. For over a century, from the National Policy of 1879 until the 1989 Free Trade Agreement with the United States, the industry was protected from outside competition by tariffs ranging up to 30 per cent.

The Post-war Boom Alters the Market

The two world wars and the resulting material shortages would have an impact on the insular, somewhat complacent, furniture industry. Unlike the consumer electronics industry, furniture manufacturers didn't play a major role in wartime production. The exception was a few southwestern

PREVIOUS SPREAD: 6.1 Reclining on aluminum garden furniture in the fifties.

Ontario manufacturers like Electrohome/Deilcraft, Imperial Furniture and Knechtel Furniture Company, Hanover, Ontario.[5] Electrohome, for example, converted its entire furniture production line to make wooden aircraft components, principally for the Cornell Trainer, which was used by Commonwealth and U.S. forces alike. By the end of the war, more than one thousand pairs of wings and eight thousand fins had been made.[6]

The lure of furniture manufacturing was irresistible. Massive post-war family formation had created a huge demand. Meeting the need, at prices consumers could afford, required ingenuity. One solution was to manufacture less rarefied designs: inexpensive component furniture that could be added to the household piece by piece, over time and often on credit. Conceptually simple, it signalled a fundamental change in the way furniture was designed, produced and sold.

Rather than being laboured over by individual craftspeople, furniture was cut, formed and assembled on production lines by factory workers. To produce more furniture faster and at a lower cost, each person performed one simple task, a process that wrested control from the artisan and placed it in the front office. Production managers, sales and marketing staff and eventually industrial designers became the key forces driving the industry.

Educational institutions were slow to adapt, often electing to train classic cabinetmakers rather than factory workers. Montreal's École du Meuble (established in 1930) gained a reputation as one of the best cabinetmaking schools in North America. The Ontario College of Art (established as the Ontario School of Art in 1876 and now known as the Ontario College of Art and Design) added design and other applied arts to its curriculum in 1945, although it described the training as craft and hobby related. Toronto's Ryerson Institute of Technology (founded in 1948 and now called Ryerson Polytechnic University) was a rare exception. Its furniture and cabinetmaking course, introduced in the late forties, emphasized industrial production.

For the most part, the manufacturing sector had to look elsewhere for industrial design talent. A few companies turned to licensing American and European styles. By taking advantage of existing research and development, manufacturing templates and advertising, the companies could shorten the new product introduction cycle and eliminate some of the risk.

The practice of licensing designs to Canadian manufacturers was already well established. One of the most notable examples in the thirties was the American Russel Wright's authorizing Imperial Furniture to produce a line. In the fifties, Canadian firms licensed designs from American companies like Widdicomb Furniture, Grand Rapids, Michigan (residential case goods) and Laverne International, New York (plastic furniture). Other licensed designs included furniture for the Danish-American designer Jens Risom (Guildhall Cabinet Shop, Toronto), the Danish designer Finn Juhl and the American designer Paul McCobb (Rockland Furniture Company, Quebec).

The Designer's Role Is Enlarged

The licensing experience taught many companies the value of recognizable design. Donald Strudley, president of Imperial Furniture and a designer in his own right, hired a young Dutch designer, Jan Kuypers, to create contemporary furniture for his firm (FIGS. 6.25, 6.26). Less successfully, Eaton's recruited Magnus Werner, a young Swedish designer, but abandoned the experiment when it was still in its embryonic stage.[7]

One of the legacies of Kuypers's involvement in Canadian furniture design was the widespread introduction of modular construction and the use of standard components. Following his lead, a number of companies mass-produced furniture that was sold across the country. Snyder's, in

TOP RIGHT: 6.3 Canadian Wooden Aircraft used plywood to achieve an art deco look in the forties.

BOTTOM RIGHT: 6.4 Russell Spanner's Pasadena table, 1953, was popular with young families.

Waterloo, developed a line of wooden occasional tables that relied on interchangeable modular pieces and introduced sectional sofas to Canada (FIG. 6.16). Electrohome's Deilcraft furniture division automated its traditional furniture manufacturing processes and absorbed numerous smaller plants to achieve economies of scale. Kaufman's of Collingwood (now part of Krug) produced bedroom and dining room suites with clean contemporary lines, often by a designer named James Leithead.

The largest-volume Quebec producers, Vilas Industries and Roxton Mill of Waterloo, Quebec, amortized an investment in design across a family of products by using component pieces and recycling existing designs. Although most of their furniture was in a traditional vein, both added modernist versions to their lines in the fifties. In Toronto, Russell Spanner's success was predicated on making inexpensive mix-and-match furniture from a few standard components.[8] His designs are more sophisticated, given these rigid, inflexible parameters.

Ironically, some of the most advanced component designs and production techniques emanated from the much smaller and less capitalized studio design community. For example, the most successful production-run chair of the era was Jacques Guillon's cord chair (FIG. 6.2).

Experiments with Metals

While the efforts of these design pioneers were enthusiastically supported by the media (in particular, Margit Bennett, the editor of *Canadian Homes* magazine, and later David Piper of *Canadian Interiors*), retailers and buyers were more difficult to win over. The industry's accepted wisdom is that most residential furniture interprets historical styles—a trend that continues today.

The situation was different on the West Coast, which was in the process of reinventing itself as a modernist community. Much of the furniture created by the small shops was casual, perhaps as a result of the indoor/outdoor lifestyle afforded by the weather. With no tradition to uphold, designers like Peter Cotton (FIG. 6.17), Klaus Grabe and Robert Calvert were free to experiment with new forms and materials like metal rod.

In hindsight, it's easy to see that their choices were dictated by the lack of an established furniture manufacturing industry. Designers had to innovate using the materials and processes at hand. These limitations often restricted a company's growth, resulting in a flurry of short-lived, designer-run businesses. One reluctant Vancouver businessman, Robin Bush, would eventually head for Ontario's furniture heartland to achieve his design goals.

In Eastern Canada, a number of well-established foundries produced architectural work: railings, gates and fireplace tools. As metal rod became an acceptable material for furniture, many added lines, often designed in-house. The most successful were John Hauser Iron Works, Kitchener, and Metalsmiths, Toronto, which both continue to operate.

Design Enters the Classroom

The school and institutional furniture market, while not a hotbed of design innovation, provided steady income to support a production line, which in turn allowed companies to work in other areas. The detailed specifications inherent in institutional furniture also helped to increase manufacturing standards. Several Ontario-based firms—Imperial School Desk Company, Petrolia; ESA (Canada), Kitchener; and Canadian Office and School Furniture (COSF), Preston—produced huge volumes of school furniture, some of which survives to this day. The National Industrial Design Council awarded ESA for its stacking chair, and COSF became a subcontractor for the

LEFT: 6.5 Douglas Ball's RACE system, 1976, revolutionized office environments.

American modernist firm Herman Miller and later manufactured office and systems furniture designed by Robin Bush (FIGS. 6.32, 6.37, 6.50, 6.51).

On the East Coast, Archibald King and Balfour Swim were experimenting with other new materials. Using wood veneers and later plastics, they cornered the institutional market (FIG. 6.19).

By the end of the fifties, the gifted amateur Walter Nugent formed a successful company in Oakville built around a single sturdy chair design (FIG. 6.33). He simply created variations on one theme, altering frames and other elements to create site-specific chairs ranging from office reception chairs to retirement-home rockers.

A few companies clung to the artisan approach to furniture design and production. Swedish-born Sigrun Bülow-Hübe, along with her associate, Reinhold Koller, produced high-quality handcrafted furniture at AKA Works, Montreal (FIG. 6.28). In Toronto, John Stene upheld the painstaking craftsmanship learned in his native Norway (FIGS. 6.29, 6.30. 6.31), and Robert Kaiser, also in Toronto, designed some of the most innovative chairs to emanate from Canada (FIG. 6.24). Although the NIDC honoured many of the designs, few approached the production runs of their more pedestrian counterparts.

Contract Furniture Design

By the 1960s the role of industrial designers was well established, if not glamorous. If anyone still needed proof, Al Faux mounted a convincing argument with his radius tension drafting table. The Norman Wade Company, Toronto, built an international business based on the success of this design. The table was recognized by the prestigious Milan Triennale in 1968 for its simple but detailed design and engineering innovation. It was produced (with some alterations) for more than three decades and became a staple in corporate and educational drafting environments in North America.

The residential market was tougher to crack, so wily furniture manufacturers pursued commercial and institutional contracts. These communities, where architects and interior designers had the power to influence taste, were more sensitive to good design.[9] It helped that buyers were also accustomed to longer lead times (sometimes up to three months) when ordering furnishings from American firms like Herman Miller and Knoll. Canadian companies sometimes grabbed a toehold in the market simply by shaving three weeks off a delivery date.

Durability and suitability were often ranked higher than cost, as purchases were amortized over years of use. Also, local manufacturers were able to consistently deliver a high-quality product. In the rigorous modern environments created by the architects Mies van der Rohe and Viljo Revell, exactitude was necessary. "God is in the details," claimed Mies, and select Canadian manufacturers proved to be just as obsessive.

Three Torontonians—Stefan Siwinski (Korina Designs), Jack Dixon (Dixon Designs), and Court Noxon (Metalsmiths)—all placed product in the signature modern buildings of the decade (FIGS. 6.35, 6.38, 6.27). Using chromed flat-rolled steel, their designs blended harmoniously into the reductive International Style modern interiors. Still, designers of the era like Dixon could justifiably lament that American design filled the rooms and Canadian design "filled the empty spots in the corners."

Herman Miller and Knoll "owned" the organizations specifying corporate and institutional furniture. At best, Canadian designers created furniture for secondary areas. Frequently, their "original" designs were much more imitative than one might have hoped. Robin Bush managed

RIGHT: 6.6 Hugh Spencer's Club chair comforted Expo 67 dignitaries.

to have his modular office furniture line advertised (in Canada only) alongside the designs of such renowned Herman Miller alumni as Charles Eames, George Nelson and Eero Saarinen, at least providing a small marketing victory. He also doggedly pursued specifiers within the Department of Transportation, which became known for its support of Canadian design, and for a time seemed to control the market for airport furniture.[10]

Licensing Foreign Designs Endures

Eventually, both Herman Miller and Knoll established marketing branches in Canada, then subcontracted some component manufacturing and assembly to local suppliers. The subcontracting arrangement in turn led to a number of local companies producing the lines under licence, and the international experience became an important training ground in both manufacturing and marketing skills.

Companies like Electrohome, hardly an insignificant player, produced Knoll designs—including Florence Knoll metal-based desks, tables and credenzas. When the contract was allowed to lapse, the division's team leader, John Hauser, returned to the family foundry in Kitchener. Smaller firms like Metalsmiths and Nienkämper used the knowledge and experience acquired as licensees as a springboard to remake their companies using homegrown talent. Other firms, like Leif Jacobsen, built their businesses by specializing in the custom work that supported large-scale office makeovers: boardroom tables, reception desks and so on.

By the mid-sixties, the furniture industry benefited from considerable "Buy Canadian" advocacy. In advance of the "triumph" of Expo 67, the country cast off some of its self-doubt and looked more favourably on Canadian-made products. Designers had a much higher profile and wielded some influence beyond their field.

The institutional market (particularly post-secondary schools) remained strong throughout most of the sixties. Walter Nugent, Al Faux, John Stene and James Murray all produced versions of sleigh-based chairs, although these are lauded more for their sturdiness than their style. The more radical designs of Christen Sorensen (FIG. 6.40) and Luigi Tiengo, both in Montreal, occasionally appeared in student union lounges—perhaps as an attempt by the institutions to bridge the growing "generation gap."

Unfortunately, regional governments reacted against the out-of-province contracts and began to erect barriers to interprovincial trade. Publicly funded Ontario institutions could only buy from Ontario sources, and so on. This further fragmented a small market—again limiting opportunities to achieve economies of scale. The situation affected contract rather than residential furniture manufacturers, as that market depended on government orders.

Residential Design in the Sixties

The xenophobia did, however, give birth to a Quebec style that was popular within its stylistically more progressive home province. Owing a huge debt to Pop, H. Singer Furniture and other fashion-forward Montreal firms had fun with furniture aimed at the residential market. The youthful designs wrapped bright colours and fake-fur fabrics over sinuously curved forms, in the manner of the French designers Olivier Mourgue and Pierre Paulin.

The style was different from the austere minimalism of Toronto designers, much honoured by the EEDEE awards. Quebeckers created their own furniture award, the Trophée du Meuble to draw attention to its achievements.

RIGHT: 6.7 Scot Laughton's Juxta plastic storage modules for Umbra, 1999, suit the new loft living.

Beneath the surface, the residential furniture market, particularly high design, was flagging. The building boom of the sixties encouraged companies from seemingly unrelated sectors to enter furniture manufacturing. Firestone Tires, General Fireproofing and others produced furniture lines. The market became saturated, just as the demand for furniture declined.

By the end of the sixties, manufacturers were also trying to cope with the onslaught of mass-produced Scandinavian furniture. Working co-operatively, furniture manufacturers in Denmark, Sweden, Finland and Norway were able to deliver well-designed quality products at competitive prices. The Scandinavian style also captured the public's imagination—albeit via a well-oiled publicity machine, which organized exhibits and awards programs.

Small but influential retailers like Shelagh's of Canada in Toronto (launched in the mid-fifties) showcased Scandinavian design to the exclusion of much else. Toronto's Karelia International boutique (established in 1960) similarly stocked the best modern design from the same northern countries.[11] The Georg Jensen shop in Toronto—chock full of Danish products—became the last word in good design after opening in 1957.

A few companies in Montreal, such as Punch Designs and Huber Manufacturing, gamely reproduced mid-priced Scandinavian style. Their designs were marketed as pseudo-Danish teak by major department stores—principally on the basis of lower cost. One of the few female designers of the period, Elizabeth Honderich, Montreal, earned an NIDC award in 1964 for a Scandinavian-style teak buffet-sideboard.

Some firms counteracted the onslaught by producing design alternatives in steel tubing. Kinetics Furniture, Toronto (FIG. 6.55), and Amisco, Montreal, worked with designers to create tubular chairs, stools and loungers, some of which are still in production. The highest-profile designer of the sixties and seventies, though, was Thomas Lamb. His first major project was metal outdoor furniture for the Bunting Furniture Company in Philadelphia.

Contract Furniture Looks South

In the commercial sector, systems furniture was beginning to take hold. Championed by professionally trained and thoughtful designers like Douglas Ball, it would emerge as a Canadian success. With better marketing and distribution, Canadian firms like Sunar Industries penetrated the U.S. market. It was a psychological victory, but more important, it created enough production volume to fund more research and development, which led to new and better designs.

In the eighties, both Jonathan Crinion and Manfred Petri would build on the framework established by Ball, and both would achieve considerable success in the international market. Crinion's systems furniture clients include Knoll in the U.S., Tecno in Italy and Teknion in Toronto. Petri designed office case goods and systems for Toronto-based Geiger International (formerly IIL) and helped to lead the firm into the U.S. It now operates as Geiger Brickel from Atlanta, Georgia, and continues to produce some of Petri's lines.

With the advent of the FTA, any remaining tariff protection was eliminated in a five-year process culminating in 1993. In the absence of protection, nimble companies focused on niche markets such as case goods or upholstered furniture, then sold the specialized items in Canada, the U.S. and even offshore.

Toronto-based executive furniture companies like Keilhauer and Nienkämper, as well as systems furniture companies like Reff, Toronto, and SMED International, Calgary (now owned by Haworth, in Holland, Michigan), have prospered under that strategy. It demands skilled shop-floor workers, intensive automation and the ability to change over the production line at a moment's notice. Other companies, like Brunswick Contract Furniture in Toronto (a continuation of the firm founded by John Stene), chose to concentrate on local markets and continues to make custom boardroom tables, credenzas and other cabinetry.[12]

Specialization permitted the companies to invest in design, research and development, and marketing—oft-cited Canadian corporate weaknesses. By gaining experience in a particular market and continuously expanding it, the best Canadian furniture manufacturers have finally achieved a measure of equilibrium in what remains a highly competitive field.

BELOW: 6.8 Baronet stakes their territory in the bedroom and updates classics with designs by Martin de Blois in 1999.

6.9, 6.10 **WACLAW CZERWINSKI AND HILARY STYKOLT**

Dining chairs

In 1946 this laminated-wood-and-moulded-plywood chair (with matching table) was shown three times in *Canadian Homes and Gardens*, twice in featured residences. Three years later, it reappears in support of a National Gallery exhibition, *Better Designs in Canadian Products*. By 1950 *Canadian Art* magazine included it in its pages, adding to the cachet. Comfortable, functional and attractive, a set in original condition (unpainted) is a sought-after collectible. Alternative versions sometimes appear, including an example with back strapping, shown at right.

1946

MANUFACTURER: Canadian Wooden Aircraft, Stratford, Ontario
MATERIALS: moulded plywood, bent laminated wood
DIMENSIONS: h 84.7 x w 41.2 x d 83.2 cm
(h 33¼ x w 16¼ x d 32¾ in.)
MARKINGS: sometimes stamped Aero club

6.11, 6.12 **WACLAW CZERWINSKI AND HILARY STYKOLT**

Lounge chairs

Waclaw Czerwinski and Hilary Stykolt were inspired by the Finnish architect-designer Alvar Aalto's 1932 cantilevered lounger, commonly known as the Springleaf armchair. Their designs, similarly made from moulded plywood and bent laminated wood, may be a less sophisticated handling of the material, but the commitment to modern design represents a remarkable attempt by a company formed to produce war-plane parts to ease into domestic production. Although few were manufactured, collector interest in moulded plywood furniture ensures a strong demand for any examples of the lounge chair that are discovered.

1946
MANUFACTURER: Canadian Wooden Aircraft, Stratford, Ontario
MATERIALS: moulded plywood, bent laminated wood
DIMENSIONS: cantilevered lounge: h70 x w68 x d66 cm (h27 ½ x w27 x d26 in.); lounge: h72 x w67 x d72 cm (h28 x w26 x d37 in.)
MARKINGS: none

6.17 **PETER COTTON**

Spring-back dining chair and high-back armchair

Peter Cotton produced economical furniture that displayed the formal qualities of mass, line, texture and proportion. His versatile spring-back chair was easy to make and, at only fourteen pounds (6.4 kg), inexpensive to ship. While it's commonly assumed that the frame is wrought iron, it is in fact electrically welded steel rod, matte finished with black lacquer. The chair was selected for Toronto's edition of the highly publicized Trend House series of modernist show homes in 1952. Fleury Arthur and Calvert (architects for Trend House in Toronto) also ordered the furniture for their Wymilwood residence at Victoria College.[16]

It was shown in the Ottawa Design Centre's inaugural exhibition *Canadian Design for Living* (1953) and, along with Cotton's desk, appeared in London's *Decorative Arts* annual (1954–55). Cotton adapted the design into a high-back model with wooden or upholstered armrests and won an NIDC award in 1953. A "pub" model replaced the upholstered back with half-height, semicircular moulded plywood. The chairs sold for $30, $35, $75 and $60 respectively. Hardwood-veneered plywood frames were available in walnut, oak, Primavera, eastern birch or Honduras mahogany.

Spring-back, 1951; high-back, 1953
MANUFACTURER: Pion Ornamental Iron Company for Perpetua Furniture, Vancouver
MATERIALS: black lacquered steel rod, hardwood veneered plywood, foam rubber, upholstery
DIMENSIONS: spring-back chair: h 71 x w 35.5 x d 39 cm (h 29 x w 22 x d 19 in.); dimensions for high-back chair unknown
MARKINGS: unknown

6.18 **JAMES DONAHUE AND DOUGLAS SIMPSON**

Plastic chair

Influenced by the fibreglass and glues used in Second World War fighter planes, the architects Douglas Simpson and James Donahue formed ten layers of fibreglass into a one-piece moulded plastic chair in 1945–46. This was four years prior to the popular fibreglass chairs of the DAR series by Charles Eames (originally designed and presented with metal shells and legs) and sixteen years before the first mass-produced one-piece plastic school chair by Marco Zanuso and Richard Sapper in Italy.[17]

The lightweight stacking auditorium chair had no joints or attachments and was made from glass-fibre-reinforced cotton and synthetic-resin adhesives that were then baked at 350 degrees Celsius. The National Research Council, Ottawa, provided technical support. The chair was exhibited in *Design in Industry*, a crucial launching pad for Canadian products, but was never produced.

1946
MANUFACTURER: National Research Council, Ottawa
MATERIALS: moulded fibreglass
DIMENSIONS: unknown
MARKINGS: unknown

6.19 **ARCHIBALD KING AND BALFOUR SWIM**

Coastline chair

The post-war demand for site-specific furniture resulted in a number of functional designs that are remarkable for their resilience. Archibald King and Balfour Swim's multi-purpose desk chair (designed with the assistance of the National Research Council) was ubiquitous in schools and institutions. The designers maximized the potential for the Coastline chair via frequent reincarnations: it was outfitted with arms, shrunk to children's size and built into a school desk. Production runs were as high as forty thousand chairs a year. Later versions had leatherette upholstered seats in brown, maroon, red, blue, and light and dark green.

Although the chair owes a debt to Alvar Aalto, there's something Canadian about a natural wood (rather than plastic) auditorium chair. The Coastline chair (and its successors, Scotialine and Roseway) were in production for decades. Ven-Rez Products now makes steel and plastic furniture, but a faithful customer will occasionally stimulate the original 1948 veneer press into action. Early versions of the Coastline often turn up at rummage sales and are dragged out for active duty—a testament to a smart, rugged little chair.[18]

1947; manufactured 1948 onward
MANUFACTURER: Ven-Rez Products, Shelburne, Nova Scotia
MATERIALS: native birch veneers, natural lacquer finish
DIMENSIONS: h81.7 x w41.6 x d51.8 cm (h32¼ x w16¼ x d20½ in.)
MARKINGS: stamped on underside of seat: Ven-Rez, Shelburne, NS

6.20 **JAMES DONAHUE**

Lounge chair

Affectionately known as the Canadian Coconut, in homage to George Nelson's 1955 Coconut chair, or alternatively the Winnipeg chair, Donahue's lounger was produced in his basement with the help of his architectural students from the University of Manitoba.[19] This resulted in some product variations like differences between the compound curves of its plywood shell, handmade wooden or rubber shock mounts and, occasionally, ungainly welds on the metal rod legs. There is also a version that features a Scandinavian-style wooden base.

About two dozen examples of what has become known as the Winnipeg chair exist, suggesting total production volume of approximately two hundred units. It was upholstered in fabrics in colours like mustard, orange and lime green. When the Hudson's Bay Company and Morgan's department stores began selling the chair, it was priced at about $35. Today, in mint condition, it can command thousands.

Late 1940s, early 1950s
MANUFACTURER: James Donahue, Winnipeg
MATERIALS: ¾" bent fir plywood, ⅝" metal rod, 4" foam; upholstery
DIMENSIONS: h76 x w87 x d79 cm (h30 x w34¼ x d31 in.)
MARKINGS: none

6.21 **HUGH DODDS**

The Dodds stacking chair

Hugh Dodds's "workmanlike" institutional chair won third prize in the second annual NIDC Wood and Aluminum Competition, 1952, losing out to Lawrie McIntosh's prototype (BELOW). Nonetheless, Dodds created a vibrant design, stylishly positioning exposed screws and introducing a dynamic bull-nose curve to the seat front as well as the rear legs. The design almost solves the problem of the all-wood stacking chair, but its reinforcement slats under the seat and back actually make it difficult to stack. The Dodds chair went into mass production, first by Aero Marine Industries and later by Curvply of Orono, Ontario. The architect George Englesmith played a significant role in facilitating the production of the chair, specifying them (with arms) for the Toronto typographers Cooper & Beatty. The installation was illustrated in *Canadian Architect*, July 1956.

1952
MANUFACTURER: Aero Marine Industries, Oakville, Ontario
MATERIALS: laminated plywood
DIMENSIONS: h77 x w38 x d41 cm (h30¼ x w15 x d16¼ in.)
MARKINGS: stamped under seat: The DODDS Stacking Chair Made in Canada by Aero Marine Industries, Oakville—Ontario

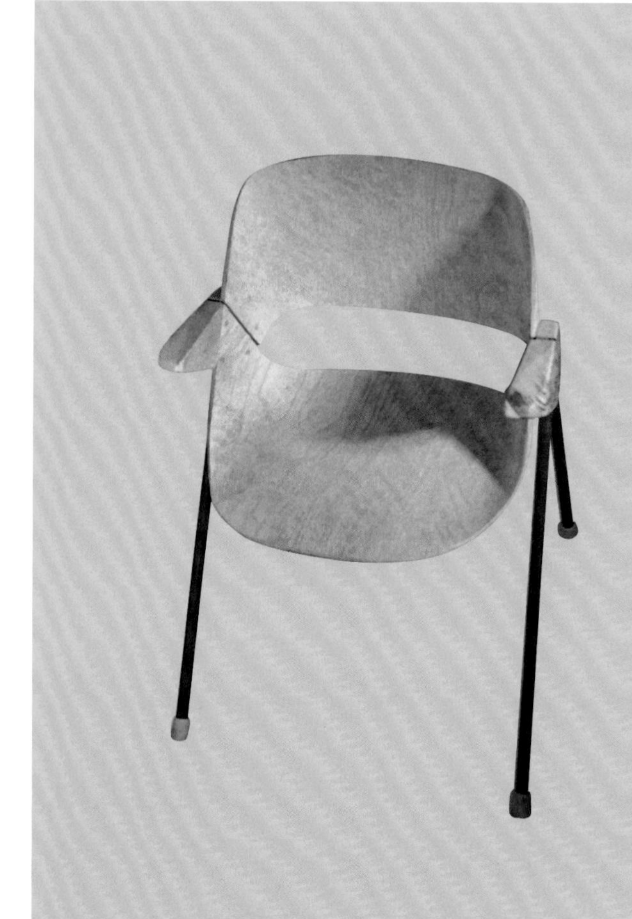

6.22 **LAWRIE MCINTOSH**

Chair

For most of his career, Lawrie McIntosh ran a busy industrial design practice specializing in product design rather than furniture, but he first made a name for himself with his moulded plywood pull-up chair. The stylish chair features distinct curves to its back and seat that in profile form a basket. The design captured top prize of $1,000 at the second annual NIDC Wood and Aluminum Competition in 1952. The jury believed that "its simple structural system...should result in easy fabrication and reasonable price." It did not. A $450 grant to build the dies went to Aero Marine Industries (ironically owned by the competition's third-place winner, Hugh Dodds), but the design never went beyond a hundred-unit production run. The majority went to a single client, the Toronto retailer Dolci Shoes, whose owner had seen the chair in *Design*, a British magazine. George Nelson of Herman Miller sat on the NIDC jury and subsequently included the design in his influential 1953 book *Chairs*.[20]

1952
MANUFACTURER: Aero Marine Industries, Oakville, Ontario
MATERIALS: moulded plywood seat, solid wood arms, steel tube legs
DIMENSIONS: h74 x w53 x d41 cm (h29½ x w20¾ x d16¼ in.)
MARKINGS: stamped underneath seat: Aero Marine Industries, Oakville—Ontario

6.23 **JULIEN HÉBERT**

Contour chair

The sculptor, designer and philosopher Julien Hébert was a brilliant manipulator of aluminum, as evidenced by this indoor-outdoor chaise longue. It consists of two undulating tubular forms resting on a triangular base that also functions as an armrest. It is stable in two positions: balanced on its base or lowered with its foot on the ground. Nylon or canvas covers were available in red, green, royal blue and gold. The lounge was exhibited at the 1954 Milan Triennale and that same year appeared in Milan's prestigious *Domus* (November 1954) magazine and London's *Decorative Arts* annual (1954–55).

Purportedly the design evolved after Siegmund-Werner, the company that manufactured aluminum ski poles, hired Hébert to design furniture to help the firm bridge several bad snow seasons. Their collaboration resulted in a complete line of garden chairs and five NIDC awards. More than a hundred thousand units of garden furniture were made. The Contour chair is the perfect synthesis of Hébert's design philosophy: inexpensive, practical and production ready but with a sculptor's eye to magnificent form. The Royal Ontario Museum has an example in its collection.

1951
MANUFACTURER: Sun-Lite Indoor-Outdoor Furniture, Siegmund-Werner, Montreal
MATERIALS: aluminum tube frame, canvas cover
DIMENSIONS: h86.4 x w56.5 x length 153.7 cm (h34 x w22 ¼ x length 60 ½ in.)
MARKINGS: none

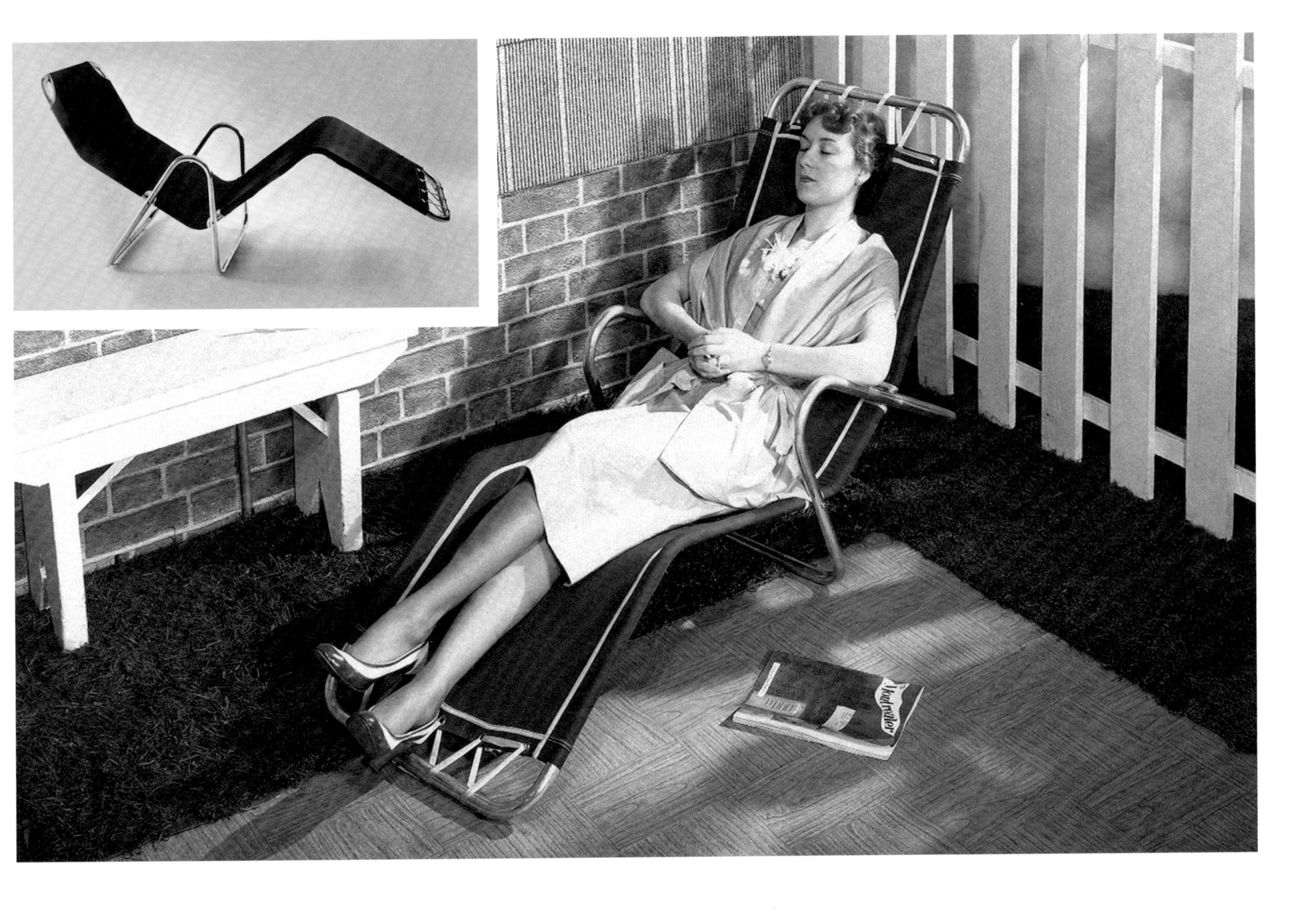

6.24 **ROBERT KAISER**
Armchair

This steel-rod-and-turned-walnut armchair reflects Robert Kaiser's interest in Scandinavian modernism and his training in art. Contrasting wood and metal give it stature as well as the flexibility to inhabit both traditional and contemporary settings. It appeared in *Canadian Architect* in Imperial Oil's head office (1957), in *Canadian Homes and Gardens* in the designer's own residence (1958) and in the *Decorative Arts* annual (1957–58). It also won an NIDC award in 1956, appeared at the Milan Triennale in 1957 and was exhibited at the Brussels World's Fair in 1958.

Kaiser produced more than a hundred designs, though many didn't go into production. Sought by collectors, his armchair and a similarly styled dining chair rarely appear in the market, as production runs were small.[21] The Royal Ontario Museum holds one in its collection.

1956
MANUFACTURER: Primavera Design Group, Toronto
MATERIALS: steel rod, turned walnut, wool upholstery, canvas webbing
DIMENSIONS: h 76.8 x w 66.5 x d 72.4 cm (h 30¼ x w 26¼ x d 28½ in.)
MARKINGS: none

6.25 **JAN KUYPERS**

Nipigon armchair

Kuypers briefly abandoned his penchant for minimalism and introduced a playful note with this armchair for Imperial's Contemporary line. Legs angled like the points of a compass were all the rage in the late fifties. Kuypers's version consisted of a birch frame, moulded plywood seat and back (in either birch or walnut veneer) and stubby arms, making it somewhat similar to Lawrie McIntosh's 1952 chair (FIG. 6.22). It was available with or without arms and sold for $40. It won an NIDC award in 1957, was featured in the Canada Pavilion at the 1958 Brussels World's Fair and appeared in an issue of *Canadian Art* in 1959.

1956
MANUFACTURER: Imperial Furniture Manufacturing Company, Stratford, Ontario
MATERIALS: solid yellow birch frame, moulded plywood seat and back
DIMENSIONS: h76 x w54 x d43.5 cm (h30 x w21 x d17 in.)
MARKINGS: metal label on frame underneath seat: Imperial Contemporary

6.26 **JAN KUYPERS**

Helsinki vanity/desk

Jan Kuypers had a special affinity for storage units. Many of his residential design series had multiple variations of matching buffets, cabinets and shelves, but those for the bedroom were especially notable, in part because "it has few, if any, inherited items," giving buyers the freedom to purchase modern objects. For growing postwar families crowded into small spaces, he designed night tables with drawers, headboards with bookshelves, double-duty desk/vanity tables and long, lithe chests of drawers. This vanity/desk, with optional mirror, was part of the Helsinki bedroom group, which included a bed, five dresser variations and a bedside table. Made from yellow birch, the pieces were largely unadorned, save for an angled, stepped motif on the front surface. The open-stock line of small-scale furniture was well received, and Kuypers followed up with two similarly styled bedroom and dining room groupings: Oslo and Stockholm.[22] Donald Strudley, the firm's president, designed the chair.

1952
MANUFACTURER: Imperial Furniture Manufacturing Company
MATERIALS: yellow birch
DIMENSIONS: h76 x w132.5 x d42 cm (h30 x w52 x d16½ in.)
MARKINGS: stamped on back: Imperial Contemporary m3568

6.27 **COURT NOXON**

Sun lounge 2750

Court Noxon used angles to create a low-slung lounge that first appeared in Toronto's Four Seasons Motor Hotel (and later at the Inn on the Park).[23] The dramatic lounge's solid redwood seat and back are supported by black squared steel tube frames. The design showcases a "second generation's" skill at manipulating metal to create high-quality furniture. For comfort, the seat and back are curved, creating a strikingly modern silhouette. The line, which included a matching chair and ottoman, was exhibited at the 1964 Milan Triennale.

1954
MANUFACTURER: Metalsmiths, Toronto
MATERIALS: square steel tube and redwood
DIMENSIONS: h40.6 x w60.9 x length 182.8 cm (h16 x w24 x length 72 in.)
MARKINGS: none

6.28 **SIGRUN BÜLOW-HÜBE**

Cocktail table No. 1244

With her horn-rimmed glasses and prim dresses, the formidable Sigrun Bülow-Hübe cultivated an image that was anti-style. In the September 1967 issue of *Canadian Homes* she claimed her work was merely "an expression of simple, dependable materials used to serve a practical function," though many of her designs distinctly reflect the aesthetics of the time. This low-slung coffee table shares a design sensibility with the work of her Scandinavian compatriot Grete Jalk, but it also capitalizes on the popularity of surfboard iconography and uses contrasting light and dark woods to emphasize its geometry. It was included in the 1956 NIDC *Design Index*. Bülow-Hübe's contribution to her craft was her insistence on high-quality traditional manufacturing methods: hand-turned structural elements, well-engineered assembly and hand-rubbed, natural oil finishes.[24]

1955
MANUFACTURER: AKA Works, Montreal
MATERIALS: alternate laminates of mahogany, oak or walnut with natural maple, solid wood frame
DIMENSIONS: h45 x w30 x length 213 cm (h18 x w15 x length 84 in.)
MARKINGS: unknown

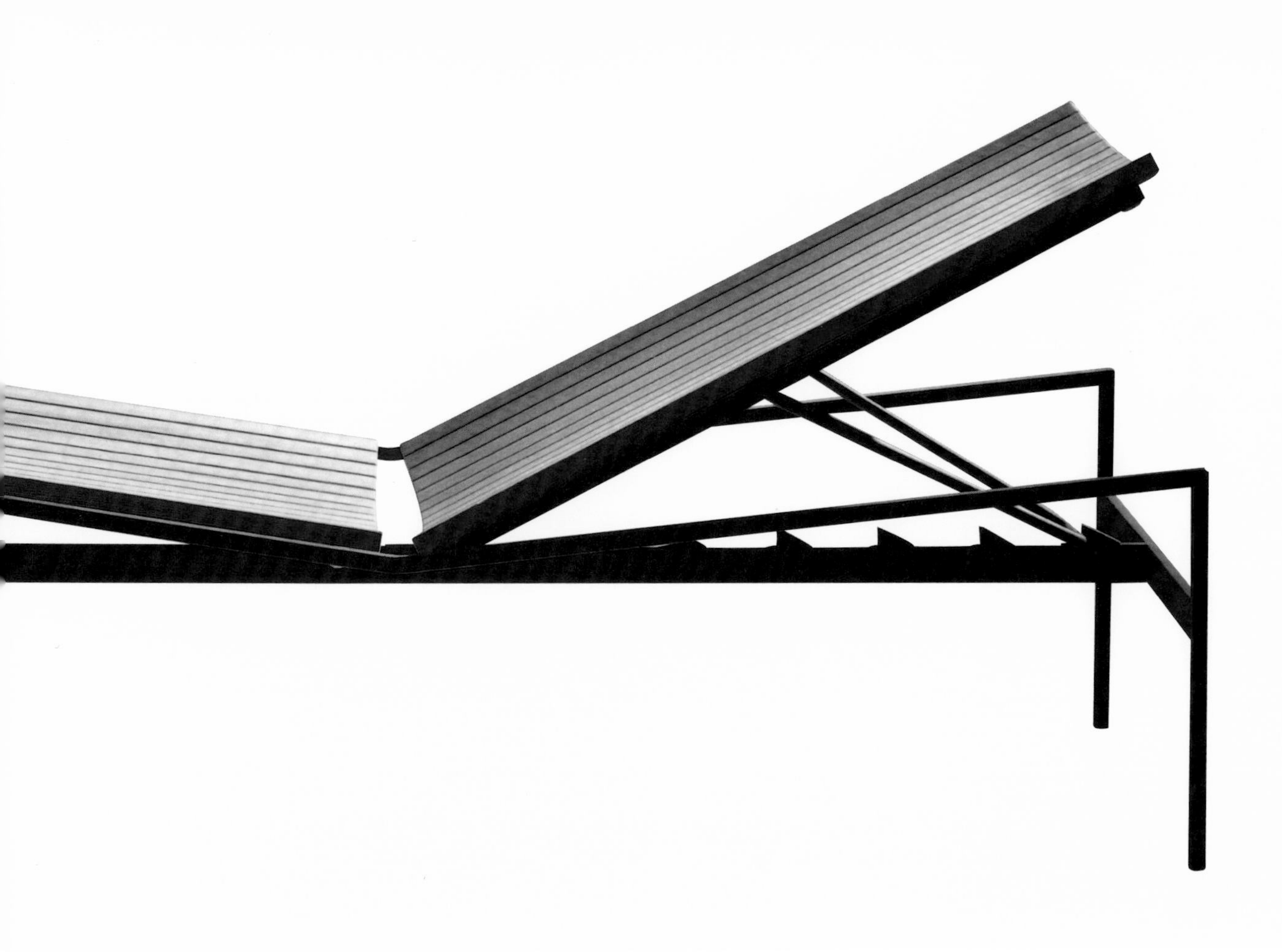

6.29, 6.30 **JOHN STENE**

Dining chair and stool

John Stene launched his firm, Brunswick Manufacturing Company, with this chair design, which won an NIDC award the following year (1958). Both the oiled walnut frame and the woven seat reflect its designer's northern heritage. Surprisingly, the "rope" is twisted Kraft paper twine. Stene taught his staff the simple weaving technique and the traditional method of securing the rope on nails on the underside of the seat.[25] Later versions feature rope wrapped continuously around the seat and back frame, resulting in a sturdier chair.

Stene also made a children's version of the chair, with legs that were three inches (7.6 cm) longer. When the child grew, the legs could be trimmed to match the adult chairs. The line also included a vanity/piano stool (that employs the same paper twine but is made from stained oak) and a thirty-inch-high (76 cm) barstool. Occasionally, an original chair resurfaces at the firm (which remains in operation), requiring only minor repairs. The company continues to make the chair on request.[26] Later versions, made from moulded plywood, have larger seats and backs.

Chair, *1957*; stool, *1958*
MANUFACTURER: Brunswick Manufacturing Company, Toronto
MATERIALS: chair: solid oiled walnut and Kraft paper twine; stool: stained oak and Kraft paper twine
DIMENSIONS: chair: h81 x w46 x d43 cm (h31¾ x w18 x d17 in.); vanity/piano stool: h46.9 x w49 x d33 cm (h18½ x w19¼ x d13 in.)
MARKINGS: none

6.31 **JOHN STENE**

Dining table

This oiled walnut table is characteristic of the handcrafted, elegant designs by Stene. He collaborated with Rudolph Rataj, a master woodworker trained in Germany, on this and most other designs. Stene plotted the "cuttings" to eliminate waste, thus allowing the firm to use solid rather than veneered walnut in the frame. It won an NIDC award in 1958.

1957

MANUFACTURER: Brunswick Manufacturing Company, Toronto
MATERIALS: solid oiled walnut frame, veneered walnut top
DIMENSIONS: h 71 x w 157 x d 81 cm (h 28 x w 62 x d 32 in.)
MARKINGS: brass label nailed to underside of table: Brunswick Mfg. Co. Ltd. Toronto

6.32 **ROBIN BUSH**

Prismasteel modular office furniture, 700 series

Robin Bush described his Prismasteel series as a Meccano set because its thirty-three metal-frame interchangeable components can be configured into 120 pieces of furniture, including desks, credenzas, seating and dividers. With its simple rectangular shapes and bright colours, it's indebted to George Nelson's Executive Office Group for the American manufacturer Herman Miller. It stayed in production for over ten years and appeared in many airports, including Ottawa's Uplands (1960) and Toronto International (1962). The entire furniture group won an NIDC award in 1958.[27] The steel frame is either white or charcoal and the steel panels either white, blue, orange, brown or Formica wood grain.

1957

MANUFACTURER: Canadian Office and School Furniture, Preston, Ontario, for Herman Miller Furniture Company, Zeeland, Michigan
MATERIALS: porcelain-enamelled steel frame, Formica walnut tops, enamelled steel panels, natural walnut drawers
DIMENSIONS: various
MARKINGS: unknown

6.33, 6.34 **WALTER NUGENT**

#22 sprung-steel chair and pedestal table

Walter Nugent patented his design for a sprung-steel chair seat and back in Canada and the U.S. The sweeping curve of the one-piece unit is supported internally by tempered steel rod. A heavy canvas sling and two-inch (5 cm) foam rubber wrapped in vinyl completes the seat. The chair was eventually available in up to seventy-five colours upholstered in either flat or ribbed vinyl, and buyers could select from a variety of bases. Pictured is Nugent's second version (#22), which introduced a polished, chromed steel base to replace the #11's Scandinavian-style oiled walnut base.

Most of these chairs could be bolted together to form multiple units, complete with side tables, for public seating. Pedestal coffee tables were available in a variety of sizes (round, square or oval), with tops made from either glass or Arborite (in either black or antique white). The chromed steel base, offered in various heights, featured three or four legs sharply angled to support the tabletop.

The #22 chair was exhibited in cities throughout North America, Australia, Japan and Europe by the federal government's Canadian Exhibition Commission and was catalogued by the Design Council of England. This low-backed model (which was also available with arms and swivel tilt) was later complemented with a high-backed version that could be outfitted with rockers for hospital use. In its many incarnations, the chair was in production for nearly fifteen years and appeared in numerous international airports (Edmonton, Winnipeg and Toronto), as well as at McGill University, the Hunt Club, Toronto, and Jasper Park Lodge. Versions won NIDC awards in both 1960 and 1967.[28]

Chair and table 1957–59

MANUFACTURER: Walter Nugent Designs, Oakville, Ontario

MATERIALS: chair: ⅜ in. tempered steel rod, ribbed vinyl, polyether foam, canvas, chromed steel; table: glass, chromed steel

DIMENSIONS: chair: h81 x w69 x d58 cm (h32 x w27 x d23 in.); table: w76 x d76 cm (w30 x d30 in.)

MARKINGS: typically, metal label: Walter Nugent Designs, Oakville, Ontario. Made in Canada Patents pending. Reg'd 1960.

6.35 **STEFAN SIWINSKI**

Lounge chair 100–1

Beautifully detailed, Siwinski's 1959 lounge chair represents the confidence of a designer who knows his craft and his market. His meticulous eye is particularly evident in the foam-rubber-and-vinyl upholstery, which had to follow the shape of the chair without creating noticeable seams. He used a combination of heat and hand-stitching to tightly "mould" the covers to their shells. The chair's strikingly thin silhouette has graced many lobbies, frequently acting as a local substitute for Mies van der Rohe's classic Barcelona chair. It was featured in the 1962 Praeger resource guide *New Furniture*. Two years later, a government architect for the Toronto International Airport saw the chair at an NIDC display and specified it for the new facility's departure lobby. Workers at Korina Designs produced five hundred chairs in ten weeks—a challenge for a firm that prided itself on hand craftsmanship.[29]

1959
MANUFACTURER: Korina Designs, Toronto
MATERIALS: stainless steel frame, seat and back of moulded plywood covered in foam rubber and upholstered in leather
DIMENSIONS: h76 x w76 x d74 cm (h30 x w30 x d29 in.)
MARKINGS: none

6.36 **STEFAN SIWINSKI**

Three-legged dining chair

Stefan Siwinski's three-legged dining chair is his most original and playful design. A hybrid, it combines Scandinavian organic modern with Pop imagery. The geometric "stick-man" chair was purchased in quantity for the executive cafeteria at Union Carbide's head office and the Blue Cross Hospital, both in Toronto. Siwinski offered the chair in a black-and-brown colour scheme. Steel rods and exposed wooden joints link its "lollipop" moulded plywood back to its apron-shaped, ribbed vinyl seat. Sculptural and modern, the chair was exhibited at the 1964 Milan Triennale.[30]

1958
MANUFACTURER: Korina Designs, Toronto
MATERIALS: steel rod, bent laminated wood, vinyl
DIMENSIONS: h80 x w46 x d40.5 cm (h32 x w18 x d16 in.)
MARKINGS: none

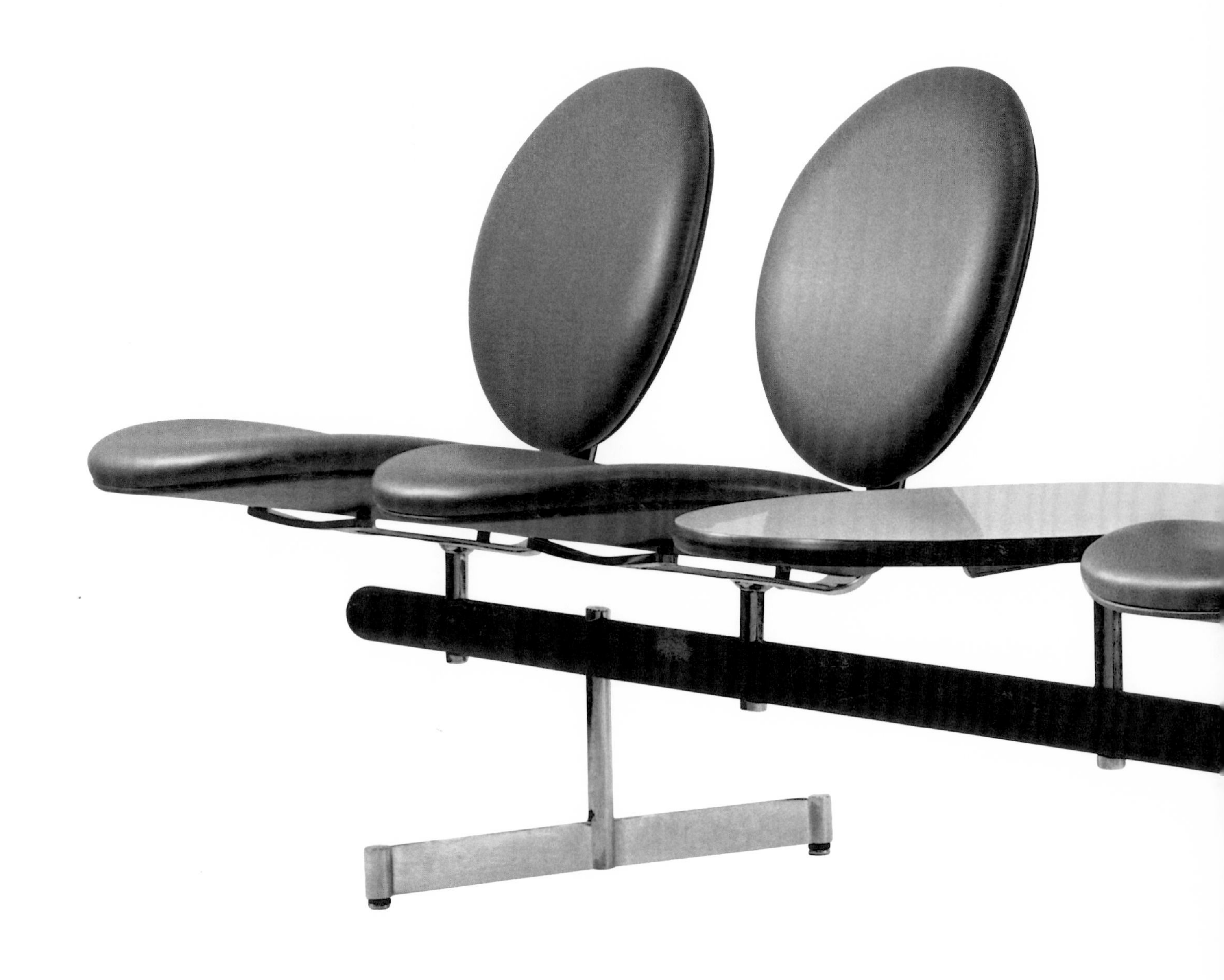

6.37 **ROBIN BUSH**

Component public seating group Lollipop

Lollipop seating earns its name from its simple round forms that share the same design vocabulary and whimsical nomenclature as George Nelson's 1956 Marshmallow sofa for Herman Miller. Its seats and backs, gently curved to fit the body comfortably, are connected by a beautifully crafted metal T-spine attached to a steel tube base. The same T-spine was employed in Bush's Preston Executive Group furniture, also manufactured by Canadian Office and School Furniture. The modular, component seating system can accommodate any number of seats as well as small circular side tables. Over twelve hundred units were produced for the initial run for Toronto's new airport terminal, and the design was eventually used throughout the building for over thirty years.[31] Robin Bush appeared in advertisements promoting Lollipop's apropriateness in modern settings.

1960

MANUFACTURER: Canadian Office and School Furniture, Preston, Ontario
MATERIALS: steel tube and hardware, black vinyl upholstery, moulded plywood and foam cushion
DIMENSIONS: four-seat component without table: h 80 x w 56 x length 236 cm (h 31 ½ x w 22 x length 93 in.)
MARKINGS: none

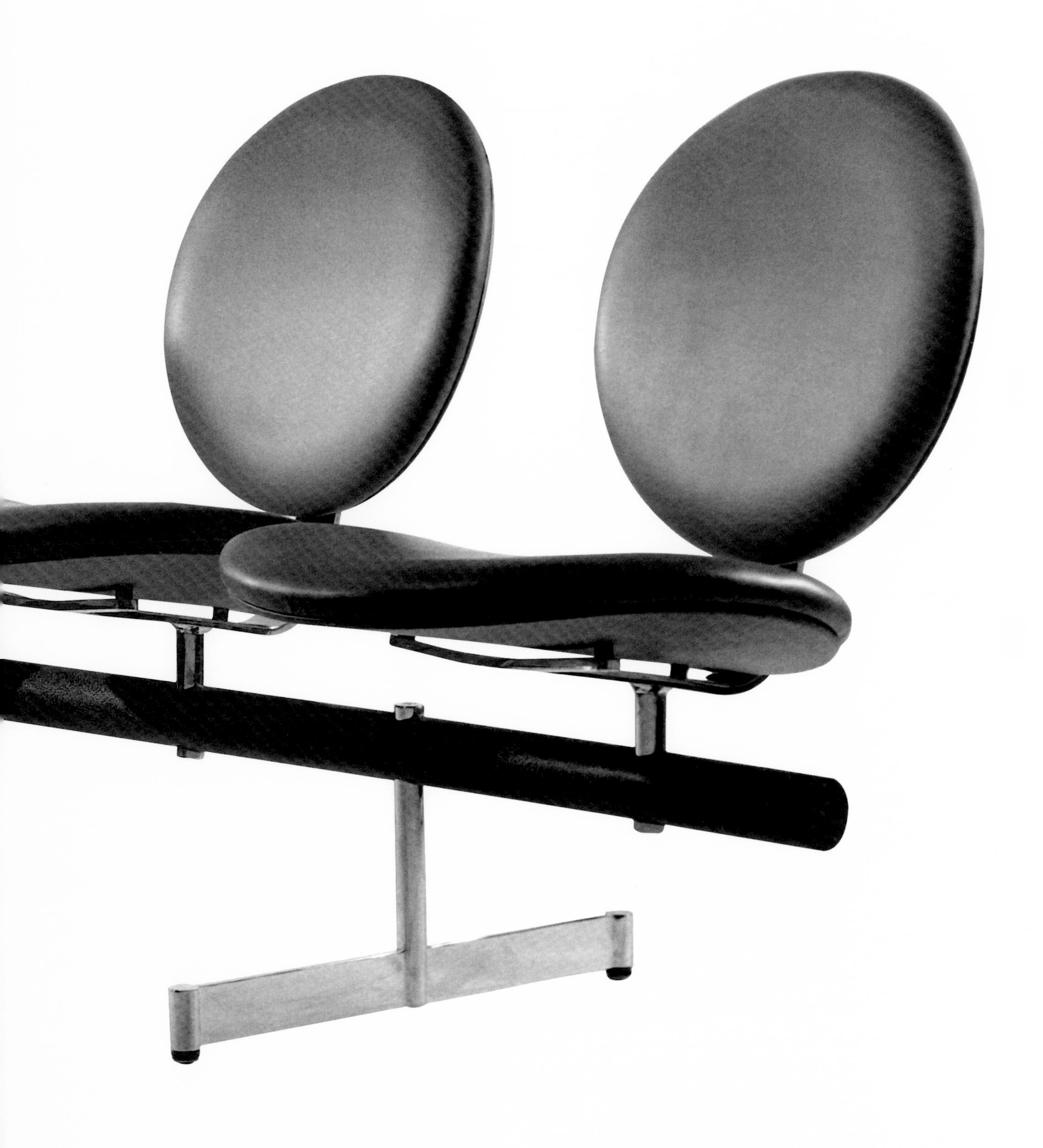

6.38 **JACK DIXON**

LG series

Jack Dixon's most successful design was the LG series, consisting of a lounge chair, ottoman, two-and three-seat sofas and a circular side chair (CSC-F), which all feature a similar sprung steel base. The upholstered seats (in vinyl, fabric or leather) are cradled within oiled wood shells. Influenced by the Barcelona chair, it appeared in the Design Collection at Toronto's Design Centre. The production run for the series was about five hundred units.

A 1960 graduate of the Ontario College of Art, Dixon formed his own company, Dixon Designs, to produce high-quality furniture for architects and designers. Envisaged to cross over into the residential market, the series landed in commercial settings such as the Canadian embassy in New Delhi, the observation lounge at the top of the Toronto Dominion Centre and at Expo 67 and Habitat, Montreal. Throughout the sixties, the firm competed against Stefan Siwinski, Walter Nugent and Court Noxon for short-run, high-end furniture. By the late sixties, Dixon began teaching part-time at OCA, and between 1975 and 1996 he was a full-time faculty member, rarely creating his own designs.[32]

1963
MANUFACTURER: Dixon Designs, Toronto
MATERIALS: chromed steel base, oiled wood frame, reversible foam-rubber cushions, leather upholstery
DIMENSIONS: h74 x w67 x d76 cm (h29 x w26½ x d30 in.)
MARKINGS: black vinyl label 5 x 2 cm (2 x ¾ in.), with logo, reads: Rd. 1964 Dixon Designs Showroom 82 Bloor St. W. Toronto, Ont. 925-2777

6.39 **SVEND NIELSEN/LEIF JACOBSEN**

Lounge chair No. 145

Svend Nielsen's lounge chair is clearly influenced by the Barcelona chair, which has been continuously manufactured by Knoll International. Nielsen was familiar with the chair, as Leif Jacobsen produced Knoll designs in Canada for nearly ten years. In the sixties, there was a cottage industry in redeveloping the famous chair.

His distinctive "improvements" to the classic design include a sleigh base. He replaced the tufted upholstery of the original with more casual pillows of the sixties in either leather or terylene fabric. It became a big seller and spawned the Canti 146-1, which featured a more exaggerated coat-hanger-style tubular steel base. Encouraged, Nielsen reworked the cantilevered chair by employing semicircular chromed-steel legs. Although cited by *Canadian Architect* magazine in 1969 as "classic and commercial," it didn't entice buyers—proving that imitation is not only flattering, it's often more profitable.[33]

1967–68
MANUFACTURER: Leif Jacobsen, Toronto
MATERIALS: chromed flat-rolled steel, laminated and solid wood frame, rubber webbing, foam and leather upholstery
DIMENSIONS: h71 x w74 x d74 cm (h28 x w29 x d29 in.) (estimated by S. Nielsen)
MARKINGS: possibly has metal label: Leif Jacobsen Limited, Toronto, as well as its logo

6.40 **CHRISTEN SORENSEN**

1+1 series

Christen Sorensen says he designed this modular seating series to be "without power symbols": to erase the distinctions between class and gender. He wanted to revolt against the rigidity that makes furniture function "like police."[34] The various foam modules, with scooped backs and seats, could be opened and combined to create chairs and sofas, or closed to make ottomans or daybeds. Lightweight, the units could be rearranged with ease. Designed during the miniskirt era, the cube's twelve-inch (30cm) distance from the floor made some women and conservatives balk, but the University of Alberta in Edmonton ordered a number for their Student Union Building. The series eventually included a low-slung table, denser cubes and a more conventional matching armchair. Lauded more for its underlying idea than for its design, the 1+1 series embodies the egalitarianism of an uninhibited era.[35]

1964
MANUFACTURER: Christen Sorensen, Montreal
MATERIALS: foam, zippered fabric slipcovers, leather connecting straps
DIMENSIONS: basic: h61 x w61 x d30 cm (h24 x w24 x d12 in.);
enlarged: h91 x w61 x d30 cm (h36 x w24 x d12 in.);
cube: h91 x w91 x d30 cm (h36 x w36 x d12 in.)
MARKINGS: none

6.41 **JACQUES GUILLON & ASSOCIATES**

Alumna desk and secretary unit

Jacques Guillon & Associates developed the Alumna desk series for Alcan's former headquarters at Place Ville-Marie in Montreal. The series offered a sophisticated patented (1968) system in which an interlocking cloverleaf extrusion connected the frame and fluted aluminum legs with wood-panel inserts. Alcan initially specified over twelve hundred desks for executives and general staff, convincing Art Woodwork, its manufacturer, to invest in tooling to produce the furniture commercially. Advancing the aluminum component industry for the furniture market, the innovative series boosted confidence in Canadian design know-how. Sunar Industries bought the manufacturer in the mid-sixties, offering the design with woodgrain tabletops for another decade. There is an example in the collection of the Musée du Québec.[36]

1961

MANUFACTURER: Art Woodwork, Montreal
MATERIALS: aluminum frame, wood, plastic
DIMENSIONS: desk and secretary: h 74 x w 152 x d 167.4 cm (h 29 x w 60 x d 65 in.)
MARKINGS: sticker: Compagnie de l'exposition universelle canadienne No. 512

6.42 **COURT NOXON**

Model 1148 tub chair

For many designers, a three-legged chair is a rite of passage. Noxon combined the horseshoe back of a traditional tub chair with a contemporary encircling steel support. The design is distinguished by its exposed fasteners, which efficiently clip together the polished chromed steel elements. Introduced in 1963, the tub chair appeared in the VIP lounge at Montreal's Dorval International Airport, and continues to sell more than thirty years later. It was available in fabric or leather. Current models have larger-diameter steel tubing.[37]

1963
MANUFACTURER: Metalsmiths, Toronto
MATERIALS: polished chromed steel tube, fabric-upholstered steel shell
DIMENSIONS: h 74 x w 69 x d 69 cm (h 29 x w 27 x d 27 in.)
MARKINGS: some may have a small metallic self-adhesive label with Metalsmiths in black letters, applied to the frame

6.43 **DONALD LAPP**

Hystron Collection

Hyman Singer, president of H. Singer Furniture, Montreal, had adventurous taste, liked modern design and invested in research and development to capture that market, says Donald Lapp, who was an in-house and contract designer for the firm for many years. In 1968 Singer introduced to the Canadian market contemporary sculptural furniture by employing moulded rigid urethane. The collaborative effort, led by Lapp, involved a sculptor, mould makers and engineers, as well as experiments to test the plastic's properties (density and cure time) and handling (mould capacity and clamping strength).[38]

Each chair consisted of a one-piece blown-foam form covered with specially developed nylon knits or jersey-backed vinyls that could stretch over the rigid urethane. Inserts in the base accommodated a variety of leg styles made from either wood or metal. The series, which included loveseats and sofas, high- and low-back swivel chairs and ottomans, was intended for both the residential and the contract market. Lapp, best known as vice-president of design for Roxton Mill, Waterloo, Quebec, which primarily made traditional Early American country furniture, scaled the chairs for "smaller, modern living spaces." The line, produced for about three years, was introduced to the American market by the Canadian consulate in New York.

1968
MANUFACTURER: H. Singer Furniture, Montreal
MATERIALS: rigid urethane interior shells, foam padding, nylon upholstery, wood base
DIMENSIONS: h79 x w76 x d91 cm (h31 x w30 x d36 in.) (estimate by D. Lapp)
MARKINGS: none

6.44 **JAMES MURRAY**

GT-3A armchair

James Murray says he was inspired by nature to give his furniture strength where needed and slenderness where appropriate. Feet that are slightly larger in the back emphasize the thinness of the steel frame. Similar to Robert Kaiser's armchair (FIG. 6.24), Murray's has softened the metal with oiled wood armrests in walnut, Burma teak or rosewood. It was available in many configurations, including the model shown and as a swivel tilt armchair with rolling base (GT-6T) and was upholstered in fabric, vinyl or leather.[39]

1967
MANUFACTURER: James Murray Furniture, Ayr, Ontario
MATERIALS: chrome-plated welded steel bar stock, moulded plywood shell, foam rubber, fabric upholstery, oiled walnut armrests
DIMENSIONS: h58 x w58 x d61 cm (h23 x w23 x d24 in.)
MARKINGS: furniture destined for export was affixed with a decal: JAMES MURRAY FURNITURE LIMITED, AYR, ONTARIO Made in Canada

6.45 **DOUGLAS BALL**

System S desk

Following the success of the F system, a wooden general office system, Douglas Ball created the S (for steel) system. It consists of self-supporting steel components such as work walls, tables, movable pedestals and drawers. The desk (a table outfitted with drawers) could be complemented with a tube table or a secretarial "return" with bin storage and a tambour door. The pedestals and matching desk drawers are beautifully crafted and from a distance resemble seamless moulded plastic, a material Ball originally intended to use. A year's worth of research and development failed to produce a solution in plastic. A toolmaker with twenty-five years' experience at Sunar Industries remembered making a steel baby carriage component, with the metal formed over the radius corners to hide all its edges, and the steel S drawer was born. (The technique was also used in the glove compartment of Porsche sports cars.)

The S system pieces were modular and integrated into all Sunar systems furniture. The line was in production (with alterations) until 1990. Compared with Herman Miller's Action Office System, it is more flexible.[40]

1969

MANUFACTURER: Sunar Industries, Waterloo, Ontario
MATERIALS: chromed steel, enamelled steel, plastic laminate surfaces, steel tube table legs
DIMENSIONS: variety of sizes; desk: (shown) is h 76.2 x w 152.4 cm (h 30 x w 60 in.)
MARKINGS: adhesive label inside drawer: Sunar Industries Waterloo, Ontario

6.46 **STEFAN SIWINSKI**

Plastic chair

Siwinski's experiments with moulded plastics in the mid-sixties resulted in a series of hemispheric fibreglass or Plexiglas chairs with either pedestal or drum bases. While the designs are typical of the period, they are remarkable for their execution, which Siwinski controlled from moulding to upholstery. Fewer than fifty chairs were made, but they were widely photographed and appeared in publications like *The New York Times Magazine* in July 1967.

1965
MANUFACTURER: Korina Designs, Toronto
MATERIALS: fibreglass
DIMENSIONS: h71 x w81 cm (h28 x w32 in.)
MARKINGS: none

6.47 **JERRY ADAMSON/KAN INDUSTRIAL DESIGN**
Habitat chair and ottoman

The Habitat chair was the first commercially produced all-plastic Canadian chair. It was moulded using the new capital- and labour-saving British technology of rotational casting: powdered polyethylene was placed in a sealed tin mould, heated and rotated. The boxy indoor/outdoor chair was part of a modular series that included a bed made from fibreglass and reinforced plastic (FRP), tables from sheet polycarbonate and dining chairs from moulded fibreglass. Designed to complement Expo 67's cube-shaped apartment building, Habitat, the prototypes drew considerable media attention from publications such as The *New York Times Magazine*, July 1967, *House and Garden* (U.K.), July/August 1967 and *Canadian Interiors*, November 1967.

John Geiger of Interiors International Limited put the chair into production, and it sold in Canada and the U.S. A few thousand were made over a ten-year period, including sales to Toronto's George Brown College and to Trent University, Peterborough. The chair was offered in either translucent or coloured opaque resins and could be outfitted with detachable padded or fully upholstered seats in vinyl or leather. Today, good examples are rare, as linear polyethylene becomes brittle and cracks with age. KAN, where the chair's designer, Jerry Adamson, was a partner, translated their experience with FRP into a successful line of street furniture in the seventies.[41]

1967
MANUFACTURER: Interiors International Limited, Toronto
MATERIALS: rotational cast linear polyethylene, fabric upholstery
DIMENSIONS: chair: h 64 x w 58 x d 53 cm (h 24 ¼ x w 22 ¾ x d 21 in.); ottoman: w 63.5 x d 63.5 cm (w 25 x d 25 in.)
MARKINGS: paper label: Interiors International Limited Made in Canada, Toronto, Ontario

6.48 **KEITH MULLER AND MICHAEL STEWART**

MS-SC stacking chair

The architect Keith Wagland urged the young designers Keith Muller and Michael Stewart to compete for the interior design contract for Conestoga College in Kitchener. They won, and from this project was born the sixties classic MS-SC stacking chair. Inspired by Alvar Aalto, Muller and Stewart made an efficient design that illustrated the modern principle of "honesty in construction"; the seat, backrest and front legs are made from one piece of ply, while the two back legs, attached with exposed screws, serve as support.

The designers introduced a cut-out in the seat back that serves as a natural handle to help when stacking. For fifteen years, the company exported these institutional stacking chairs (in natural stain or bright colours of red, green and purple) to schools and colleges across the continent. Marlin Superior Products in Gravenhurst, a manufacturer of plywood aircraft and boats, made the first production run. Curvply Wood Products of Orono later fabricated the chair for Ambiant Systems. The designs now turn up in twentieth-century antique stores for moderate prices.

1968

MANUFACTURER: Ambiant Systems, Toronto
MATERIALS: stained red and green moulded plywood, exposed bolts
DIMENSIONS: h61.5 x w82 x d75 cm (h24 ¼ x w32 ¼ x d29 ½ in.)
MARKINGS: none

6.49 **KEITH MULLER AND MICHAEL STEWART**
JACK DIAMOND (CONCEPT)

Image Series armchair

Casual modernism best describes the Image Series armchair. The padded upholstery makes the boxy geometric frame informal and inviting. Assembled on-site with an Allen key (a wrench), the chair could be shipped flat. The Image Series was Muller and Stewart's first critically acclaimed design and the most commercially successful of their wooden products. The knock-down chair was originally designed for the home of the prominent Toronto architect Jack Diamond (who collaborated on the concept).

Muller and Stewart produced their first run (manufactured by Precision Craftwood in Toronto) for Trent University. Max Magder of the Toronto manufacturer Du Barry Furniture saw the Image Series at the EEDEE awards program, where it won best of show. Magder bought the licence to produce the furniture in the U.S. and Canada. He introduced a corner table to the line and standardized the finished plywood to one inch (2.5 cm) thick.

In addition to the table, the Image Series included a sofa, armchair and ottoman in blue, black, natural and white, the most popular colour. The upholstery came in corduroy, "totem" linen, leather and a synthetic leather.[42]

The Image Series sold to institutions across North America but was also popular in the residential market. Du Barry distributed to Macy's and Bloomingdale's, and *Canadian Homes* photographed it for several profiles of contemporary designed interiors. After 1985 Ambiant Systems offered an irregular production of the Image Series, supplying replacement orders.

1967
MANUFACTURER: Precision Craftwood, Toronto
MATERIALS: birch plywood, painted lacquer finish, Lackawana leather, exposed bolts
DIMENSIONS: h61.5 x w82 x d75 cm (h24½ x w32 x d29½ in.)
MARKINGS: none

6.50 **ROBIN BUSH**

Radial Office System

In the sixties, Bush valiantly tried to convince Canadian Office and School Furniture to prototype a radical secretarial system. It could be configured as a single or twinned workstation and included a motorized drum-shaped filing unit and movable side table. The individual units could be grouped in an open area, with each occupying as little as thirty-two square feet (3 m^2). Its playful support pillars and sensuous curves would have made an attractive alternative to the rectilinear monotony of typical office furniture. Bush designed a house following this spherical scheme, but it similarly was never built.[43]

1962

MANUFACTURER: Canadian Office and School Furniture, Preston, Ontario
MATERIALS: unknown
DIMENSIONS: h 76 x w 162.5 (h 30 x w 64 in.);
drum filing unit: dia 76 cm (30 in.)
MARKINGS: unknown

6.51 **ROBIN BUSH**

Steel Shell chair

Bush claimed in a 1962 *Time* magazine article that he wasn't "interested in what people think they want, or what style sells the most. We're interested in the right solution. If it happens to be a good-looking piece, it's purely accidental." The happy accident of his 1963 swivelling Steel Shell chair was short-lived. Made from crackled steel, it was heavy, difficult to adjust and challenged the production capabilities of its manufacturer, Canadian Office and School Furniture. Only one production run was made.[44] Bush included the design as part of the Simpson's contract department proposal to furnish Toronto's New City Hall. More than a decade earlier, the same challenge (a steel shell chair) had stumped the noted American designer Charles Eames.

1963
MANUFACTURER: Canadian Office and School Furniture, Preston, Ontario
MATERIALS: steel frame and shell, upholstered seat
DIMENSIONS: unknown
MARKINGS: unknown

6.52 **HUGH SPENCER**

Slinger lounge (steel frame) 706S

Hugh Spencer not only offered his designs in a variety of materials and finishes but would also alter dimensions to suit a customer's needs. The "sling" of this lounger was designed to conform to the user's posture, could be easily changed or reversed and was available in cowhide, fabric or tapestry canvas. Aimed at both the commercial and residential markets, it came with a steel or wood frame in walnut, maple, rosewood or teak lacquered in black or Japanese red. The stationary steel base is the most appealing. A similar design, outfitted with a rolling swivel base, won a Canadian Furniture Mart award in 1967. Combined with a Spencer-designed desk, the Slinger became part of a fledgling "Canadian collection" at the Toronto outlet of the venerable Danish retailer Georg Jensen.[45]

1967
MANUFACTURER: Opus International, Toronto
MATERIALS: chromed-steel base and hand-stitched eight-ounce leather sling seat
DIMENSIONS: unknown
MARKINGS: none

6.53 **PAUL BOULVA**

Lotus chair

The indoor/outdoor design, which resembles Jerry Adamson's Habitat chair (FIG. 6.47), was intended for the Montreal Olympics in 1976 but was not ready in time. However, the stackable plastic chair with metal sleigh base was later produced in tens of thousands. The plastic shell came in off-white, orange, brown and yellow, and the seat could be upholstered. A plastic-laminate side table with steel frame accompanied the line, which was also sold in the U.S. and Britain. Originally named Chair 2000, it represents a sophisticated handling of materials.

Paul Boulva, who graduated in industrial design from the Université de Montréal in 1972, was the in-house designer for Artopex between 1972 and 1982. Artopex was the result of the amalgamation of three Quebec manufacturers, Artena, Opus and Bonnex. Boulva also designed the Cobra chair for Artopex (1974), which featured the same tempered-steel inner frame as Walter Nugent's patented furniture (over which there was a patent dispute).[46] Artopex Plus reopened in Granby, Quebec, under new management but no longer produces the design. Boulva currently runs Art Design International (ADI), an ergonomic office furniture manufacturer that specializes in patented steel frames.

1976

MANUFACTURER: Artopex, Laval, Quebec
MATERIALS: injection-moulded polypropylene, steel tube frame
DIMENSIONS: h 72.3 x w 45.7 x d 58.4 cm (h 28 ½ x w 18 x d 23 in.)
MARKINGS: moulded under seat: Artopex 80 série Lotus series Artopex Ltée/Ltd. 2121 Berlier Laval, Que. FABRIQUE AU CANADA—MADE IN CANADA

6.54 **BOB FORREST**
2001 chair

Inspired by the sphere-shaped chairs of the era, Bob Forrest positioned an acrylic hemisphere on an upright base, thus making his own version. It was originally a custom design for the well-known Toronto advertising executive Jerry Goodis but captured the imagination of the avant-garde. Over a fifteen-year period, about three hundred chairs were made.[47] It appeared in numerous print and television advertisements in its day.

The forty-five-degree cantilevered central support (a two-inch [5 cm] square tube) is an accidental but compelling design feature. The chair was unstable when the central support was attached to its perimeter ring at a ninety-degree angle. A special swivel mechanism ensures that the chair always returns to its original position. The metal frame was available in either a bright chrome or a bronze finish.

The clear acrylic sphere was the largest that could be drawn by local plastics firms at that time and required meticulous quality control. Initially, the acrylic work was contracted out to Hickey Plastics and, later, Dunlea Plastics (both of Toronto), but it was eventually brought in-house. The cushions were offered in a variety of textured fabrics, wool and psychedelic prints. Today, rediscovered by the current generation of media and entertainment tastemakers, a 2001 chair in excellent condition can command thousands.

1970
MANUFACTURER: L'image Design, Toronto
MATERIALS: ⁵⁄₁₆ x 2 in. bronzed bar stock, 2 x 2 in. bronzed tubing, ½ x 1 in. bronzed tubing, return swivel, ½ in. clear acrylic shell, feather- and down-filled wool cushions
DIMENSIONS: h66 x w84 x d71 cm (h26 x w33 x d28 in.)
MARKINGS: later versions may have a clear sticker bearing the L'image Design name and address

6.55 **PHILIP SALMON AND HUGH HAMILTON**
Stools

The patented angular stool is Philip Salmon and Hugh Hamilton's most successful design of their three-year partnership, formed in 1969. One continuous tube is machine bent four times to form the cantilevered base and footstool. The residential and contract stools, with their early application of powder coating known as Kincoat, came in yellow, red, black, white or polished chrome. Other tubular products in this line included a low stool without footrest and the Zee chair and bookcase. The stool was originally manufactured by Bentube and marketed by Form Canada, but the designers sold the licence to Kinetics Furniture in the early seventies. Kinetics improved the production quality and introduced a bull-nose cap where the leg terminates, rather than a flat disk. Haworth in Holland, Michigan, now owns Kinetics and produces the line in the original Rexdale, Ontario, plant. The design is unaltered and includes the Haworth label under the seat. Amisco, the furniture manufacturer in Quebec, also produced a version, but the bend in the leg reveals a dimple and the bull-nose cap is rubber.[48]

c. 1969
MANUFACTURER: Kinetics Furniture, Toronto
MATERIALS: powdered-coated steel tube
DIMENSIONS: h71 x w43 x seat dia 31 cm (h28 x w17 x seat dia 12½ in.)
MARKINGS: none

6.56 **BILL LISHMAN**
Rocker

When his wife, Paula, needed a rocking chair in which to knit, Bill Lishman, who sculpts in metal, bent a piece of steel rod to fluidly create a template for a wooden rocker. As the design took shape, Lishman preferred the rocker in metal. Through constant refinement, he developed "sculpture you can sit in." It is an update of the classic bentwood chair made by the German manufacturer Thonet. Patented in both Canada and the U.S., the rocker appeared on the cover of *Canadian Interiors*. It was also a prize on the American television game show *The Price Is Right*.

When local manufacturers were stymied by the chair's compound curves, Lishman designed the jigs and dies and produced the chairs himself, with the aid of assistants. The suspended "tractor" seat is based on a traditional design; Lishman created a new matching backrest. The first twenty-five chairs were made of steel rod, the next hundred from three-quarter-inch steel tubing and the remainder from one-inch tubing, for a total production run of fewer than three hundred to date. The earlier versions are coated with white PVC and later with white nylon, and some appear in stainless steel. The chair occupies a large area and therefore is used more frequently outdoors. As for Paula's knitting, it became a patented process for turning fur pelts into yarn.[49]

1976
MANUFACTURER: Bill Lishman, Blackstock, Ontario
MATERIALS: one-inch steel tubing, nylon coating, back and seat cast aluminum
DIMENSIONS: h 94 x w 81 x d 137 cm (h 37 x w 32 x d 54 in.)
MARKINGS: serial number stamped on loops that connect the seat back to the chair frame

6.57 **THOMAS LAMB**

Steamer chaise longue

To help revitalize the Dominion Chair Company, an established Nova Scotia–based furniture manufacturer specializing in steam bending of wood, Thomas Lamb updated the classic design of an ocean liner lounge chair. He simplified steam bending to a single angle and reduced the number of solid maple slats to three standard lengths. The genius of his design is that these organic wavelike slats can be assembled into a family of indoor/outdoor knock-down furniture, including a chaise longue, pull-up chairs and tables.[50]

While the project failed as a renewal strategy for Dominion Chair, Lamb switched the production to Ontario-based manufacturers (Du Barry and Ambiant) and then produced the design himself at a factory in Malaysia.[51] Simple yet elegant, it became the first Canadian furniture design to be selected for MOMA's Study Collection, and the honour vindicated the design profession in Canada.[52] Now a part of the canon of Canadian chair classics, the design has been featured on a 1998 Canadian postage stamp and still sells in the MOMA shop.

1977
MANUFACTURER: Steamer Furniture Company, Malaysia
MATERIALS: steam-bent plywood
DIMENSIONS: h85 x w60 x length 175 cm (h34 x w24 x length 70 in.)
MARKINGS: none

6.58 **MICHAEL FORTUNE**

Number One chair

The Number One chair is Michael Fortune's signature design. Since 1981 he has produced more than two hundred examples of the chair. The bent vertical slats recall Thomas Lamb's Steamer chair (FIG. 6.57), but Fortune adds elements of Frank Lloyd Wright and Charles Rennie Mackintosh to the design by dropping the seat back to the foot rail. Using the technology of steam bending, hot pipes and hydraulic machines, Fortune moves away from dovetailing and peg-and-dowel construction, standard in traditional craft furniture. He spent five months making the first Number One, but today the production time is considerably shorter. To prepare for each production cycle, Fortune's studio follows his instruction manual and reviews a training film, then employs thirty-eight sequenced jigs to fabricate the chair. The Number One chair won the Virtu 2 competition.

original design *1981*, designed *1989*
MANUFACTURER: Michael Fortune, Peterborough, Ontario
MATERIALS: Macassar ebony, silver inlay, wool upholstery
DIMENSIONS: h85 x w63 x d63 cm (h34 x w25 x d25 in.)
MARKINGS: stamped: Michael C. Fortune studio, back of front stretcher

6.59 **PAUL EPP**

Nexus chair

Paul Epp's Nexus chair follows in the tradition of Thomas Lamb's classic Steamer collection (FIG. 6.57), demonstrating a superb use of a few bent plywood components. Epp arranges each bent ply—the nexus or link—in opposing directions to form the frame, the back legs and the seat supports, which extend into the arms. With a second bend, he makes the seat and backrest. This chair reflects the eighties penchant for increased ornament and neo-classicism. It has bold sweeping arms and a square pattern of nine bored holes on the seat back. Ambiant Systems produced approximately one thousand Nexus chairs in eight variations—with or without arms, fully or partly upholstered or solid seat—until the company closed in 1989. The Canadian Museum of Civilization includes a chair in its collection. It won a silver from Design Canada in 1984.

1983
MANUFACTURER: Ambiant Systems, Toronto
MATERIALS: moulded bleached ash plywood, grey lacquered armrest, cane seat
DIMENSIONS: h 83 x w 46 x d 38 cm (h 32¾ x w 18 x d 15 in.)
MARKINGS: none

6.60 **VELLO HUBEL**

Clover Leaf nesting tables

In the early 1990s, CSL, a manufacturer of metal closet organizers, took advantage of "cocooning" and expanded its housewares line. Vello Hubel, the venerable designer and chairman of the OCA industrial design department, joined his former student Scot Laughton and other young designers, like Max Leser and Tom Deacon, to design the Clover Leaf nesting tables. They became the most memorable product of CSL's Steelform Collection.

Hubel's design features a stylized cloverleaf tabletop in colourful MDF (medium-density fibreboard) supported by jet-black metal rod legs. Sold in sets of three, the retro nesting tables came in black, mahogany, cognac, teal blue and cherry. CSL also made a version with a metal tabletop. They were popular in France and Japan, and Crate & Barrel's and Bloomingdale's mail order catalogues also sold them in the U.S. The tables were selected for *ID* magazine's 1993 annual.[53]

1992

MANUFACTURER: CSL (Creative Space Ltd.), Scarborough, Ontario
MATERIALS: stained medium-density fibreboard, steel rod
DIMENSIONS: h 47 x dia 41 cm (h 18 ½ x dia 16 in.)
MARKINGS: none

6.61 **JONATHAN CRINION**

Gazelle chair

Inspired by animal imagery, Jonathan Crinion's indoor/outdoor Gazelle chair features sweeping, hornlike arms that flow gracefully into its legs. The design echoes earlier zoomorphic chairs such as the British designer Ernest Race's 1951 Antelope and the Danish designer Arne Jacobsen's 1958 Swan. *ID* magazine included it in its 1988 annual product review, and institutions such as Disneyland, the Art Gallery of Ontario and the American broadcasters ABC and CNN own Gazelles.

Lee Jacobson, the original co-owner of the Toronto furniture manufacturer AREA, saw a prototype of the "spec" chair in Crinion's studio and encouraged its production by arranging start-up money for retooling from Umbra. Over fifty thousand were produced at a steel-forming plant in northern Italy, but the chair was assembled in Toronto.[54] It was available in black, white or green.

1987
MANUFACTURER: AREA, Toronto
MATERIALS: steel, powder-coated epoxy finish
DIMENSIONS: h72 x w56 x d53 cm (h32 x w22 x d21 in.)
MARKINGS: adhesive label: AREA

6.62 **DOUGLAS BALL**

Clipper CS-1

Clipper is a self-contained virtual office designed for ergonomically correct computer use. The "cockpit" of the capsule can be a fully equipped workstation, complete with seating, lighting, air supply and storage. Its multi-position side wings and tables offer stages of privacy, including completely enclosed, causing one clever techie to remark, "Someone finally invented the perfect workstation. When you die, they bury you in it."

Aimed at computer specialists working long shifts, Clipper's combination seat and footrest aligns the buttocks, back and neck in a comfortable position and places the keyboard almost in a person's lap, a position considered ideal. Fewer than a hundred have been produced, but it has attracted considerable attention. A unit is in the Conran Foundation Collection in London, England, and it was published in *Conran on Design* (1996).

1993
MANUFACTURER: New Space, Fort Worth, Texas
MATERIALS: moulded plywood, translucent plastic, Lexan, powder-coated steel frame, fibreglass-and-foam seat, leather upholstery
DIMENSIONS: h 1.22 x w 2 x d 1.22 m (h 4 x w 7 x d 4 ft.)
MARKINGS: none

6.63 **MARK MÜLLER**

Parabola Tangent shelving

The horizontal planes of these minimalist wood shelves appear to float freely on the wall, with no visible means of support. In fact, a long, slim wall-mounted extruded aluminum bracket bites into the underside of the shelf, holding it in place. For greater load capacity, the shelves can be fitted into a matching wall panel. The sculptural shelves come with straight or curved edges. This design reflects Mark Müller's efforts, as design director of Nienkämper, to move the company away from hand finishing into machine tooling, without decreasing luxury or style. Veneers are wrapped around a solid MDF core, using a patented process from Italy, to create a wide, curved profile. White aluminum with light-coloured wood is another feature introduced by Müller, and this, along with the round-edged wood veneers, epitomized the look of Nienkämper's nineties furniture collection. The shelving was selected in the 1998 *ID* annual review and Virtu 1998.[55]

1997
MANUFACTURER: Nienkämper, Toronto
MATERIALS: clear-finished medium-density fibreboard, aluminum brackets
DIMENSIONS: shelves 1 in (2.5 cm) thick; variable lengths
MARKINGS: label on the aluminum bracket

6.64 **TOM DEACON**

Tom chair Model 9563

By puncturing the thinly curved back with an elegant grid of holes, Tom Deacon invokes the classic lattice patterns of the Austrian designer Josef Hoffmann. He further distinguishes the design's lower back by adding a "bustle," a long vertical box, which hides the adjustment mechanism.

Keilhauer spent two years developing this product, investing over $2 million to create new tooling to lower production costs. The result is a plastic component frame that is 20 per cent less expensive to produce than metal. The Tom was the company's first high-performance chair—its seat, back and arms lift and tilt. The investment paid off: since its launch, more than sixty-five thousand chairs have been produced, and the Tom has become the company's best-seller. Contributing to the design's commercial success is the witty marketing campaign created by Concrete Design Communications. The Tom chair won a silver award at the 1997 NeoCon World's Trade Fair and also captured the 1998 Design Effectiveness award.[56]

1997

MANUFACTURER: Keilhauer, Scarborough, Ontario
MATERIALS: steel-and-plastic frame, blue wool jersey, foam
DIMENSIONS: h 99 x w 50 x d 47 cm (h 39 x w 19¾ x d 18½ in.)
MARKINGS: stamped on seat back: Keilhauer; paper label and stamp under seat

7.1 **WILLIAM TROTT**

Sunspot floor lamp

William Trott's refined floor lamp is an example of post-war modern "industrial craft." Its severe aluminum composition contrasts with the delicate placement of the concave shade on the spun cone. Bored holes encircle the cone to make a necklace of light when the lamp is on, revealing Trott's careful attention to detail. According to his son Nicholas, Trott produced fewer than five hundred lamps. This one, together with his Sunspot gooseneck desk lamp (illustrated in *Canadian Art*, summer 1951), represent his brief foray into residential lighting. It appeared in the NIDC's *Design Index* in 1949.[1]

1948
MANUFACTURER: Lighting Materials, Winnipeg
MATERIALS: spun aluminum with iron core in base, patented alumlite finish (tarnish-free)
DIMENSIONS: h 158 x base dia 30.7 x shade dia 46 cm (h 62 ½ x base dia 12 x shade dia 17 ½ in.)
MARKINGS: none

7 Lighting

Lighting Industry Becomes Design Aware

The relatively small indoor lighting industry in Canada has been notably responsive to design and innovation.[2] Since the war, like the Canadian furniture industry, it has been divided according to purpose, producing residential lighting and standard as well as contract (custom) fixtures for institutional, corporate and commercial architecture.

In the fifties, the contract lighting industry, larger and more modern than its residential counterpart, expanded to complement the new architecture and style of Canadian institutional buildings. Fluorescent tubes, with metal egg-crate boxes and plastic sheaths, largely prevailed.[3] Winnipeg-based Lighting Materials, operated by William Trott, produced most of the institutional lighting for Western Canada, including for the landmark B.C. Electric building in Vancouver.

In Montreal, Modulite, also known as Rameck Supplies, furnished the Quebec market, and its contemporary and traditional-style lighting included some notable steel pendant and gooseneck lamps by the architect Norman Slater. Modulite's owner, Harry Gottlieb, was sensitive to modern design, and participated in NIDC design award programs.[4] J.A. Wilson Lighting and Display and John C. Virden Lighting, operating from the forties to seventies, produced modern lighting for the Toronto area, with its product and marketing manager, Donald McCormack, introducing a number of his designs.[5] Frank Reed joined Virden in 1953 and launched the Ringmaster lamp (FIG. 7.2).[6]

For the home, modern fifties lighting consisted of bent-tube floor, pendant and sconce lights. These designs reflected the influence of the Swedish-American designer Kurt Versen, whose work was sold through Simpson's.[7] The Crown Electrical Manufacturing Company (also called Crown Electric), a subsidiary of a company in St. Charles, Illinois, was one of the first electric lamp manufacturers in the country. The company produced stained-glass shades with metal and brass fixtures, distributed through Eaton's. Founded in 1910, it was a decade later the largest lighting fixture plant in Canada. In 1936, following bankruptcy, it was taken over by the powerful Verity family (connected with Massey-Ferguson Industries), and it closed in the eighties.[8] William Campbell, a former Crown Electrical manager, founded his eponymous company in 1933 and launched competitive product lines (FIG. 7.11). It was acquired by Deilcraft, the furniture division of Electrohome, in 1955, and the plant operated until 1969. It became the largest residential lighting manufacturer in Canada.

Ceramics and Plastics Compete with Metal

After the Second World War, the Canadian pottery companies Medalta and Medicine Hat Potteries made ceramic lamp bases for the industry in baroque colours and traditional shapes to satisfy

OVERLEAF, TOP: 7.2 Frank Reed's Ringmaster pendant lamp, 1953, outfitted schools and institutions.

OVERLEAF, BOTTOM: 7.3 Michael Baldwin put some Pop into Electrohome's 1970 Mushroom lights.

war-rationed markets. Medalta, which produced over ninety-three styles in more than twenty-four colours, worked with lighting manufacturers to provide custom designs for consumers.[9]

By the mid-fifties, simpler Scandinavian ceramic lighting emerged as an alternative. Bostlund Industries was one of the first companies in Canada to offer earth-toned stoneware lamps in the mid-fifties (FIG. 7.9). A decade later, Céramique de Beauce in Quebec introduced organic Scandinavian-style lamp bases to its existing lighting collection under the direction of the studio potter Jacques Garnier.[10]

With the advent of space-age Pop styling, in the late sixties, the lighting industry adopted brightly coloured plastics, globular light bulbs and chrome in expressive and playful shapes. Electrohome/Deilcraft's principal lighting designer, Michael Baldwin, produced an array of styles from traditional to Pop (FIG. 7.3) to complement the company's furniture and consumer electronics line.[11] Origina Canada and Galaxi Lighting, both in Toronto, mixed plastic with chrome (FIG. 7.5).

Halogen Re-energizes the Market

As the eighties approached, new Quebec manufacturers began making inexpensive contemporary lighting for home use. Bazz, Sverige and Clique supplied the province's blossoming home furnishing boutiques. Capitalizing on earlier Danish styles, especially Poul Henningsen's classic metal pendant designs for Louis Poulsen & Company, Bazz and Sverige created simple suspension lights in colourful painted metals, often favouring aluminum (FIG. 7.5). By the mid-eighties, they switched to halogen. This technology was first popularized in the early seventies by Richard Sapper's Tizio lamp. Halogen, powered by an independent transformer, allowed minuscule bulbs to throw strong light, thereby liberating the designer from the constraints of cumbersome shades. Michel Morelli, the industrial designer for Sverige, helped to introduce sleek black "Eurostyling." His designs represent some of the earliest mass-produced halogen lights in Canada (FIG. 7.4).[12]

The Ontario branch of the Canadian Standards Association (CSA) was more stringent than its Quebec counterpart. Wary of the new technology, it initially imposed restrictions that slowed the conversion to halogen. The first halogen designs in Ontario came in limited editions from independent designer manufacturers like the brothers Aldo and Francesco Piccaluga, who created lights for the retail store Italinteriors, Toronto, and Portico (FIG. 7.17).

Lighting Design Matures

In the nineties, young designers and studio manufacturers were attracted to lighting because of the easy access and availability of CSA-approved sockets, wiring and tubes. Using a handful of inexpensive off-the-shelf parts, they fashioned high-design lighting. Andrew Jones in Toronto, Addison Lanier (FIG. 7.7) in Vancouver and Bakery Group in Ottawa participated in this trend.

Designers also contract their lights offshore. Stephan Copeland's Tango desk lamp (FIG. 7.6) was produced by Arteluce, a leading Italian manufacturer, the first North American design for this company.[13] Kirk Mosna, based in Hamilton, worked with Egoluce, a Milanese company, to manufacture his Dragonfly spot lamp (FIG. 7.20).

For the most part, the lighting industry still caters to local markets, although there are some notable exceptions. Axis Lighting in Montreal works with in-house industrial designer Dirk Zylstra (who collaborates with Koen de Winter) to fabricate linear pendant lighting that plugs directly into office systems furniture (rather than the ceiling or wall). For Calgary-based SMED International, the company developed Spectrum, a horizontal winged canopy made from stretched Lycra.

PREVIOUS PAGE, TOP: 7.5 Simon Ben Ghozi's pendant lamp for for Bazz, 1981, employed colourful metals.

PREVIOUS PAGE, BOTTOM: 7.4 Michel Morelli's halogen Eurostyle Rapolla lamp for Sverige, 1989.

In Langley, British Columbia, Peter Murphy established Ledalite Architectural Products to outfit Expo 86 in Vancouver, and he has since built a reputation for high-quality, well-engineered fluorescent lighting. Now American owned, the company has collaborated with the Vancouver architect Peter Busby on energy-saving lighting.

The direct lighting industry (wire, track, recessed and halogen) is more regionally based. MP Lighting of Vancouver specializes in metal track halogen lighting systems, designed by the owner, Mirek Pospisil. Eureka and Systemalux run decorative lighting companies in Montreal and make custom products for interior designers as well as their own lighting lines. Systemalux employs the industrial designer Michel Foti. His Ring lamp, designed with Marie-Josée Dufresne, has received considerable attention and furnishes the offices of Daniel Langlois, founder of the animation company Softimage, in Montreal.

BELOW LEFT: 7.6 Stephan Copeland's zoomorphic Tango desk lamp, 1989, features a flexible double arm.
BELOW RIGHT: 7.7 Addison Lanier's sinuous Peruse/Illumine flow lamp, 1997, is an example of studio manufacturing.

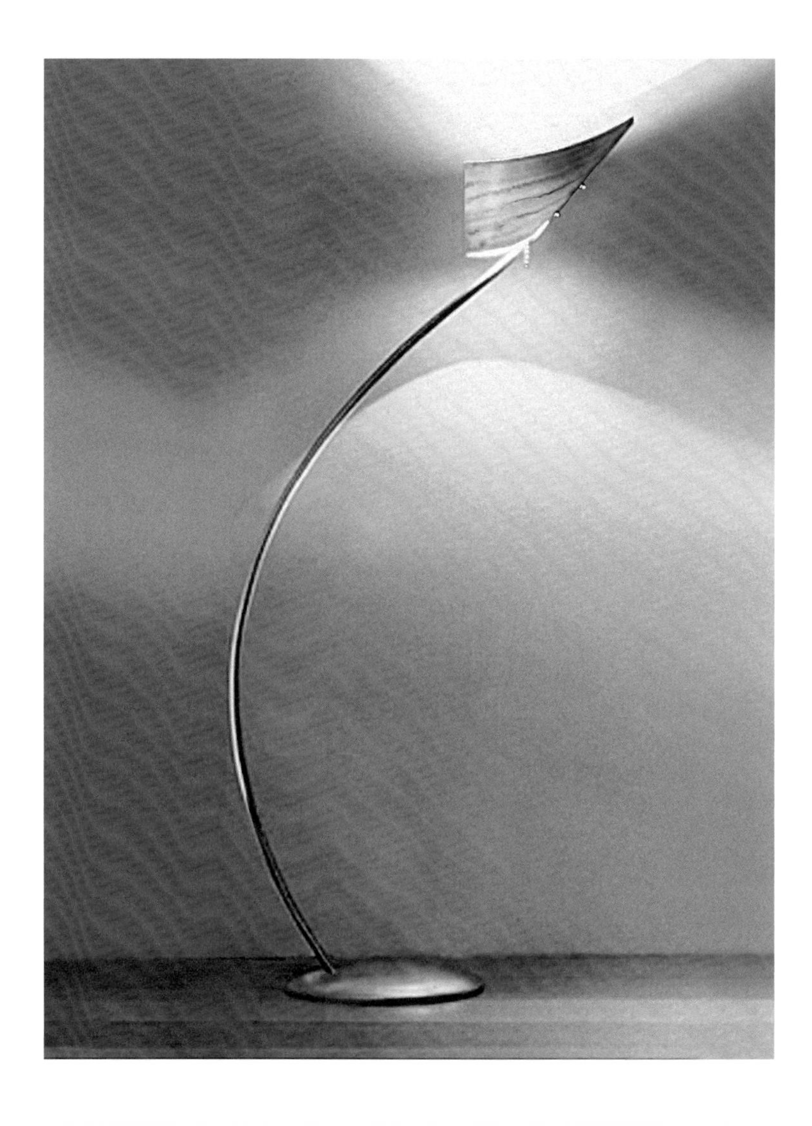

BOSTLUND
OAK RIDGES ONTARIO

7.8, 7.9 **LOTTE BOSTLUND**

Table lamp #900

In the sixties, Canadian interior designers like Murray Oliver and J & J Brook admired Lotte Bostlund's genuine Danish styling. Scandinavian design shops across the continent sold Bostlund lamps, including the prestigious Georg Jensen department store, which advertised a Lotte design in *The New Yorker* of September 1, 1962. Bostlund lamps are characterized by their tall and thin slip-cast ceramic bases glazed in muted colours of brown, cream and robin's egg blue. Lotte signed only designs that she had hand-carved and hand-painted (harlequin, diamond and stripe [LEFT]), usually placing her Lotte signature above the electrical cord. The tapered fibreglass shades echo the shape of the ceramic base, while the wound string and jute serve as textural contrast to the smooth clay.

The family-run Bostlund Industries collaborated with the Eaton's contract design department, supplying lighting to Canadian embassies and government buildings. Much to the horror of the family, pacifists all, Prime Minister Pierre Trudeau announced the War Measures Act in 1970 on live television sitting next to a Bostlund lamp.

The #900 lamp, introduced in 1965, is distinctive for its squared-off base and Chinese Tenmoku brown glaze. It was illustrated on the cover of the Bostlund Industries catalogue.[14]

1964
MANUFACTURER: Bostlund Industries, Oak Ridges, Ontario
MATERIALS: base: cast-moulded stoneware, with semi-gloss glaze finish, felt bottom; shade: filament-wound fibreglass with impregnated yarn
DIMENSIONS: base: h 44 x dia 18.5 cm (h 17 ¼ x dia 7 ¼ in.); shade: h 41 x dia 37 cm (h 16 x dia 18 ½ in.)
MARKINGS: stamp on shade's metal ring: Bostlund Industries

7.10 **PETER COTTON**

Tripod lamp

Peter Cotton's tripod lamp was hand-forged from steel and featured a silk shade. Seeking economy in materials and manufacturing, he followed it up with a simpler three-legged "lug" lamp that replaced the steel with standard bent rod. He showcased the lamps with the cord wrapped around a single leg, artfully disguising the problem of a dangling cord.

1950
MANUFACTURER: Perpetua Furniture, Vancouver
MATERIALS: hand-forged black-lacquered steel and silk shade
DIMENSIONS: h 63.5 x dia 28 cm (h 25 x dia 11 in.);
MARKINGS: unknown

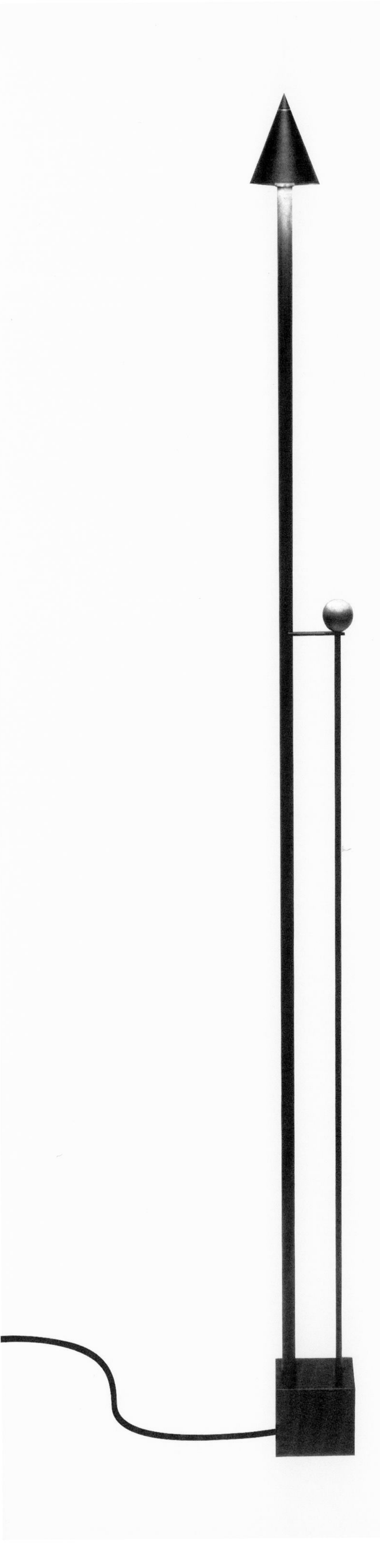

7.17 **SCOT LAUGHTON WITH TOM DEACON**

Strala lamp

Time magazine cited the Strala in its "best design of 1988" issue. Strala (or "ray of light") explores the symbolism of light within a decidedly postmodern framework—a black steel totem is lit by turning its base. Scot Laughton designed the lamp while he was still a student at the Ontario College of Art and collaborated with Tom Deacon on a limited edition of forty, which they signed and numbered. Later, Laughton produced several hundred more through Portico, the Toronto-based studio furniture company. Arteluce, the Italian lighting manufacturer, made a prototype but did not put it into production. Design awards include a citation from the 1987 *Progressive Architecture* annual furniture competition and from Virtu 2.

1986
MANUFACTURER: Portico, Toronto
MATERIALS: spun aluminum cone, epoxy coated steel tubing, patina bronze sphere/switch, cast bronze base
DIMENSIONS: h 175 x w 102 x d 12.5 cm (h 69 x w 4 x d 5 in.)
MARKINGS: signed under base

7.18 **ALDO AND FRANCESCO PICCALUGA**
Aztec wall lamp

In the seventies, well-designed inexpensive spot lighting was not widely available. Aldo and Francesco Piccaluga developed their own, based on a ribbed metal canopy that functioned as a pendant, sconce and recessed ceiling light. They adopted the economical metal-spinning process employed in their Synthesis furniture to accommodate the small production runs (from twenty to two hundred). Like the furniture line, the lighting fixtures used the new electrostatic-charged powdered epoxy paint to add colour. Gary Sonnenberg of Precision Craftwood produced the Opus architectural lighting collection in 1977, and the Italian lighting company I Guzzini (which at that time operated a manufacturing plant in Sherbrooke, Quebec) fabricated the design in polystyrene. Currently Systemalux offers the lighting, renamed Aztec, in spun aluminum with a sheet metal bracket.[15]

1997; original design 1977
MANUFACTURER: Systemalux, Toronto and Montreal
MATERIALS: aluminum, sheet metal
DIMENSIONS: h14 x w37 cm (h5½ x w14½ in.)
MARKINGS: none

7.19 **JEAN-FRANÇOIS JACQUES**

Zenith lamp

The design was sold in New York City and Montreal in the early eighties, and about fifteen hundred were produced. Zenith represented a whimsical handling of Canadian designers' favourite material, aluminum.

1987
MANUFACTURER: Météore Design, Montreal
MATERIALS: embossed spun anodized aluminum
DIMENSIONS: h28 x dia25 cm (h11 x dia10 in.)
MARKINGS: label inside shade: Météore Design

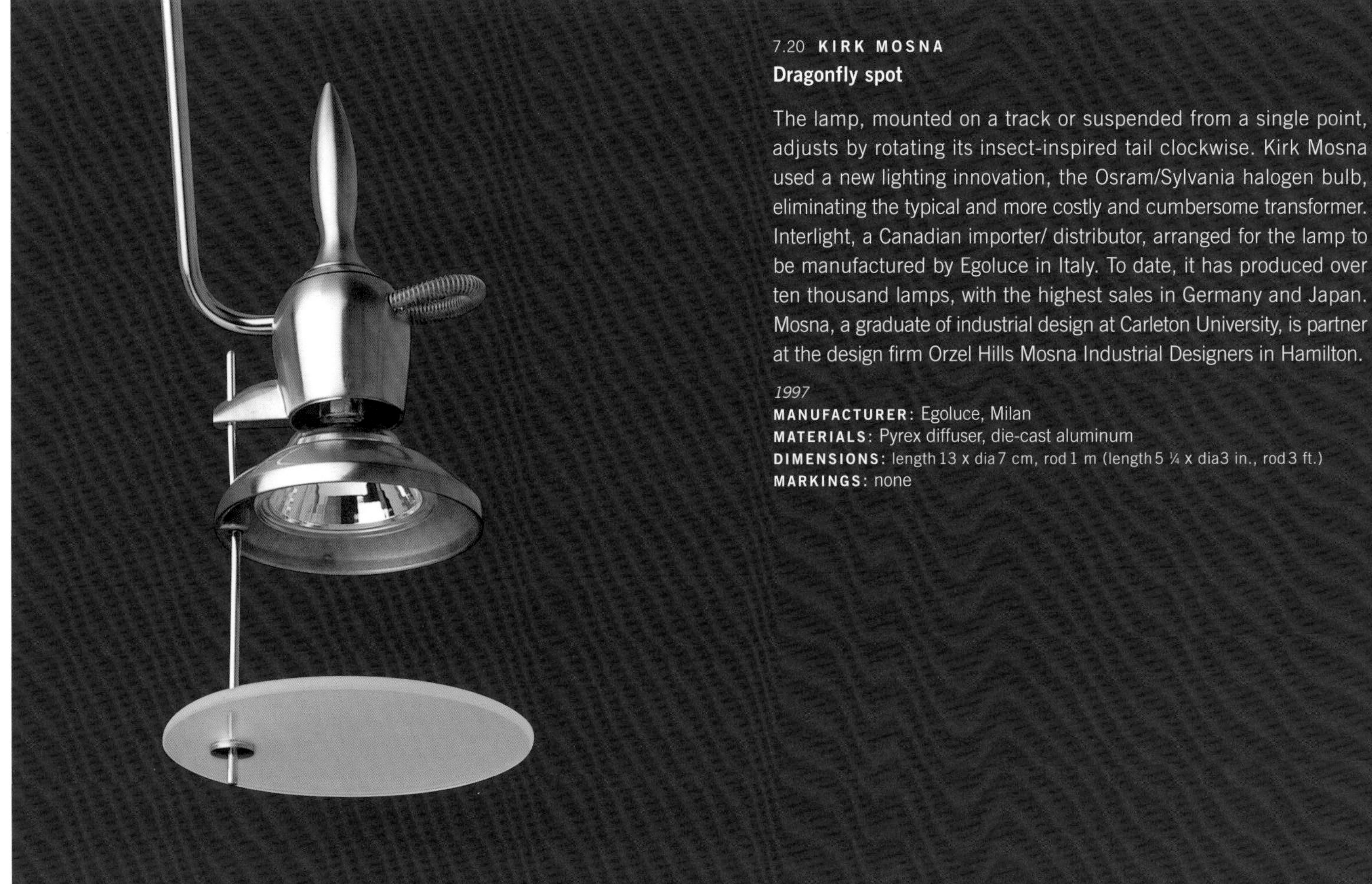

7.20 **KIRK MOSNA**

Dragonfly spot

The lamp, mounted on a track or suspended from a single point, adjusts by rotating its insect-inspired tail clockwise. Kirk Mosna used a new lighting innovation, the Osram/Sylvania halogen bulb, eliminating the typical and more costly and cumbersome transformer. Interlight, a Canadian importer/ distributor, arranged for the lamp to be manufactured by Egoluce in Italy. To date, it has produced over ten thousand lamps, with the highest sales in Germany and Japan. Mosna, a graduate of industrial design at Carleton University, is partner at the design firm Orzel Hills Mosna Industrial Designers in Hamilton.

1997
MANUFACTURER: Egoluce, Milan
MATERIALS: Pyrex diffuser, die-cast aluminum
DIMENSIONS: length 13 x dia 7 cm, rod 1 m (length 5 ¼ x dia3 in., rod 3 ft.)
MARKINGS: none

8.1 **STANLEY COSGROVE**

Trees with Leaves textile

Original screen-print designs for textiles are rare in Canada, particularly by fine artists such as Stanley Cosgrove. His flat, abstracted autumn landscape features clusters of trees among falling leaves. Sponsored by the Quebec government, Trees with Leaves was part of the Canadart Print Collection, which included designs by other well-known artists. Intended for curtains, lampshades and upholstery as well as apparel, the designs were available by the yard in three background colours in twelve combinations. They sold exclusively at Morgan's in Montreal and Simpson's in Toronto. The fabrics were promoted in *Canadian Homes and Gardens*, *The Globe and Mail* and *Canadian Art* in 1950, and the publicity caused a stir. A sample of Cosgrove's design is in the Musée des Beaux-Arts, Montreal.[1]

c. 1949

MANUFACTURER: Canadart Print Company, Montreal

MATERIALS: screen-printed cotton

DIMENSIONS: w120 cm, repeat 86 cm (w47 ¼ in., repeat 34 in.)

MARKINGS: printed on the selvage: A Canadart PRINT BY STANLEY COSGROVE NO. 47

8 Textiles

Textile Design Prospects Improve

Canada's commercial textile industry, based in Quebec and Ontario, has generally catered to the more lucrative local garment industry (where styles change each season) and the industrial product market rather than home furnishings. Industrialists who opened textile mills in the nineteenth century did not cultivate a local design tradition.[2] Dominion Textiles of Drummondville, Quebec, eventually dominated the Canadian industry for most of the next century. A 1948 report on Canadian textiles organized by the NIDC observed that industry treated designers like "technicians" who were expected to borrow and adopt designs from abroad.[3] The newly opened post-war schools teaching textiles, like the École du Meuble and the Ontario College of Art, favoured an artistic approach, while the École du Textile in St-Hyacinthe encouraged a more commercial approach. This pedagogy split textiles into limited-production high-design and generic commercial production. A decade later, when the NIDC enlisted Courtauld's Canada, a synthetic fibre company in Cornwall, to sponsor a competition to promote printed fabric for upholstery,[4] it received only eight submissions.

In the sixties, the building industry and contract furniture market boomed, which improved prospects for Canadian-designed textiles. In 1963 Gary Smith founded Unifab in Montreal, initially to import fabrics from Europe, but it soon expanded into manufacturing. Now operating in Toronto, the business continues under his son Steve Smith. Norm Proud, a former employee of Unifab, founded Tandem Fabrics in 1969, and since 1975 has run its mill in Moncton. Both Unifab and Tandem work with interior designers to create custom patterns for commissioned furniture for corporate offices. In Nova Scotia, the small company Designcraft opened in 1976 and commissioned designs by Suzanne Swannie, who later became a noted textile artist. Her woven upholstery patterns caught the eye of Thomas Lamb when he worked for the Dominion Chair Company.[5] By the eighties, however, there were few professional designers working exclusively for the textile and wallpaper industries. Rather it is the small handicraft industry scattered across the country that has achieved critical acclaim, producing printed and woven textiles in limited quantities.

Weaving Attracts Interior Designers

In the thirties and forties, an economic renewal program and the arrival of portable hand looms, like the Thackerey, made the craft relatively inexpensive and easily adaptable to home life.[6] Through drive and determination, Karen Bulow in Montreal ran one of the few firms to produce fabrics for the interior design market, distinguishing herself from the hobbyist (FIG. 8.4). Similarly, Velta Vilsons (FIG. 8.5), who had worked for Bulow, founded the Toronto Handweaving Studio in 1956 to provide woven fabrics for interiors.

Across the country, Helme Ehasalu in Montreal, Honey Hooser in Vancouver and Dorothy and Harold Burnham in Toronto contributed to the handicraft weaving industry. From their studio in Jordan, Ontario, the Burnhams produced yardage for apparel, tablecloths and other linens.[7] Since the sixties, however, the studio craft movement inspired professionals to adopt a more artistic approach.

Production handicraft weavers virtually disappeared in 1989, with the closing of both Karen Bulow Ltd. (owned by Margareta Steeves since 1960) and Sheridan College's weaving department. Weaving studios continue to exist, such as the Silo Weavers in St. Jacobs, Ontario, but they typically create piecework for table linen and apparel rather than drapery and upholstery fabric.

Block Printing Becomes Economical

Post-war designers have favoured silkscreening and block printing designs on blank fabrics, a relatively simple and inexpensive technique, which provides immediate results. Interior designers often produced custom designs to enliven modern interiors and introduced contemporary styling. Henri Beaulac, a professor at the École du Meuble, designed flamboyant print fabrics in the fifties. On the West Coast, Marianne McCrea McClain, a graduate of the Vancouver School of Art, created custom-silkscreened drapery material under the trade name Marian.[8] The self-taught designers Joanne and John Brook, in Toronto, used printing to create all-over linear-patterned draperies to complement their modern furniture (FIG. 8.9).

Occasionally printed fabrics served as a means for artists to reach wider audiences. Peter Freygood in Montreal founded Canadart in 1949 to reproduce in fabric the work of renowned artists like Stanley Cosgrove (FIG. 8.1), Alfred Pellan and Paul-Émile Borduas. The collection received considerable media attention and remained in production for several years, although Borduas later disassociated himself from the project.[9]

The craft artist Thor Hansen employed Canadian imagery and wildlife in his textiles (FIGS. 8.6, 8.7). When his long-time employer, the British American Oil Company, asked him to co-ordinate the arts and crafts for its new interiors, he also commissioned designs from artisans like Elizabeth Wilkes Hoey.[10] Based in Ontario, Wilkes Hoey designed custom patterns for local clients, including Canadian Westinghouse Company and Appleby College.[11] Like Hansen, she built a respectable career printing her own line of Canadiana fabrics depicting abstract renditions of spinning wheels, birch trees and maple leaves.

In the hands of professionals and amateurs alike, printed textiles in limited editions continued to flourish and reflected the styles of the day. In the mid-sixties, Sven Sandin, who ran Samo Textiles, Montreal, as a fabric import business for over thirty years, commissioned silkscreen designs from Jacques Corriveau, Philip Salmon and Michel Ducharme. The latter's bold supergraphics were notable. In the eighties the short-lived Design Cooperative, Toronto, founded by graduates of Sheridan College, introduced whimsy and coloured fabrics to studio furniture, and Paula MacMillan explored her retro-modern motifs for nineties home accessories. The Montrealers Robert Lamarre and Monique Beauregard are long-established textile designers who have carved out a niche market balancing craft and industry (FIGS. 8.12, 8.11).

Reflecting the nineties phenomenon of designers catering to export markets, Creative Matters in Toronto make prints strictly for carpets. Creative Matters was founded in 1988 by Carol Sebert (who studied textiles at Sheridan College), Luba Huzan (a graduate of graphic design at Ontario College of Art) and Donna Hastings (fashion at Ryerson Polytechnical Institute) and has collaborated with the interior designers Brian Gluckstein in Toronto and Bill Solfied and the Rockwell Group in New York City.

TOP RIGHT: 8.2 Lucia Kinghorn and Liz Crawford's Dead Guy felt area rug capitalized on the nineties sense of irony.

BOTTOM RIGHT: 8.3 Karen Bulow Ltd. supplied textured blinds for corporate interiors in the seventies and eighties.

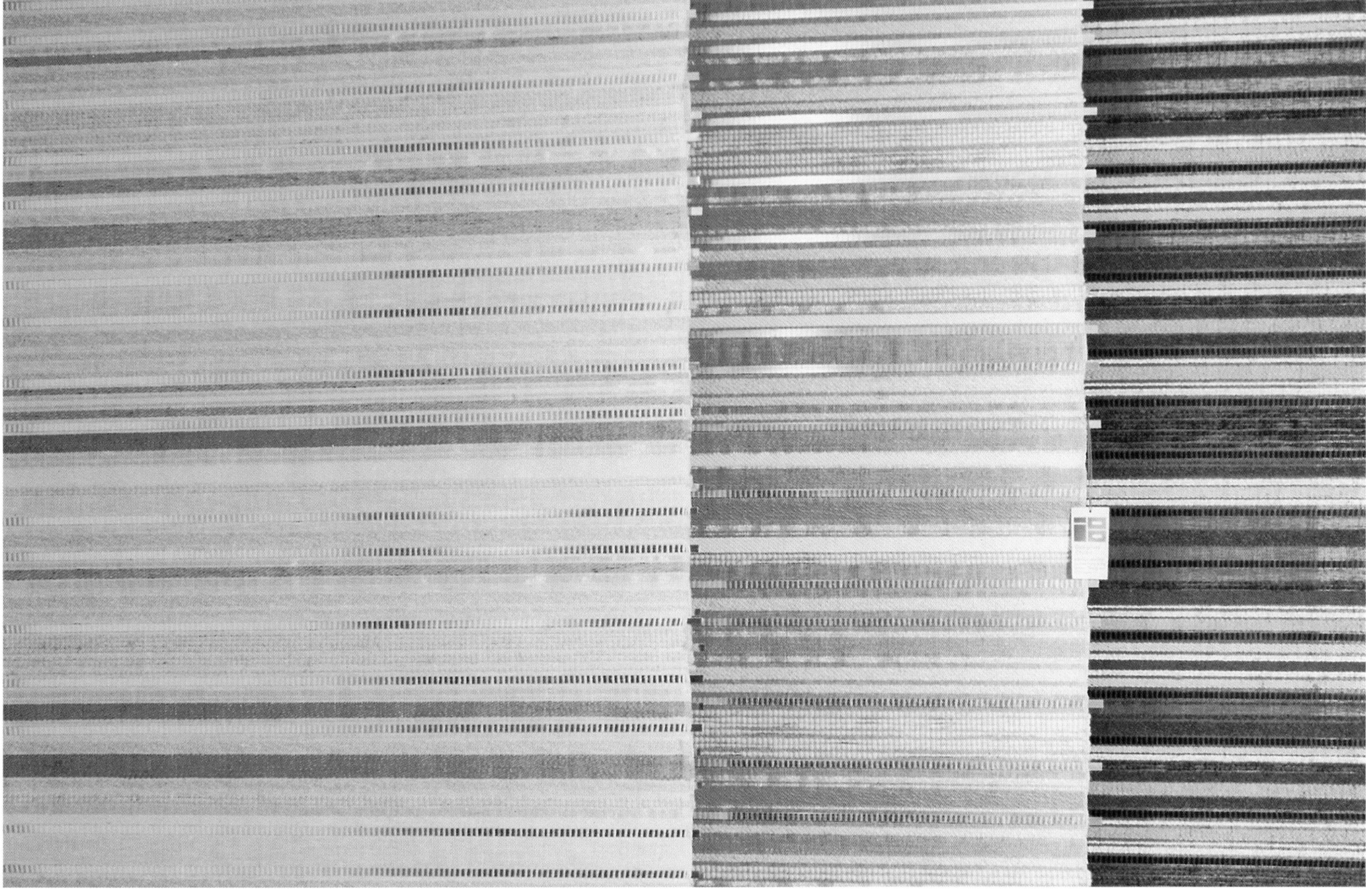

8.4 **KAREN BULOW**

Fabric samples

Karen Bulow collaborated closely with designers on custom fabrics. Typically, she designed on the spot with the interior decorator sitting next to her at the loom, catching the shuttle while they "sketched." She was also known for creating design samples in the countryside with her portable table loom. Bulow preferred subdued palettes, which she achieved by mixing coloured yarns with strands of grey. Focusing on the texture of the material, she kept pattern to a minimum.

1954
MANUFACTURER: Canadian Homespuns, Montreal
MATERIALS: wool, gold thread
DIMENSIONS: grey: w 122 cm, repeat 42 cm (w 18 in., repeat 16 ½ in.); blue: w 127 cm, repeat 25 cm (w 50 in., repeat 10 in.)
MARKINGS: none

8.5 **VELTA VILSONS**

Drapery fabric

Velta Vilsons synthesizes craft and industry with her handsome woolen weaves.

1961
MANUFACTURER: Vilsons Weaving, Toronto
MATERIALS: cotton, wool, linen
DIMENSIONS: w114 cm (w45 in.)
MARKINGS: none

8.6 **THOR HANSEN**
Jack-in-the-Pulpit textile

8.7 **THOR HANSEN**
Sunridge/Geese in Flight textile

The Danish émigré Thor Hansen originally developed the Sunridge/Geese in Flight (ABOVE) design for a tapestry (hooked in Cape Breton) to be displayed in the British American Oil building in Toronto. Inspired by the Group of Seven, Hansen was a craft advocate and believed in cultivating a national style. The success of this design led him to produce a series of fabrics for drapery and upholstery based on Canadian flora and fauna. Jack-in-the-Pulpit (LEFT) and Blue Heron were among the favourites. Hansen's subtle contemplative colour scheme of yellows, greys and maroons made the fabrics ideal for fifties-era ranch-style houses. The Huronia Museum in Midland, Ontario, holds Hansen's collection of fabric samples, linoleum panels and metalwork designed for the BA Oil buildings.

c. 1950s
MANUFACTURER: A. B. Caya, Kitchener; screen-printed by Montreal Fast Print, Montreal
MATERIALS: viscose
DIMENSIONS: w108 cm, repeat 56 cm (w42 in., repeat 22 in.)
MARKINGS: none

8.8 **JOHN GALLOP**

Meadow

John Gallop ran the interior design department at the Toronto-based architectural firm John. B. Parkin Associates. With his "flower power" pattern, he moved away from modernism and into Pop.

1960
MANUFACTURER: Contemporary Distribution, Toronto
MATERIALS: silkscreened Peru linen
DIMENSIONS: w 27 cm, repeat 60 cm (w 50 in., repeat 24 in.)
MARKINGS: paper label: Contemporary Distribution; printed on the selvage: Contemporary Distribution Meadow by John Gallop

8.9 **JOANNE AND JOHN BROOK**

Tobacco Leaf

Tobacco Leaf has a bold and free-spirited style that is characteristic of the work of the husband-and-wife team operating as J & J Brook, whose draperies were admired for inventive design. Catering to the architecture and interior design trade, the company made patterns in any combination of twenty-six colours. The designers used a variety of sources to produce their work, including Elizabeth Wilkes Hoey in Bronte, Ontario, and Jim Farquhar in Toronto. Tobacco Leaf was included in the *Design Index*, 1954.

1952
MANUFACTURER: Contemporary Distribution, Toronto
MATERIALS: hand-blocked Peru linen
DIMENSIONS: w122 cm, repeat 60 cm (w48 in., repeat 24 in.)
MARKINGS: paper label: Contemporary Distribution

8.10 **FRANÇOIS DALLEGRET**
KiiK black-and-white fabric

Dallegret sold the KiiK pattern to the American manufacturer Knoll, who reproduced it as Lines, a large-scale pattern for drapery. It features slightly curved opaque parallel stripes on a sheer background. The pattern of rippling stripes—similar to the lines made by a rake drawn through sand—hang vertically or horizontally. The fabric was launched in black and white. Six additional colours, each with white, were introduced in 1973 (camel, brass, stone, grey, charcoal and copper). Black was discontinued in 1974, but the others were in production until 1978.[12]

1969
MANUFACTURER: Knoll, New York City
MATERIALS: silkscreened polyester
DIMENSIONS: w122 cm, repeat 76 cm (w48 in., repeat 31 in.)
MARKINGS: paper label: Knoll Textiles 745 Fifth Avenue, New York 10022

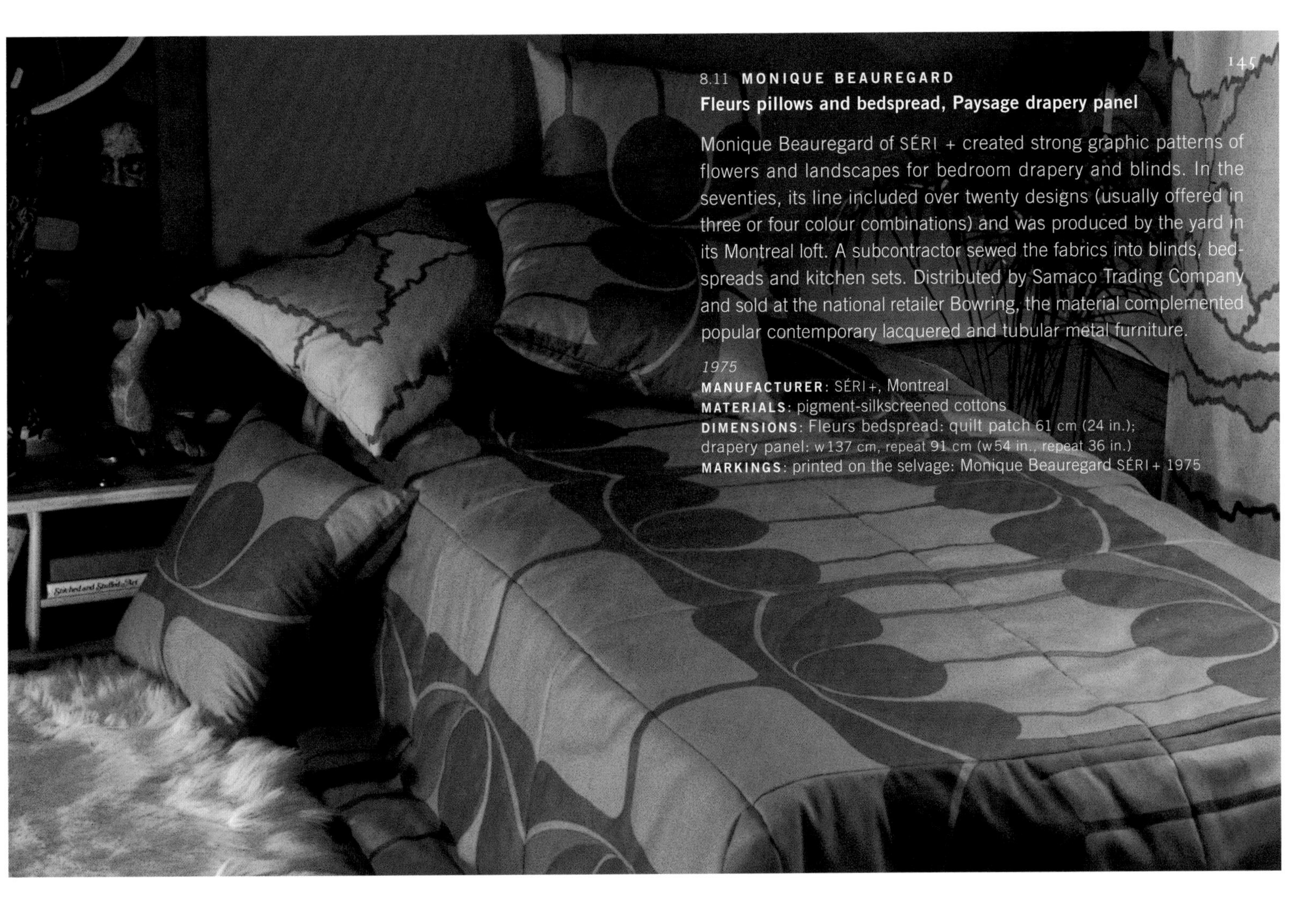

8.11 **MONIQUE BEAUREGARD**

Fleurs pillows and bedspread, Paysage drapery panel

Monique Beauregard of SÉRI + created strong graphic patterns of flowers and landscapes for bedroom drapery and blinds. In the seventies, its line included over twenty designs (usually offered in three or four colour combinations) and was produced by the yard in its Montreal loft. A subcontractor sewed the fabrics into blinds, bed-spreads and kitchen sets. Distributed by Samaco Trading Company and sold at the national retailer Bowring, the material complemented popular contemporary lacquered and tubular metal furniture.

1975
MANUFACTURER: SÉRI +, Montreal
MATERIALS: pigment-silkscreened cottons
DIMENSIONS: Fleurs bedspread: quilt patch 61 cm (24 in.); drapery panel: w137 cm, repeat 91 cm (w54 in., repeat 36 in.)
MARKINGS: printed on the selvage: Monique Beauregard SÉRI + 1975

8.12 **ROBERT LAMARRE**

Oxide yardage

Robert Lamarre and Monique Beauregard opened Éditextil in 1997 and equipped the studio with high-tech machinery to pursue new design techniques. In Oxide, Lamarre achieved lustrous colours by scanning a photograph of an oxidized metal door, then heat-transferring the design to fabric. Known as dye sublimation, the technique requires no water and therefore creates no rinse-off pollution. The process is ideal for creating one-off art pieces or for developing prototypes for industry.

1997
MANUFACTURER: Éditextil, Montreal
MATERIALS: PES charmeuse, dye sublimation
DIMENSIONS: 132 cm, repeat 33 cm (w52 in., repeat 13 in.)
MARKINGS: printed on the selvage: Robert Lamarre 1997 Éditextil

9.1 **GORDON DUERN**

701 stereo

Taking its cue from the cantilevered aluminum bases of Herman Miller's Action Office line, the 701 looks more like a storage unit than a stereo. Spare and elegant, it was promoted as a pedestal for objets d'art with room inside for seventy-five records or a hideaway bar. Its fully equipped stereo seemed almost incidental. A rolltop door preserved the unit's long, low silhouette. In 1978 Electrohome updated the design with Tempo, a centre-pedestal stereo with a similarly low-slung cabinet and sliding tambour cover. About six thousand of the 701 units were sold.[1]

Marketed 1967

MANUFACTURER: Electrohome, Kitchener, Ontario

MATERIALS: natural walnut, aluminum finish

DIMENSIONS: h65 x w160 x d61 cm (h25 ½ x w63 x d24 in.)

MARKINGS: paper label, rear left side:

"Circa" 75 model 701 stereo—serial # 00000.

Made in Canada by Electrohome Limited, Kitchener, Ontario

9 Consumer Electronics

Radio Enters the Living Room

Building on a series of inventions by scientists from Europe and North America, Reginald Fessenden, born in East Bolton, Quebec, became the first person to transmit the sound of a human voice without the use of wires. The event occurred in 1900, a year before Guglielmo Marconi transmitted a Morse code signal across the Atlantic. A new century of communications was born.

Initially, radio sets were mere boxes, designed to house cumbersome and often inflammable components. As the technology progressed, it became important to convince homeowners that these ungainly contraptions should inhabit their living rooms. To make them more acceptable, radios were placed within cabinets that mimicked popular furniture styles.

Alternating current tubes allowed radios to operate on household electrical currents, thus eliminating the need for bulky batteries and aerials. In 1925 Toronto's Edward Rogers made the tubes more reliable, which in turn popularized radio.[2] In quick succession, indigenous companies such as Standard Radio Manufacturing (Rogers) and Electrohome, as well as a host of branch-plant firms like Canadian General Electric Company (CGE), Canadian Westinghouse Company and Canadian Marconi Company—most established shortly after the turn of the century—began manufacturing radios in earnest. These companies formed a consortium to permit members to use one another's patents (on a pro rata basis), thus streamlining access to new developments and, to an extent, restricting competition.[3]

Plastics Make Design Viable

The widespread introduction of plastic changed the way radios were perceived and marketed. Within two decades of electrification, radios became decorative objects, sought after principally for their cabinet design. The material's debut, however, was inauspicious. The American manufacturer Crosley built the world's first moulded plastic radio cabinet in the early thirties, but it was clumsy and unappealing. The lingering effects of the Depression slowed the development of both plastics and radios, although in 1934 Wells Coates designed his now famous round Bakelite plastic radio in England.

When the Second World War broke out, many manufacturers, including CGE, Northern Electric Company and a newcomer, Addison Industries, began making military communications equipment such as navigational aids and walkie-talkies. These "military converts" also tinkered with consumer-oriented radios on the side, although material shortages stalled the designs in the prototype stage. By war's end, the former fledgling firms had become professional assembly-line operations, with excess capacity to be absorbed. The miniaturization of electronic components for military use, coupled with the development of new plastic compounds and handling techniques, would have a profound effect on the consumer electronics industry.

Electric Goods Market Chaos

After the war, most communications-oriented firms began manufacturing other electrical goods. Northern Electric, which had long produced telephone systems, branched into products for the movie and broadcasting industries and more. Addison Industries added large household appliances like refrigerators and stoves to its manufacturing mix. And CGE made just about anything with a two-pronged plug.

This led to a chaotic market, with almost daily announcements of new companies, followed by mergers, consolidations and failures. Brands traded hands like baseball cards as manufacturers sought to create their most efficient lineup. Companies even made entire products on behalf of their competitors—if the favour was reciprocated.

These byzantine networks and arrangements were fuelled by enormous pent-up consumer demand. After years of caution and penny-pinching, people wanted consumer goods, although they were still forced to choose between a new washer and dryer or a combination radio/record player, which often created conflict at the manufacturing level. Radio was the glamour product, and soon new models were appearing at the rate of forty to sixty a year.[4] Appliances, more profitable and less buffeted by trends, became the bread-and-butter products. Many designers of vacuum cleaners and floor polishers justifiably felt overlooked as resources and esteem followed those working with radio.

This is not to say that Canadian radio design was a hotbed of innovation but simply that radio became a marketing tool to increase the appeal of an entire brand. A snazzy radio created a "halo" effect over a manufacturer's more conventional output. As a result, many manufacturers prominently featured their most popular radios in multiple-product advertisements, which in turn gave radios incredible, if somewhat undeserved, attention.

Design as a Marketing Tool

The first breakout Canadian radio designs were the Addison Model 5 (FIG. 9.7) and, to a lesser extent, Northern Electric's Model 5000 Baby Champ (FIG. 9.8), both introduced immediately after the war. The design motifs were borrowed from the genteel and luxurious art deco period of the 1920s: modern, but certainly familiar to buyers.

Most other Canadian manufacturers were even more conservative in their approach. Liberated from the initial need to match household furniture, most radios merely imitated other familiar objects like luggage, music boxes and sports equipment. One of the first modest design trends was the introduction of sportsmen's radios. By the late forties, Rogers, Northern Electric and Marconi launched models that operated on either batteries or electricity and looked appropriate alongside a fishing-tackle box.

The sameness may have resulted from the co-operative nature of the business. At one time or another, 174 companies were making radios in Canada,[5] yet there were scant differences between brands. Not only did companies buy the coils, capacitors and tubes from the same manufacturers (often a competitor in the finished goods sector), but they also used the same subcontractors. Speakers and grille cloths came from a short list of suppliers, and a single factory stamped out many of the chassis.

Eventually, newly formed industrial design divisions of companies began to flex their creative muscles. In the mid-sixties, the Montrealer André Morin helped to introduce innovations within

TOP RIGHT: 9.2 Eric Winter created a modular radio with six face plates for Rogers-Majestic in 1951.

BOTTOM RIGHT: 9.3 The G3 stereo of the mid-sixties by Clairtone was the last and most portable of the G line.

Rogers Majestic

the two-man design department at Marconi, then later at RCA Victor Company, within its 150-person engineering department. RCA was independent of its U.S. counterpart in design, administration and manufacturing—an aberration for its time.[6]

Just as designers began to have an impact on the market, radio ceased being a commodity item and entered the pantheon of popular culture. This phenomenon can make the inherent value or even function of the object immaterial. True to form, when Addison Model 2 radios (FIG. 9.6) became "hot," buyers and sellers overlooked their poor-quality sound reproduction and cracked plastic cabinets. In a memo of surprising candour, a sales manager at Eaton's asked the company's purchasing department for more colourful plastic radios while admitting product returns were at all-time highs.[7]

The Influence of Automotive Styling

One of the reasons for the crossover into popular culture is that radio design began to draw its inspiration from the automotive industry. The trendiest radios soon sported sleek aerodynamic styling. Dials and grilles were borrowed from a dashboard or headlight and balanced by a decorative panel in the shape of, say, a Cadillac fin. Developments in plastics and pigments made colours more expressive and vibrant: bright reds, yellows and greens, reflecting the trend in automobiles. Lettering and stripes became an obsession.

The reality for most Canadian manufacturers, however, was that the market wasn't large or sophisticated enough to support constant innovation. The trick was to find an inexpensive method to give the appearance of responding to trends. The styling solution was to create a "family" of radios with distinct personalities aimed at specific market segments that used the same cabinet and chassis to reduce tooling and manufacturing costs. Only the faceplate, the knobs and a few other cosmetic elements changed.

This strategy, of course, required a high degree of communication and collaboration among the design, engineering, production and even marketing departments. Marconi, RCA and Rogers-Majestic (so named after Rogers purchased the U.S. company that made the Majestic radio brand) all created successful modular radio families (FIG. 9.2). But planning and co-ordination was a time-consuming and daunting task, so the exercise was rarely repeated. It did teach manufacturers in Canada and elsewhere how to squeeze out small efficiencies with one-off radio styles.

The stiffest competition, however, wasn't from outside the country but rather from the branch plants operating within Canada. They had the advantage of eliminating expensive development costs by borrowing moulds and tool-and-die equipment to stamp out the latest American model. Often identified as "Made in Canada," virtually all the output of companies such as CGE, Westinghouse and Philco were produced in this manner. The numbers tell the story. A typical production run in the U.S. was fifty thousand units; in Canada, a five-thousand-unit run earned you a promotion.[8]

Despite the advantages of catering to a much larger market, American engineers frequently made fact-finding missions to Canada. The Canadian Standards Association safety code was more stringent, resulting in better-quality radios, while American manufacturers were plagued with sparking models that could result in death.

A Mature Market Demands Innovation

By the early 1950s, the radio market was already showing signs of maturing. Of the 3.9 million households in Canada, most owned two radios, for a total of 6.2 million units.[9] Manufacturers began to employ the age-old strategy of creating "product extensions": putting designers to work creating radios for the kitchen, cottage or teenager's bedroom. Radios doubling as lamps were popular, as was the clock/radio combination. Radios were also coupled with record players, amplifiers and televisions. Although it created a dizzying volume, the emphasis on function over form decreased the design content.

The advent of transistors, which made even smaller components possible, introduced to the market radios in novelty shapes like telephones, teddy bears and coffee mugs, reflecting popular culture rather than design quality. Marconi developed its own transistors, the earliest versions of which were plug-in models (as the company didn't know whether transistors would have to be replaced).[10]

By the mid-sixties, most companies in both Canada and the U.S. ceased making radios—all swept away by the Asian juggernaut of "boxed" components. Many purists, however, remain attached to tube radios and consider the quality of sound far superior. Collectors must wrestle with a fluid and sometimes unfathomable market, where objects are valued for their plastic cabinets, electronics, nostalgia and even popularity and rarity. Or sometimes all of the above.

Designers Remake Audio

As with radios, audio equipment manufacturing began in Canada at the turn of the century when companies like Berliner Gramophone established phonograph and disc manufacturing facilities. Once again, there was a plethora of competition from American companies like Concertphone, Crescent Silver Tone and Musicphone until the market began to rationalize, between the two world wars.

After the Second World War, many of the same branch-plant operations that dominated radio and television manufacturing also controlled the console audio market. These companies included Admiral, CGE, Westinghouse, Philco, Philips, RCA, Sylvania and Zenith. Only three—Clairtone Sound Corporation, Electrohome and Fleetwood Electrical Products—were wholly indigenous, although CGE, RCA and Westinghouse established plants early in the century and made significant investments in Canada. Others followed the route chosen by foreign automobile manufacturers. When threatened with potential trade tariffs, these companies elected to manufacture (but not design and develop) products locally.

Although collectively the Big Ten employed about thirty-five industrial designers,[11] until the sixties their output, like that of their radio colleagues, consisted primarily of modifying traditional cabinet designs to accommodate the technical requirements of audio components (amplifier, tuner, record player and optional tape recorder). The fact that this country has any memorable designs in the area of electronics is the result of a few determined Canadians: Peter Munk and David Gilmour of Clairtone; Carl Pollock, the son of the founder of Electrohome; and Floyd Toole, the chief acoustical consultant at the National Research Council in Ottawa.

Under Toole's leadership, the unique testing and measuring facilities at the NRC made it possible to measure sound performance, thus supporting (or refuting) the claims of manufacturers and audiophiles. It helped to spawn a small community of Canadian specialty manufacturers dedicated to perfect sound reproduction. Edward Manning in Toronto operated one of the first "boutique" audio shops. His Manning Combination high-end audio components of the early fifties (which appeared in the industrialist E.P. Taylor's residence in north Toronto) were developed with the aid of the NRC.

It would be another decade before the engineer Munk and the self-taught designer Gilmour founded Clairtone to build audio equipment for aficionados, including the world's first stereo consoles that completely dispensed with tubes and used transistors instead, in July 1963. That same

year, the company's groundbreaking Project G (FIG. 9.15) challenged Electrohome. George Eitel, a twenty-year veteran of Electrohome's design department, had publicly announced, "I hate gingerbread,"[12] and it had long been Pollock's mission to re-energize his father's company by introducing modern design. The company's Circa series of 1965 (FIGS. 9.17, 9.18) was intended to establish a progressive image and be the company's standard-bearer for years to come.

Both corporate design departments were influenced by furniture styles emanating from Scandinavia and the U.S., in particular the American inclination for combining wooden cabinetry with metal bases as expressed by the designers Charles Eames, George Nelson and, to a lesser extent, Florence Knoll.

Many consumers weren't prepared to place a sleek, unabashedly contemporary stereo in the middle of their living room—at least not when such high style cost as much as $1,500. A chief designer at one of the multinationals, where the output was 90 per cent original design, explained the volume difficulties. Every year his department tracked buying trends in stereos: for example, 30 per cent Colonial, 15 per cent French Provincial, and so on. It plotted its production goals based on style, price and expected sales volume. All the selection, however, militated against achieving economies of scale.

Modern design became an aberration. Both Clairtone and Electrohome depended almost entirely on conservative styling to make their sales targets. One exception to the "forward design/underperforming sales" rule was Electrohome's Circa 711/Apollo series (FIGS. 9.19, 9.20). By the end of the decade, André Morin had even more success with his Forma Collection for RCA (FIGS. 9.12, 9.13, 9.14). It briefly became the best-selling RCA stereo model in Canada.[13]

Before any company could seriously mount a design-driven assault against pedestrian stereo cabinet design, the market changed. Components began to appear in unimaginative metal boxes that flaunted technology rather than style. In the absence of a necessity for local designs to match components to indigenous furniture, companies around the globe withdrew from the market. The "Golden Age" of stereo design in Canada didn't outlast the sixties. A handful of international companies came to dominate mainstream audio.

Throughout the latter part of the twentieth century, a few Canadian companies, undaunted by the onslaught of offshore product, served niches at the high end of the market. Most depended on the testing resources of the NRC. Chris Russell of Peterborough created the rugged Bryston power amp, Montreal's Ed Meitner designed a CD player and Dave Reich of Classe Audio, Lachine, Quebec, made amplifiers. For the most part, these achievements were based on technology rather than design, although most rock 'n' roll roadies have a story that illustrates the sturdiness of a Bryston amp, surely a compliment to its design.

In 1982 Marcel Riendeau of Oracle Audio Corporation, Sherbrooke, Quebec, received a Design Canada Award for Excellence for his Oracle Turntable—Premier Series II. It was cited for both its dramatic good looks and its functionality.[14] As the century ended, many Canadian boutique manufacturers had strongholds in micro sectors of the market such as reintroducing vacuum tube amplifiers or building speakers. Designing and manufacturing consumer electronics for the average family, however, was no longer feasible for most industrialized countries around the world.

Television Enslaves a Generation

Many of the developments that made television possible occurred before the turn of the century, but most agree that the medium's wide-scale introduction to North American consumers was at

LEFT: 9.4 Hugh Spencer designed the iconic Project G stereo for Clairtone, 1963.

CLAIRTONE COLOR

the 1939 New York World's Fair. General Electric, RCA and Westinghouse had working models, and a rudimentary transmission system was soon established in major urban centres in the U.S. Later that same summer, television was showcased at the Canadian National Exhibition in Toronto. But it would be more than a decade before regular television transmission was available in major cities, and even later the farther north you travelled from the forty-ninth parallel.[15]

By the time television became widely available in Toronto and Montreal (1952 and 1953 respectively), it was a commodity item, dominated by the same Big Ten firms that controlled radios, stereos and a host of other electrical products. Almost 60 per cent of Canadian households within range of a transmitter owned a television by 1956.[16]

Canadian manufacturers initially sought to serve the market with upscale console televisions made from real wood, mimicking the furniture-style introduction of radio and stereo. These models quickly decreased in size as television invaded suppertime, leisure time and family time. It became a case of homes (and people) adapting to television rather than the other way around. Technological advances (colour, remote controls and larger screen size) rather than design drove the market. More design content went into producing TV tables, swivel supports, rolling carts and other accessories than into the televisions themselves. Consumers had to accept the medium's ungainliness and work their world around it. Any of the smaller manufacturers that had hoped to make the leap from radios to televisions disappeared or contracted the "majors" to produce a model under their brand. Even Northern Electric, already a substantial corporation, hired Sylvania to make their televisions.

As with radio, many television components were purchased from a limited number of sources, reducing the manufacturer to an assembler. Components like cathode ray tubes imposed limits on creativity as they were large, heavy and required careful venting. Compared with the development of radio, miniaturization of television (with no attendant loss of picture quality) was a gradual process.

Colour television was introduced in the U.S. in 1966 and a year later in Canada. By then, about eighty thousand televisions were sold annually in this country. To have a presence in each sector (portables, black-and-white and colour, console models in various styles and so on), each of the manufacturers made up to twenty "new" televisions a year. Adding original design content only complicated the process. Designers often had to satisfy themselves by adding stripes, knobs or other inexpensive surface decoration.

Initially, buyers didn't want to convert to colour television. Clairtone, which marketed colour televisions before the signals were widely available, faced an uphill battle, as did its much larger American competitors, like Zenith. Capitalizing on the cachet of its Project G stereo, Clairtone launched the GTV (FIG. 9.5). Ironically, this consumer resistance occurred during an era when Canada was one of the leaders in broadcast transmission technology as a result of the Anik A1 satellite and other innovations.

By the end of the sixties, offshore manufacturers controlled the market and launched models targeted at specific market segments, often using space-age imagery. Thirty years later, the Victoria-based designer Peter Andringa revisited these motifs. His line of retro-styled televisions, assembled by his own firm, Mercury 7 Manufacturing, exhibit a "Jetsons" brand of exuberance about the future. The recent introduction of LED and gas plasma television screens, and the encroachment of new transmission formats like high-definition and digital television, should create an opportunity for determined designers to provide alternatives to the ubiquitous box.

LEFT: 9.5 Clairtone's GTV, 1965, by Al Faux and Anthony Mann, added style to the ubiquitous box.

9.6 **ADDISON**

Model 2, R5

Dubbed "the waterfall" because of its cascading grille, this small plastic radio has made a surprising journey since its introduction to the market in 1945. Initially a hard sell (Eaton's 1950–51 catalogue shows a four-tube model with battery pack for just $28.95), it has now become one of the most popular and expensive radios in the international collectibles field. Its striking yellow-and-red model appears in numerous radio publications, often labelled as originating from the U.S. Although it has never been attributed to a single designer, it was likely created in the Toronto plant in collaboration with a plastic moulding firm during the Second World War and put into production only after the war ended.

Five-tube AC/DC models, sold by Eaton's in 1946–47, came in just three colours: maroon or mottled green with ivory trim, or ivory with red trim. In various incarnations, Model 2 eventually became known and collected for its fifteen (or more) dramatic colour combinations. It's also noted for its pinwheel knobs, stacked base and gothic window dial.

Slightly altered, the radio was also known as the R5 (with various subsets) and offered in either a wooden, Catalin or Plaskon plastic cabinet. The lighter-weight Plaskon, an amino plastic similar to Bakelite, was moulded rather than cast, has less heft and feels soft to the touch. The similar L-2 edition was cast from the heavy dark plastic, Bakelite. The short-run, slow-baked and hand-polished Catalin models are preferred.

Catalin colours identified include yellow and red, yellow and maroon, marbleized green and yellow, marbleized black and white with white, maroon and white, white and vermilion, brown and white, turquoise (pistachio) and white. Bakelite was available in various hues of dark, solid colours and was sometimes painted. With the Plaskon versions, many of the colours have faded or yellowed, and many radios were assembled from miscellaneous parts in the factory, making it nearly impossible to authenticate all the colour combinations.[17]

c. 1940

MANUFACTURER: Addison Industries, Toronto

MATERIALS: Catalin plastic

DIMENSIONS: h 15 x w 27 x d 14 cm (h 6 x w 10½ x d 5½ in.)

MARKINGS: Addison logo type below tuning dial; paper label on bottom: Addison Industries Limited Toronto Canada; also paper label: CSA approved; metal label screwed inside back: Chassis R5A1 serial #00000.

9.7 **ADDISON**

Model 5

Apparently introduced simultaneously with the Model 2, this substantial tabletop radio exhibits a different personality, albeit also drawn from the art deco period. Many collectors attribute its stepped shoulders and five-column grille to the Mayan or Egyptian temples that had captured the public's imagination in the 1920s. Others claim it reflects the opulence of cinema palaces.

It first appeared in trade journals in 1945 and was available in walnut, mahogany or bleached wood. Two years later, it was showcased to the public in Catalin plastic. At nearly $60, it was twice the price of its competitors. Fully featured, it had five tubes, could pick up long-or short-wave radio and functioned on AC or DC currents, all of which contributed to its higher price. Its mottled striations (in dark green, maroon or ivory) are well conceived and appropriate to its design.

c. 1940
MANUFACTURER: Addison Industries, Toronto
MATERIALS: Catalin plastic
DIMENSIONS: h24 x w32 x d18 cm (h9½ x w12½ x d7 in.)
MARKINGS: Addison logo type above speaker grille

9.8 **NORTHERN ELECTRIC**

The Baby Champ, Model 5000 series

Commonly referred to as the "rainbow" model because of its distinctively shaped grille, this art deco design appears earlier in a 1939 compression-moulded Bakelite cabinet for Northern Electric's office intercom loudspeaker, the Magnaphone. In 1946 the company's internal publication, *The Northern Circuit*, featured four similarly styled Baby Champ radios being presented to a military hospital in Halifax. By autumn, Eaton's catalogue showcased an inexpensive four-tube Bakelite radio. The following spring, Eaton's labelled it Model B4100 and increased its price by 10 per cent. Later in 1947, a five-tube version in one of six colours was offered. This is generally believed to be the Model 5000, although the design underwent numerous modifications (like reducing or changing knobs) with each production run, ending with the 5100 series. The 5000 series purportedly established a Canadian record for units produced.

Collectors like the radio for its streamlined machine-age three-tiered banding and attention to detail; for example, the tuning dial on earlier models is a miniature propeller. Although its Bakelite housing is sturdy, the painted finish, chosen to compete against the newer brightly coloured soft plastics, is less durable. The Baby Champ name continued with the 5200 series, which bears little resemblance to its predecessor. Larger and more utilitarian, if technologically advanced, it won an NIDC award in 1948. The 5400 Baby Champs as well as the 5900 series of 1953 were plain rectangular radios enlivened by prominent dials and needles.

Baby Champs featured a paper label on the back with an image of a baby posing as a boxing champion. The initial Bakelite models were painted walnut or ivory. Later models came in six colours: carnation red, red/gold, blue, green, ivory and brown. Two companies, Engineering Products of Canada, Montreal (January 1948), and Peerless Engineering, Toronto (April 1948), advertised in *Canadian Plastics* magazine that they made the moulds.

First appears as Magnaphone cabinet in 1939, as radio cabinet in 1946
MANUFACTURER: Northern Electric Company, Montreal
MATERIALS: Bakelite plastic
DIMENSIONS: h16 x w27 x d16 cm (h6½ x w10½ x d6½ in.)
MARKINGS: Northern Electric logo type on bottom of tuning dial; silkscreened on back: Patented 1932 1933 & 1937 MADE IN CANADA NORTHERN ELECTRIC COMPANY LIMITED MONTREAL CANADA; Baby Champ paper label

9.9 **NORTHERN ELECTRIC**

Midge Model 5708

The design evolution of the Midge was the opposite of Northern Electric's Baby Champ series: each new version was increasingly more radical. Introduced in 1949 as a big-featured but lower-cost and smaller competitor to the Baby Champ, it underwent various styling changes, culminating with the 5700 series.

Launched under the Midge moniker in 1949, it began as a simple, front-grille box with rounded shoulders, brightened by snappy colour choices such as yellow, red and green. The design department first added a fin to its dial. The diagonal motif (5400 series) appears a year later in its grille, which wraps around its front and side panel. This edition also marks the introduction of a round tuning dial and was also made by Farnsworth in Indiana. The squared shoulders typical of a Midge have become rounder, as if flattened. With the 5700 series, developed at the Belleville plant, it has been squashed into a bulletlike shape, with two dramatic "headlights" for its grille and dial. Colour choices were red, green, blue, brown, ivory or white.

Midge name introduced 1949; 5700 series Midge dated 1953
MANUFACTURER: Northern Electric Company, Belleville, Ontario
MATERIALS: Bakelite plastic
DIMENSIONS: h 15 x w 31 x d 15 cm (h 6 x w 12 ¾ x d 5 ½ in.)
MARKINGS: Midge plus Northern Electric logo type on faceplate; printed on cardboard on back: Northern Electric Model 5708

9.10 **MAX DUCHARME**
Bean radio

This transistor radio was purportedly the last radio made in Canada and takes its name from its bean-like shape. The model was a rare departure for the risk-averse company. Its tooling alone costing $200,000. Luckily, buyers appreciated its quirky charm and bought forty thousand units—a remarkable volume in a mature and fragmented market. After this radio, Philips and all other Canadian manufacturers conceded the transistor market to offshore competitors.

In 1961 Interplas, an international plastics show in London, England, named the radio the best Canadian design. (Germany was awarded for a design by Max Braun.) The radio dial hints at the three-pointed-star logo of Mercedes Benz.

1960
MANUFACTURER: Philips Electronics Industries, Toronto
MATERIALS: injection-moulded Styrene
DIMENSIONS: h 11 x w 30 x d 12 cm (h 4 ½ x w 12 x d 5 in.)
MARKINGS: on faceplate: Philips, plus logo; on back: blue maple leaf decal and label: Union Made

9.11 **CANADIAN WESTINGHOUSE COMPANY**

Personality, Model 501, and Personality Plus, Model 502

This jaunty little radio (ABOVE LEFT) was initially named Personality but has become known as the five-way radio because it functions resting on any of its surfaces. It can also be affixed to a wall. It first appeared in 1948 and, although superseded by the Personality Plus in 1950, remained in production until 1951. Both models are compression-moulded from Plaskon, which is more translucent than Bakelite, permitting pigments to be added to the resin itself. Initially offered in seven colours, it has been seen in as many as ten—reflecting its four-year production run. Its plastic shell, though warmer and softer to the touch than Bakelite, is less stable when subjected to heat and humidity. Somewhat surprisingly, the radio is popular with U.S. collectors (they collect all ten colours), although it never did in appear that country.[18] Personality came in black, grey, blue, green, maroon, ivory and walnut; later, additional colours were white, yellow and turquoise; Personality Plus was brown, ivory, blue, green, maroon or grey.

Personality, 1948; Personality Plus, 1950
MANUFACTURER: Canadian Westinghouse Company, Hamilton.
Both were moulded by Joseph Stokes Rubber Company, Welland, Ontario
MATERIALS: both compression-moulded Plaskon ureaformaldehyde;
DIMENSIONS: 501: h22.8 x w15.2 x d12.7 cm (h9 x w6 x d5 in.);
502: h15.2 x w22.8 x d12.7 cm (h6 x w9 x d5 in.)
MARKINGS: 501: silkscreened on back in white printing: Made in Canada. Canadian Westinghouse Co. Ltd. Model 501; 502: Westinghouse logo type below the speaker

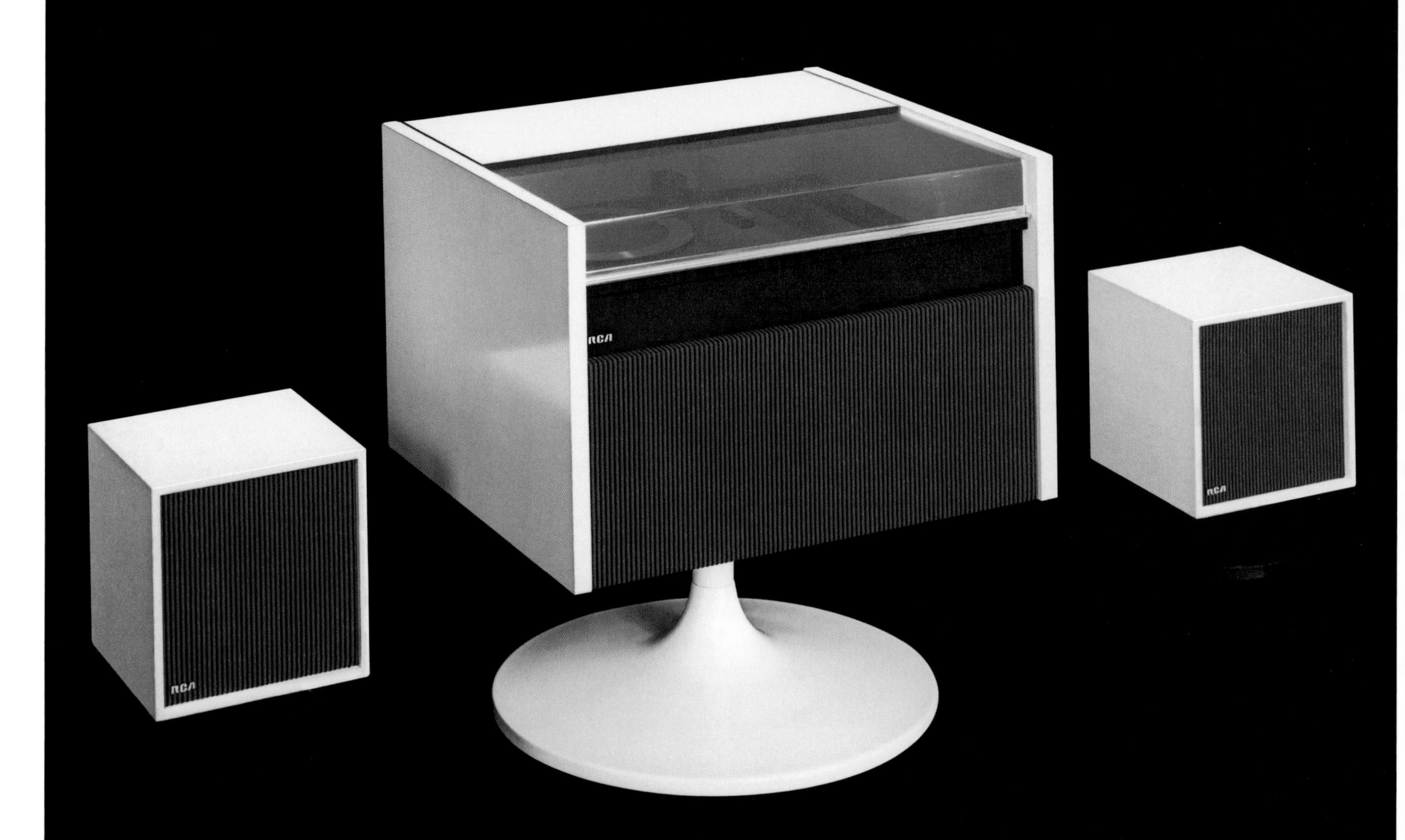
RCA
RCA
RCA

9.12, 9.13, 9.14 **ANDRÉ MORIN**

Forma Collection: Model SFA 1094, NéoForma Model SFC 1000 and MultiForma SRB 1098

When the marketing department of RCA's Canadian division issued a design brief for a stereo priced at $199, they weren't prepared for the first Forma prototype and rejected it outright. With the support of a more sympathetic corporate vice-president, André Morin convinced Eaton's (a major client of RCA's) to test-market the prototype. By three o'clock the first day, the department store had sold sixteen units and wanted to place an order for five hundred more. Somehow, RCA was persuaded to produce a first run of five thousand units, which promptly sold out within a month. In the first year of production, 165,000 units were purchased.

The Forma Collection became the company's largest-selling line in Canada and attracted the attention of the American RCA president, David Sarnoff. Even Sarnoff, however, wasn't able to talk the U.S. industrial design department into adopting something produced by their northern cousins, although he purchased a Canadian unit for his teenage children. Armed with that level of visibility, success and support, Morin designed more radical units.

The Forma Collection dispenses with the notion of a stereo cabinet as furniture and reconstitutes it as eye candy. The young hipsters gobbled it up, even though the company's brochure creaks with insecurity ("Are you daring enough...") and with apologies ("No wallflower, this stereo"). The design of the SFA 1094 is straightforward—a simple melamine cube on a plastic pedestal, with matching free-standing speakers. But it was the Life Savers colours of its plastic dust covers and speaker grilles that shook the establishment: wild magenta, intrigue blue or acid green. As a result, RCA suggested its "swinging" cube stereos might be better suited to a kitchen or bathroom!

Model SFC 1000 refines the cube's base, simultaneously providing a place for the speakers, while Model SRB 1098 deconstructs it into its most basic, stackable form. Many of the ideas suggested by the Forma Collection were taken up in more conservative units, with names such as the Sophisticate or the Modernist, which both sported tubular chrome bases. In 1972 the Forma Collection won a Canadian Award of Excellence in Industrial Design from the federal government.[19]

1967–72

MANUFACTURER: RCA Victor Company, Montreal

MATERIALS: melamine, teak or walnut veneer cube walls; melamine, plastic or aluminum bases; coloured plastic dust covers; fabric grille cloth (depending on model)

DIMENSIONS: varied, although typically based on 71 or 50 cm (28 or 19½ in.) cubes

MARKINGS: large RCA logo type on front face of stereo or smaller on faceplate of control panel

9.15 **HUGH SPENCER**
Project G stereo

In the sixties, Clairtone's Project G stereo was the epitome of "bachelor pad" cool. It transformed the traditional console box by cantilevering its speakers outside a Scandinavian-style modern cabinet. It won a silver medal at the 1964 Milan Triennale, the international design industry's most prestigious exhibition, and was adopted by progressives within the arts and entertainment community. Frank Sinatra endorsed it ("Listen to Sinatra on Project G; Sinatra does"), and G series models appeared in films such as *The Graduate* and *A Fine Madness*, with Sean Connery. Hugh Hefner reportedly bought a unit for the Playboy mansion.[20]

Constructed of rosewood with eighteen-inch (7 cm) anodized aluminum globe speakers (hence the G), it was labour-intensive to produce and retailed for as much as $1,600 U.S. With the exception of its imported Garrard turntable, the audio components were produced locally and earned five patents for sound reproduction methods. The stereo employed transistors instead of tubes, which eliminated the need for cabinet ventilation. As a result, the unit is finished on all sides and rolls on castors, allowing it to break away from its expected stationary position against a wall. Despite sales of fewer than a thousand units, the futuristic style and sexual-revolution swagger of the Project G makes it an icon of the space age. It is the most important stereo of its period and is a coveted international collectible.

1963
MANUFACTURER: Clairtone Sound Corporation, Toronto
MATERIALS: Brazilian Palisander (rosewood) cabinet, leather side panel inserts, movable matte black anodized aluminum speakers, brushed aluminum base
DIMENSIONS: h 71 x w 206 x d 47 cm (h 28 x w 83 x d 18 ½ in.)
MARKINGS: extensive paper labels underneath stereo: Clairtone Sound Corp. Ltd., with model number, finish, serial number, etc.; also, Clairtone logo on faceplate of tuner/amp; Garrard logo on turntable

9.16 **AL FAUX AND HUGH SPENCER**

G2

Although initially intended as a reworking of the Project G stereo to lower manufacturing costs, the G2 is a satisfactory design in its own right. The key feature it adds is flexibility: both the chassis and speakers can be detached from the main console to function as independent units. Its smaller, thirteen-inch (33 cm) speakers, made from cast aluminum, are less vulnerable to irreparable dents and can be rotated 360 degrees, whether housed in the main unit or plugged into optional free-standing supports (to better direct the sound).

The finicky rolltop cover of the Project G has been replaced with a removable Plexiglas dust cover. Although it's less expensive and more functional, electronically it is considered more sophisticated. While not as desirable as its antecedent, the G2 nevertheless has a strong following in the collectibles market.

1966

MANUFACTURER: Clairtone Sound Corporation, Stellarton, Nova Scotia
MATERIALS: rosewood chassis, cast aluminum speakers (painted black), Plexiglas dust cover, brushed aluminum base
DIMENSIONS: h 28 x w 198 x d 37 ½ cm (h 28 x w 78 x d 23 ¾ in.)
MARKINGS: label under base: Clairtone Sound Corporation Made in Canada

9.17 **GORDON DUERN**

Circa 75 stereo and sound chair (prototype only)

To compete with Clairtone's Project G stereo, Electrohome anticipated consumer demands ten years into the future. The resulting prototypes (of which only two were made) were aimed at the youth market.[21] The circular stereo and matching "sound" chair, with fingertip controls and speakers in its headrest, were conceived in 1965 as a "futuristic communications and entertainment nerve centre" to monitor the household, shop via television and create a music library. In addition to miniaturized stereo components, the Scandinavian-style circular unit could house a computer-driven record/playback audio device, global-time clock, videophone and colour projection television.

Advertisements trumpeted that consumers could "tune into a London theatre, the New York Stock Exchange, or a stadium in Rome to bring news and entertainment from every corner of the world." Electrohome used the prototypes (inset) as three-dimensional marketing tools to promote its new design outlook and introduced the Circa series by showcasing the units along with a film about its technological advances. In 1973 Design Canada conducted an extensive study on the product line. Both prototypes are in the company's museum collection at its head office in Kitchener.

9.18 **GORDON DUERN**

Circa 703 stereo and 704 sound chair

The Circa 703 stereo and 704 sound chair are production models of the prototypes. The circular unit is smaller than the prototype (which was too large for most living rooms) and limits itself to typical audio components rather than the global communications products envisaged for the prototype. The 703 features a tambour door and was available in walnut or teak. The 704 chair mimics the encasement-style chairs of the period and pumps the sound out through its headrest.

1967

MANUFACTURER: Electrohome, Kitchener, Ontario
MATERIALS: stereo: teak; chair: Naugahyde
DIMENSIONS: 703: h 65 x dia 91 cm (h 25 ½ x dia 36 in.);
704: h 105 x w 72 x d 69 cm (h 41 ½ x w 28 ½ x d 27 in.)
MARKINGS: paper labels on record player and on bottom of sound chair;
Circa 75 stereo (or sound chair) model #703 (or #704) Serial #00000.
Made in Canada by Electrohome Limited Kitchener, Ontario

9.19, 9.20 **GORDON DUERN AND KEITH MCQUARRIE**
Apollo 861 and Circa 711 stereos

Electrohome management believed the tabletop Apollo stereo would be an image design rather than a viable product, so its initial production run was a mere five hundred units. Its space-age styling (acrylic top and detached ball-shaped speakers) struck a chord with buyers, who kept the model in production for six years. Eventually, more than 42,500 were sold.[22] Spurred on by this success, the designers developed a pedestal version (with a similar Plexiglas top) that was marketed as "shades of 2001" (INSET). Keith McQuarrie, a graduate of Toronto's Ryerson Polytechnical Institute and the Ontario College of Art, primarily designed household products such as heaters and fans. He was responsible for creating the unit's Eero Saarinen–style spun aluminum base. Together the designers reduced the number of required component pieces from thirty-four to thirteen, resulting in a mid-range price of $300.

Known as either the Bubble or Smarties stereo, the pedestal version (INSET) went through various incarnations to eventually rack up market-leading sales of over ninety-five thousand units. The rarest editions feature coloured anodized aluminum bases in sought-after shades such as plum, tangerine, grape and lemon (hence the name Smarties).[23] Some include a cassette tape recorder. Less successful versions of the design, including some with wooden bases, were marketed under a number of department store brand names. About 10 per cent of the units were exported to the U.S., and the pedestal version was patented in the U.S. and Japan. The Circa 711 won a silver EEDEE award in 1970.

Apollo 861, 1966; Circa 711, 1970
MANUFACTURER: Electrohome, Kitchener, Ontario
MATERIALS: smoked Plexiglas bubble, brushed aluminum base
DIMENSIONS: Apollo: h 28 x dia 46 cm (h 11 x dia 18 in.); Circa 711 with lid: h 83 x dia 52 cm (h 32 ¾ x dia 20 ½ in.); speakers: h 27 x dia 20 cm (h 10 ¾ x dia 7 ¾ in.)
MARKINGS: paper label on bottom of base: Apollo stereo model #861 (or Circa 711 stereo)—Serial #00000. Made in Canada by Electrohome Limited Kitchener, Ontario

9.21 **ELECTROHOME STAFF/JOHN B. PARKIN ASSOCIATES**

Perception Modules

In 1961, with great fanfare, Electrohome launched Perception Modules at the National Association of Music Merchants Show, at the Canadian National Exhibition, and with inserts in *Canadian Architect* magazine. *Canadian Homes* featured it in layouts. The stereo was a collaboration between the company's marketing department and designers at Toronto's John B. Parkin architectural firm, which was designing Electrohome's modernist headquarters at the time.[24] The black-and-white modules, looking like something Braun's Dieter Rams might have designed, could be arranged in any configuration. Since the company's Deilcraft division specialized in fine furniture, the cubes were made from wood that was painted with polyurethane (rather than made from plastic melamine laminate). However, all the quality construction, fashion-forward design and intensive promotion didn't make the line a best-seller. Leftover cubes often ended up in storage bins in the Electrohome plant.

1968

MANUFACTURER: Electrohome, Kitchener, Ontario
MATERIALS: white polyurethane over wood
DIMENSIONS: based on modules of h43, 36 or 53 x w38 or 76 x d19 or 39 cm (h17, 14 or 21 x w15 or 30 x d7½ or 15½ in.); support bases: w114 or 152 cm (w45 or 60 in.)
MARKINGS: cubes were stamped Deilcraft; electronic units had paper labels with Made in Canada by Electrohome Limited Kitchener, Ontario

10.1 **JACQUES GARNIER, L'ATELIER L'ARGILE VIVANTE**

Teapot and warmer

The teapot is remarkable for its bold red glaze and rectilinear spout and handle. Holes and gentle imprints add to the design. It was part of a sixteen-piece tea service, which included a sugar bowl, creamer and cups and saucers. The service was displayed at the Canadian pavilion at the Milan Triennale in 1964 and awarded first place by the Tea Council of Canada that same year. Despite its critical success, the tea service had a production run of under five hundred. After Jacques Garnier left Beauce, he opened a studio known as Éstri-Céram in Magog, Quebec, creating ceramic sculpture and production pottery. He revised the tea service, introducing a larger fluted base. It was also offered in matte-finish olive green.

1964
MANUFACTURER: Céramique de Beauce, St-Joseph-de-Beauce, Quebec
MATERIALS: glazed earthenware
DIMENSIONS: teapot: h 18.2 x w 17.5 x d 13.8 cm (h 7 x w 6¾ x d 5½ in.)
MARKINGS: C/B; Argile G Vivante

10 Ceramics

Commercial Potteries Produce Tableware

Commercial potters, based primarily in Ontario, Alberta and Quebec, were generally small and privately owned but enjoyed some success after the Second World War. They produced thick and durable tableware for the hospitality market and thinner versions for residential use. Most also made other ceramics like ashtrays, lamp bases and vases. Employing machinery was standard practice, although artisans occasionally hand-painted surface decoration. More commonly, potteries imported transfer patterns (engraved copperplate impressions) or decals (multicoloured lithographs) from the U.S. and Britain. These patterns typically reflected the traditional and conventional fashions of the day rather than modern contemporary.

The Canadian commercial tableware industry developed much later than in other industrialized countries, as imports from Britain were so inexpensive that there was no incentive for local production. Only a smattering of family-owned potteries, making crockery from local red clays, operated across the country before 1900, and most disappeared with the advent of industrialization. The lone exception, working with white clays, was St. Johns Stone Chinaware Company in Quebec (1840 to 1896).[1]

A 1931 federal report on the Canadian ceramics industry documented the absence of a tableware industry that used finer white clays.[2] Two years later, Sovereign Potters opened in Hamilton and became the second pottery in Canada to make semi-porcelain white earthenware for buyers like Canadian Pacific Hotels.

In Western Canada, the tableware industry centred in Medicine Hat, near red clay deposits and railway lines. Medalta Potteries had been producing red clay crockery from the turn of the last century. In 1937 Ed Phillipson, a young engineer with the company, discovered more desirable white clay deposits in Saskatchewan, allowing the firm to develop a comprehensive line of hotelware as well as a residential line (FIG. 10.2) inspired by Fiesta, the popular American tableware produced by Homer Laughlin China Company. The following year, the rival Medicine Hat Potteries (later known as Hycroft China) opened to serve similar markets and became one of the longest-lasting potteries in Canada.

In Quebec, the Syndicat des Céramistes Paysans de la Beauce distinguished itself from other commercial potters through its synthesis of craft and design. Beauce was the only Canadian pottery to actively commission studio potters to make moulds.[3]

Two ceramic engineering schools emerged in Canada during the late 1920s, but their programs focused on technical and production issues rather than design. W.G. Worcester, head of the influential program at the University of Saskatchewan, helped his former student Phillipson locate the white clay deposits. Professor Robert Montgomery ran a department at the University of Toronto and served as technical adviser to the Guild of Potters as well as the ceramic art programs at Central Technical School and at OCA.[4]

Since skilled labour was difficult to find, commercial potteries often imported talent from Staffordshire, England, the heart of the world's ceramics industry. British-trained Thomas Hulme ran the art department at Medalta Potteries, while Sovereign Potters employed Bertram Watkin, a model maker originally from Staffordshire.[5] A shortage of expertise led to staff and even proprietors frequently moving between Alberta and Ontario.

During the Second World War, commercial potters equipped themselves with conveyor belts and tunnel kilns to create government-mandated utilityware known as "wartime cup." Ceramics replaced tin armyware, which the military needed for other applications.[6] Producing these sturdy dishes for the allied forces gave the industry valuable experience in mass-production techniques, improving speed and quality. After the war, potters were protected from imports by a government embargo and supplied "informal dinnerware" to the burgeoning home market. Commercial potters were proud to point out their "all-Canadian" pedigree in women's magazines like *Canadian Homes and Gardens* and *Mayfair*.

The removal of protective tariffs in 1953, however, damaged the industry. Without an export strategy, businesses weren't able to compete with the rejuvenated British ceramics industry. High labour costs, as well as inexpensive imports from Japan, contributed to the decline. By 1989 both Hycroft China and Céramique de Beauce had closed.

Poterie Vandesca, based in Joliette, Quebec, and founded by Bertrand Vanasse in 1947, managed to produce hospitality ware into the nineties under various owners, including Syracuse China Company of New York, Canadian Pacific Hotels and Libbey Glass of Toledo, Ohio. Blue Mountain Pottery in Collingwood, Ontario, has operated continuously since 1947. The pottery's signature turquoise glaze designs by Dennis Tupy and Alfred Dubé are in the Scandinavian style and collectible. The company stopped making tableware after the mid-eighties and now produces souvenir animal figurines. Laurentian Pottery, St-Jérôme, Quebec, founded in 1937, makes cast-moulded conventional flowerpots, lamp bases and jugs.[7]

Art Potteries Focus on Expression

The art pottery tradition began in 1900 with the Arts and Crafts movement and continues to this day. Operated by one or two people, with the help of technical assistants, the art pottery produces on a smaller scale than the commercial pottery. Typically, art potters pour liquid clay, known as slip, into porous moulds. They decorate objects with commercial glazes but prefer hand techniques to transfer patterns. The art potter's small size precludes the large production runs required of dinnerware, so the focus is on serving pieces, vases, souvenirs and artware.

George Emery, who ran Ecanada Art Pottery in Hamilton, spearheaded the art pottery tradition in Canada in the thirties and early forties. He persuaded his long-term employer, the Canadian Porcelain Company, to allow him to use its facilities on his own time to make decorative Wedgwood-style pottery.

On the West Coast, Herta Gerz, a German émigré who operated B.C. Ceramics, introduced modernism to art pottery (FIG. 10.5). Like most Canadian art potteries, her company relied on the production and sale of figurines and Canadian souvenirs to survive.

Quebec has the strongest tradition of art potters. Pierre-Aimé Normandeau, who trained at the

TOP RIGHT: 10.2 In the forties, Medicine Hat Potteries made colourful mix-and-match dinnerware, like Hatina (PLATES), while Medalta made sturdy hotel ware (FRONT).

BOTTOM RIGHT: 10.3 Staff at Hycroft China factory, Medicine Hat, Alberta, trim mugs, c.1940s.

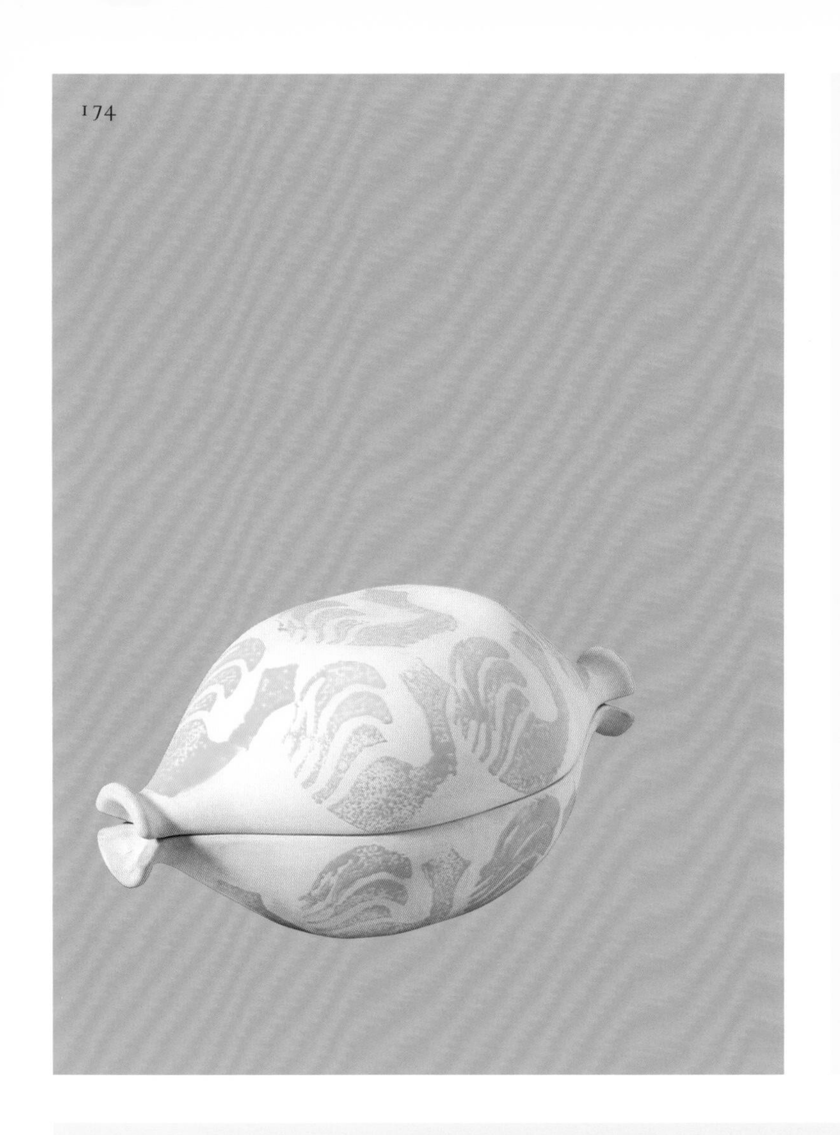

prestigious Sèvres school in France, was an influential figure, directing the province's first ceramics school in 1935, which merged with the École du Meuble in the forties. The respected graduate Gaétan Beaudin ran the small but prolific Décor pottery in Rimouski (FIG. 10.13), but discouraged with the limitations of small-scale production, he later co-founded the commercial venture Sial II, using the technique of pressure casting (FIG. 10.15). Carrying on Quebec's art pottery tradition today is Louise Bousquet. A protégée of Beaudin's, she employs similar industrial processes for her comprehensive line of white porcelain dinnerware. Denise Goyer and Alain Bonneau, also from Quebec, produce fine sculptural tableware in delicate porcelains using the technique of slip casting.[8]

Studio Potteries Produce Limited Runs

Studio potters stem from and overlap with art potters.[9] Participants in the movement consciously reject industrial practices and take pride in the individual look of objects. In the thirties, studio potters made their production work from heartier red earthenware clays, using kick wheels or building and coiling by hand. One of the first schools to teach craft pottery was Central Technical School in Toronto, where Zema (Bobs) Coghill Howarth taught between 1923 and 1962. Its graduates Tess Kidick, Nunziata D'Angelo and Bailey Leslie formed the Canadian Guild of Potters in 1936.

Securing equipment and materials presented real challenges for pioneering studio potters, as did developing recipes for colour glazes that were stable in firing and could be duplicated. Many turned to commercial companies for assistance in firing as independent kilns were prohibitively expensive.[10] Sovereign Potters generously allowed artisans to use their kilns. But outside Quebec, there was little collaboration between studio potters and industry, and studio potters tended to distance themselves from industry. The first pottery exhibition at the National Gallery of Canada, titled *Canadian Ceramics 1955*, co-organized by the Canadian Handicrafts Guild in Montreal and the Canadian Guild of Potters in Toronto, showed no examples from industry. This annual event ran until the early seventies and continued to ignore industry.

The Danish-born husband and wife Kjeld and Erica Deichmann in New Brunswick represent Canada's most important studio potters. Friendly with the studio pottery guru Bernard Leach, they combined his commitment to handcrafted pottery with Danish design (FIG. 10.9). While they never taught or took on apprentices, their critically acclaimed work was sought after, helping to raise awareness of studio pottery in Canada.[11] British-born David Lambert cultivated the studio pottery movement in Vancouver. Opening Lambert Potteries in 1944, he modelled it after a small country pottery and made wheel-thrown jugs and tableware in local clays. Known for sharing his technical expertise, he made his kilns available to colleagues and distributed pottery supplies.[12] Luke Lindoe led the studio pottery tradition in Alberta, founding the ceramic art program at the Alberta College of Art in 1947. He began his own pottery, known as Ceramic Arts, in 1957, and his wife, Vivian Lindoe, Walter Dexter and John Porter worked there until it closed in 1978.[13]

The American Ruth Gowdy McKinley, one the finest studio potters of the seventies, arrived in Toronto with her husband, Donald McKinley, a furniture-maker. She meticulously decorated her carved earth-tone and thrown vessels, raising domestic pottery to a higher standard.[14] The Five Potters Studio, Toronto, in operation since 1960, is one of the last vestiges of the Leach-inspired pottery tradition. Run by Mayta Markson, Bailey Leslie (now deceased), Marion Lewis, Annette Zakuta and Dorothy Midanik, Five Potters independently produces functional pottery for commissions and for its annual sales.

TOP LEFT: 10.4 In the seventies, Jean Cartier introduced colourful hand-sponged patterns to casserole dishes for Céramique de Beauce.

TOP RIGHT: 10.5 Vancouver's Herta Gerz designed a line of contemporary shapes, like vase 2816, decorated in Mink glaze.

BOTTOM LEFT: 10.6 A teapot by Goyer-Bonneau emphasizing line and shape was produced by Céramique de Beauce in the eighties.

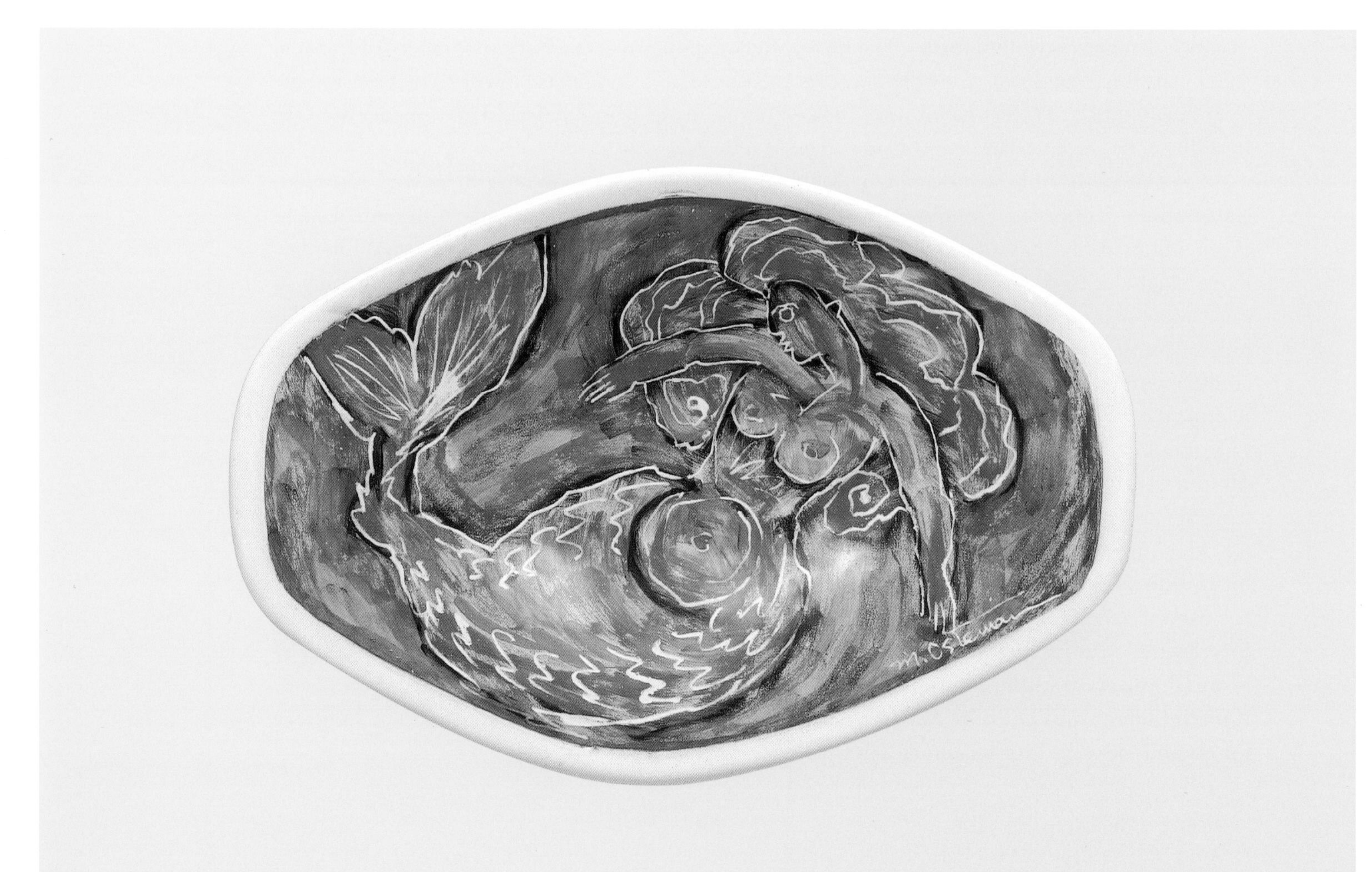

European immigrant designers broke away from the Leach tradition to reflect a more continental approach. Theo and Susan Harlander, a couple who trained at the College of Ceramics in Landshut, Germany, moved to Brooklin, Ontario, in 1952. For the next thirty years, they produced custom-ordered household ceramic items, such as tiles, lamp bases, tableware and some figurines, from a converted inn. Theo shaped forms on the wheel and Susan decorated the work with freehand incised (sgraffito) patterns (FIG. 10.7). Some of Theo's bold modern shapes reflected the impact of the European ceramist Hans Coper, while Susan's colourful palette and folk art drawings anticipated a new direction in colour and decorative imagery.[15]

In the seventies, the studio pottery movement (reflecting the general trend for all crafts) splintered between fine-art sculpture and production ware. Studio potters like Quebec's Leopold Foulem and Paul Mathieu established a movement for an intellectual approach to pottery that incorporated history and narrative metaphor.

In production pottery, colour, ornament and historicism replaced the more severe Anglo-Chinese tradition. Jan Phelan, in Hillsdale, Ontario, won a 1984 commission to design a china service for the governor general Ed Shreyer. Her delicate wheel-thrown porcelain designs, reminiscent of art nouveau, featured carved edges and swirling, lustrous glazes with mother-of-pearl and gold accents.

The new decorative sensibility also brought greater use of figurative imagery and bright colours and revived the Byzantine technique of majolica. Majolica, a glazed, low-fired earthenware, all but ignored in the Leach era, made a comeback in the eighties. Walter Ostrom became an influential force in ceramics in Atlantic Canada. Part of the wave of new teachers who came from abroad to teach the popular craft, he joined the Nova Scotia College of Art and Design in the early seventies and helped to revive this ancient decorative tradition. Another significant designer working in majolica is Matthias Ostermann in Montreal, who learned the technique from the noted British expert Alan Caiger-Smith. Since the late eighties, Ostermann has won critical acclaim for his sculptural vases exploring mythological themes. He supports his art with production tiles and serving pieces (FIG. 10.8). Scott Barnim in Dundas, Ontario, works solely as a production potter, concentrating on decorative tea and coffee services, platters and tureens. He prefers high-fired stoneware, in colourful floral and fish patterns.

TOP LEFT: 10.7 The Harlanders used incised graphics to enliven this organic ceramic vessel, 1950s.

BOTTOM LEFT: 10.8 Matthias Ostermann explored the ancient tradition of majolica in his mermaid platter, 1990s.

10.9 **ERICA AND KJELD DEICHMANN**

Soup tureen service

Drawing from ancient Chinese pottery, the design departs from the Deichmanns' spontaneous, folksy look. The traditional raised scroll decoration, which serves as a knob handle, adds to the service's formality. The Royal Ontario Museum holds the service in its collection.

1950s
MANUFACTURER: Dykelands, Moss Glen, New Brunswick
MATERIALS: red stoneware with celadon green glaze
DIMENSIONS: tureen base: h 14 cm (5 ½ in.), dia 9.5 cm (4 in.)
MARKINGS: inscribed on bottom of each bowl and tureen: Deichmann/monogram/N.B.

10.10 **SOVEREIGN POTTERS**

Dinnerware

Sovereign Potters' "all-Canadian dinnerware" design blended American and British influences. Jamboree, promoted as a "modern square design," featured an angular organic shape. Characterized by its rounded square corners, the dramatically styled line has a cream pitcher, a sugar bowl and a casserole dish. Sets were sold in solid and mottled colours, including Bermuda coral, daffodil yellow, sprout green, dove grey and pine green. Decorated patterns included Chanticleer (FRONT LEFT), a hand-painted silhouette of a rooster. Las Vegas features abstract foliage on a speckled grey-and-white background.

They were usually marked with a detailed back stamp. Sovereign also produced designs by Russel and Mary Wright, with patterns that resemble Russel's Highlight (TOP LEFT) produced by Paden City Pottery of West Virginia in the U.S.

1949
MANUFACTURER: Sovereign Potters, Hamilton
MATERIALS: glazed earthenware
DIMENSIONS: Jamboree plate: a square, 22 x 22 cm (8¾ x 8¾ in.); Chanticleer plate: 24 x 24 cm (9½ x 9½ in.); Pink bowl: h 6.5 x dia 17 cm (h 3 x dia 6¾ in.)
MARKINGS: Jamboree cup: marked Canada; Jamboree saucer: blank; Jamboree bowl: blank; Jamboree plate: blank; Chanticleer: stamped: Chanticleer Sovereign Potters earthenware made in Canada, maple leaf logo hand-painted; Pink bowl: stamped Mary & Russel Wright by Sovereign Potters

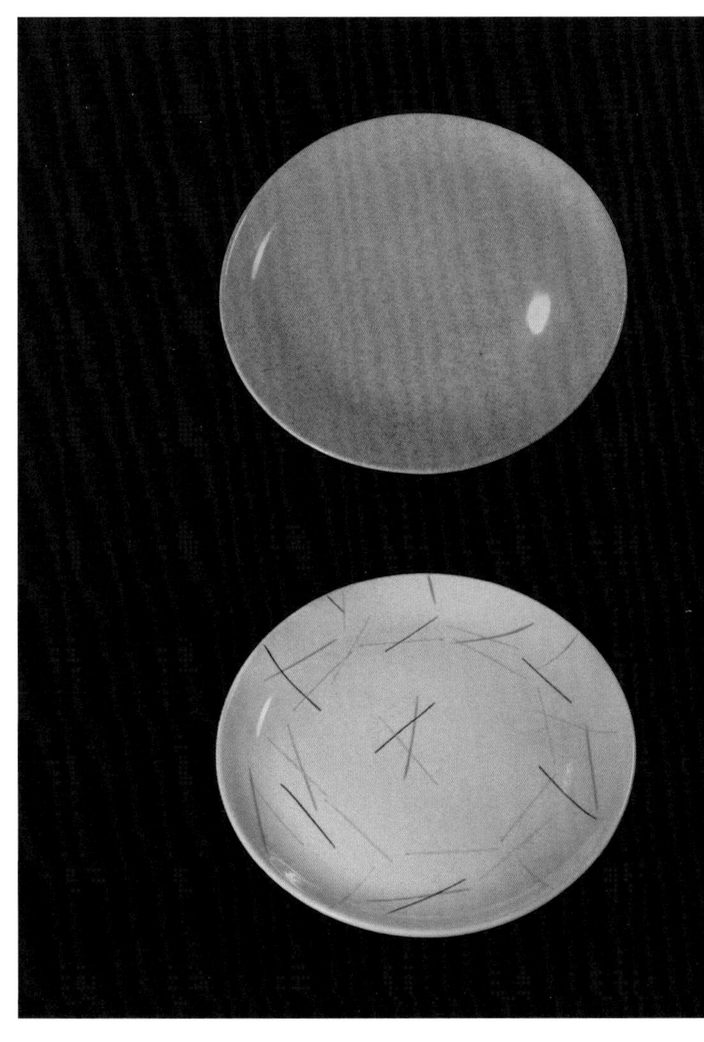

10.11 **MEDALTA POTTERIES**

Stardust plate

When William Pulkingham, the former president of Sovereign Potters, took over Medalta in 1952–53, he changed the focus to lighter and thinner dinnerware for the residential market. Under his direction, Medalta made speckled patterns called Stardust and Confetti, resembling Sovereign's contemporary lines, which were marketed to movie theatres as weekly giveaways.

1953
MANUFACTURER: Medalta Potteries, Medicine Hat, Alberta
MATERIALS: glazed earthenware
DIMENSIONS: dinner plate: dia 23.5 cm (9¼ in.)
MARKINGS: stamped: Stardust

10.12 **HYCROFT CHINA**

Jack Straws plate

Jack Straws was part of Hycroft's dinnerware line, which also included Cubes, New York, Lynn, Denim, Tartan and Calico. The hand-filled transfer patterns were available in three different shapes: round, oval or spiral fluted on the rim. The company produced one style of coffee mug for all three lines. The sets were extensive and included three plate sizes, platter, cereal bowl, teapot, creamer and sugar bowl, egg cup, serving dish, salt and pepper shakers and gravy boat stand. All patterns were sold through Woolworth's and Kresge's.

1956
MANUFACTURER: Hycroft China, Medicine Hat, Alberta
MATERIALS: glazed white earthenware, hand-painted lead foil stencil
DIMENSIONS: dinner plate: dia 24 cm (9½ in.)
MARKINGS: Hand decorated Hycroft Canada/Medicine Hat, Alberta

10.13 **GAÉTAN BEAUDIN**

Mug

An example of Gaétan Beaudin's early production ceramics, the mug (FOREGROUND) was slip-cast, as were other simple designs from his Décor pottery in Rimouski, where he taught at a local technical school. The country design has a strong sculptural quality, created by the organic inset handle. Décor produced the mug in solid colours of bone, brown, black, pink and ochre, and it appeared in NIDC's *Design Index* in 1954. It is represented in the collection of the Musée du Québec.

1950
MANUFACTURER: Décor, Rimouski, Quebec
MATERIALS: glazed earthenware
DIMENSIONS: h 16 x dia 6 cm (h 6½ x dia 2½ in.)
MARKINGS: Décor Rimouski Canada

10.14

Mug

Czech-born Victor Kominik opened Laurentian Pottery in 1937. It has been run by his son John since the sixties. The firm produced Beaudin's design well into the seventies. Back-stamped Components of Canada, the mug is shorter and heavier than Décor's.

c. 1960
MANUFACTURER: Laurentian Pottery, St-Jérôme, Quebec
MATERIALS: glazed earthenware
DIMENSIONS: h 15 x dia 6 cm (h 6 x dia 2½ in.)
MARKINGS: Components of Canada

10.15 **GAÉTAN BEAUDIN**

Oval carafe and goblets

Oval carafe service and dinnerware emulated the look and feel of country pottery, but it was pressure cast on a high-speed production line. Special ceramics, composed of 75 per cent porcelain and 25 per cent stoneware, were developed to vitrify in a single firing. Oval was offered as a twenty-piece breakfast set or a forty-five-piece dinner set and had twenty-seven complementary individual pieces, including a coffee set, goblets, decanters and covered casseroles. The dinnerware's basic shape consists of two intersecting ovals in a variety of textures and colours. Its burnt-orange background colour was offset with off-white, Tenmoku (brown), or grey or celadon green, reflecting Beaudin's travels in Japan. Despite its homemade look, it was durable, could go from oven to fridge and was dishwasher safe.

1978

MANUFACTURER: Sial II, Laval, Quebec
MATERIALS: porcelain and stoneware ceramic
DIMENSIONS: carafe: h 25.5 x w 14.6 x dia 11.7 cm (h 9 ½ x w 5 ½ x dia 4 ½ in.)
MARKINGS: stamped on base: Sial made in Canada

10.16 **KOEN DE WINTER**
Porcelaine de Chine tableware

While the original moulds for Koen de Winter's white sculptural tableware came from Quebec, the cost to produce locally was beyond reach. In the early eighties the designer went to Swatow, China, where there was a good source of white Kaolin clay and underutilized factory equipment. The region, however, lacked the requisite technical expertise, since much of the skilled labour had been displaced during the Cultural Revolution. Ironically, the thirty-seven-piece set helped renew the local tradition of commercial tableware in China rather than Canada. Much of the line remains in production but is no longer exclusive to Danesco.

1985
MANUFACTURER: Danesco, Montreal
MATERIALS: porcelain
DIMENSIONS: coffee pot with filter: h 19 x dia 11 cm (h 7 ½ x dia 4 ⅓ in.)
MARKINGS: paper label: made in China

10.17 **DENISE GOYER AND ALAIN BONNEAU**

Ocean service bowl

In the Ocean bowl, Denise Goyer and Alain Bonneau have relied on asymmetry and irregularity. One side is scooped out or scalloped and deliberately challenges our preconceptions of form and function by creating a precarious imbalance. The design remains in production.

1989
MANUFACTURER: Goyer-Bonneau, Carignan, Quebec
MATERIALS: cast porcelain, cream-and-blue glaze
DIMENSIONS: h 33 x dia 11.5 cm (h 13 x dia 4 ½ in.)
MARKINGS: hand signed

11.1 **DOMINION GLASS COMPANY**

Depression glass tableware

Dominion Glass's luncheon and dinner sets are known by today's collectors as Canadian Depression glass. One of its first patterns in the forties was Swirl (in clear glass as well as a rare cobalt blue), so named for its swirling flutes. Stippled Swirl, introduced a few years later, mostly came in clear and some fired-on ceramic colours like red, green, yellow and blue.

Dominion Glass introduced two new designs in 1945: Saguenay, a ribbed pattern in clear glass, and some fired-on pastel colours modelled after American Homespun and Hiawatha, which imitated cut glass.[1]

c.1945
MANUFACTURER: Dominion Glass Company, Wallaceburg, Ontario
MATERIALS: glass
DIMENSIONS: Swirl pitcher (BACKGROUND): h 20 cm (h 8 in.); Saguenay plate (FOREGROUND): dia 18 cm (dia 7 ⅛ in.)
MARKINGS: none

11 Glass and Miscellany

Depression Glass Reaches the Masses

Glassmaking was established in Canada in the mid-nineteenth century, when glasshouses (from small workshops to large factories) opened in Ontario, Quebec, and Nova Scotia. Following the European apprenticeship tradition, they produced hand blown domestic glass, windows and bottles. In the early twentieth century, the glass industry mechanized and underwent a series of mergers. Dominion Glass Company in Montreal and Consumers Glass in Hamilton emerged as the two leaders that dominated the market for the rest of the century. In the post-war era, only two of the country's larger commercial glassworks produced household objects like tableware and stemware. Most focused on containers (bottles, jars and so on) and architectural glass.

Dominion Glass Company, operating from its plant in Wallaceburg, Ontario (the old Sydenham glass factory), made tableware in Canada in the thirties. It concentrated on the hospitality market, with a few lines of drink ware, mixing bowls and serving dishes in colourless glass featuring popular American patterns such as Spike (FIG. 11.2) as well as designed-in-Canada motifs like maple leafs.

Shortages of ceramic dinnerware during the war years and technical advances that reduced glass manufacturing costs convinced the company to expand its line of Depression glass (machine-pressed patterned glass (FIG. 11.1)). The line was sold through Simpson's and Eaton's catalogues, often promoted along with china dinnerware by Sovereign Potters.[2] In the improved economy of the fifties, the company produced coloured and stencilled domestic ware.

Corning, the large American manufacturer that specialized in heat-resistant glass, established a plant in Toronto in 1945, becoming the second company (after Dominion) to make glass tableware (FIG. 11.3). It was fabricated using ten-year-old American moulds and appeared in two colours: cream (stamped Corex) and blue (nicknamed "Pie-crust" by collectors). It also made a "Crown" pattern from a different mould that came in blue or turquoise (stamped Pyrex). In 1950 the company began manufacturing opal glass ovenware, mixing bowls and fridge jars in similar opaque pastel colours bearing the trademark Pyrex. Production runs were relatively small, making it difficult to compete with imports. In 1954 the company focused solely on warehouse storage and distribution.[3]

Cut Glass Introduces Decoration

The cut-glass industry—the decoration of imported glass and crystal by local artisans who used hand and automated tools like wheel engraving and abrasives—developed in Canada during the late nineteenth century. Acid polishing (immersion in an acid bath) enhanced the sparkle of cut glass, fuelling its demand into the twentieth century. Small decorating and engraving glass shops

included George Phillips in Montreal, and Clapperton, in Quebec. The W. J. Hughes Company, Toronto, was widely known and specialized in one design, Corn Flower. William Hughes had apprenticed at the well-established firm Roden Brothers, in Toronto, which specialized in silver and cut lead crystal, but he created the floral pattern for his own company. Corn Flower featured a distinctive cross-hatched twelve-petal flower with long fluid stems (FIG. 11.7). The lightly cut design decorated a wide range of items, including tumblers, stemware, vases and serving dishes, and was sold only in the Canadian market.[4]

Post-War Artware Tradition Thrives

In the post-war period, a handful of small glassworks were founded, often by highly trained European immigrants, to cater to the souvenir market. John Furch, from Czechoslovakia, founded Altaglass in Medicine Hat, Alberta, in 1950. He served as master glass blower, making pressed glass (from steel moulds fabricated in England according to his specifications), mouth-blown glass (blown into wooden forms) and free-form glass. His line of candlesticks, vases, candy dishes and figurines was rarely signed, although some had diamond-shaped paper labels or, after 1967, maple leaf labels. At first, Furch made his own glass because supplies weren't widely available. His daughter and son-in-law, Margarete and Les Stagg, took over the firm in 1976. Four years later, they closed the furnace and ceased making vessels but continued to produce figurines from Pyrex rods until 1988. Joe Takash of Hungary, who worked for two years at Altaglass, founded Continental Glass in Calgary, producing similar ornamental glass between 1961 and 1974.[5]

Earl Myers, an importer of giftware, sponsored European families to come to Canada and work at Lorraine Glass Industries, established in Montreal in 1960. Mario Cimarosto, who trained in Murano (the heart of Italian glass manufacturing), was the company's glass master. He made contemporary-style stretch or free-form glass vases and dishes in bright colours like orange, lime green and cranberry. The company employed as many as thirty people and produced thousands of pieces a month. The studio-glass artist Toan Klein apprenticed with Cimarosto for several years in the early seventies (when few training programs were available), coincidentally making the company's one-millionth swan. Klein has operated his own studio-glass firm since the late seventies, first in Montreal and now in Toronto.[6]

Chalet Artistic Glass of Cornwall, Ontario, produced colourful free-form art ware, similar to Lorraine Glass but better known in the collectors' market. The designs are either signed or have an adhesive label on the base.[7] Founded in 1958 in Montreal as Industries de Verre et Miroirs, it operated briefly as Murano Glass (1960 to 1962), then became known as Chalet Artistic Glass. The Venetian brothers Angelo and Luigi Tedesco, along with Pagnin Sergio, ran it. The studio employed twenty-five artisans, including Angelo Rossi from Murano. Rossi, who purchased Chalet in 1970, had worked briefly at Lorraine Glass. Under the name Artistic Lighting, Rossi produced light fixtures, Tiffany-style glass and custom chandeliers for hotels. After Chalet closed in 1980, he opened a smaller studio, Rossi Artistic Glass, which focused on giftware. He now operates his glassworks in Niagara Falls.

On the Atlantic coast, Denis Ryan opened Novascotian Crystal (modelled after traditional Waterford crystal) in Halifax in 1996. Located on Halifax's waterfront, the "old-world" glasshouse employs two Irish master craftsmen, Jack Tebay and Philip Walsh, who produce mouth-blown hand-cut crystal and oversee fifteen artisans.

TOP LEFT: 11.2 Dominion Glass's colourful Spike kitchenware items of the fifties are now collectibles.

BOTTOM LEFT: 11.3 Corning's "Pie-crust" (FOREGROUND) and "Crown" tableware of the fifties came in pretty pastels.

11.4 **MAX LESER**

Circular scent bottle

What makes this design so striking is Max Leser's unusual choice of thick, flat bulletproof glass for a feminine perfume bottle. The geometric shape, accentuated with clean etched horizontal banding, and the frosted tube flask are typical of Leser's style. Trained as a glass sculptor, he redrilled and sandblasted the prefabricated glass in his Toronto studio. In the eighties, his collection of nine scent bottles and two vases sold in top design stores in North America. He licensed the design to Elika, a manufacturer in Los Angeles. Leser trained at the Banff School of Fine Arts in glass sculpture, and in the eighties he developed critically and commercially acclaimed functional glass products. A decade later, he embraced industrial design and opened a commercial studio in the Czech Republic.

1982
MANUFACTURER: Elika, Los Angeles
MATERIALS: 25 mm structural aquatint glass
DIMENSIONS: h11.5 x w12.5 x d2.5 cm (h4½ x w5 x d1 in.)
MARKINGS: signed Leser on underside

11.5 **JEFF GOODMAN**

Compass bowl

The Compass bowl series reflects Jeff Goodman's blending of sculptural and artistic concerns with retail savvy. He balances form, texture and colour to create a simple but functional decorative piece. With his wife, the graphic designer Mercedes Rothwell, Goodman designed the bowl with the four compass points mechanically sandblasted on the rim. In 1989 he introduced a free-form blown version, which requires considerably more labour. Goodman's studio produces the Compass series intensively in cycles for months at a time, similar to the way Michael Fortune makes his Number One chair (FIG. 6.58). To date, his studio has cast over two thousand and blown more than nine hundred Compass bowls. Goodman does all the blowing himself. This signature line is sold in stores like Barneys, New York, and William Ashley, Toronto. The Compass series provides Goodman with 40 per cent of his annual revenue and enabled him to set up a private glass studio, one of a handful in Canada. The design won the Virtu 8 competition.

1988
MANUFACTURER: Jeff Goodman Studio, Toronto
MATERIALS: depressed glass cast in a graphite mould
DIMENSIONS: h11 x dia17 cm (h4⅛ x dia6¾ in.)
MARKINGS: signed Jeff Goodman under the base

Studio Glass Achieves Recognition

Since the early seventies, there has been a revival in studio glass, with numerous artisans achieving success by producing limited runs of art glass. In addition to training at the newly founded glass-making schools, some of these artisans apprenticed with the previous generation of European-trained glassmakers. The American Robert Held (FIG. 11.9) came to Sheridan to run the ceramics program, but his interest in art glass led to his founding the first glass school in Canada.[8] Karl Schantz, another American, briefly joined Held at Sheridan and taught the "cold-work" techniques of cameo-glass and sandblast engraving. He went on to run the glass program at the Ontario College of Art from 1980 until the department closed in the late nineties, where he taught cold construction of laminated sheet glass. Daniel Crichton (known for his colourful geological vessels) has run Sheridan's program since 1979. He introduced students to the techniques of carving and splatter glass.

Elena Lee helped to raise the level of artistic practice in Canadian glass by founding Verre d'Art in 1976 in Montreal, when studio glass was still in its infancy. Her gallery promoted many new artists, including the late François Houdé, who explored artistic metaphor in glass and created the highly acclaimed Ming dynasty horse series. In 1988 Houdé founded the Centre des Métiers du Verre du Québec in Montreal, a school devoted exclusively to glass.

In Manitoba, Ione Thorkelsson, a former architecture student and self-taught glass artist, is an important studio glassmaker. She creates animal-like sculptures as well as a trademark line of wine goblets decorated with specks of colour that are sold across Canada. The third generation of studio glassmakers includes Jeff Goodman (FIG. 11.5) in Toronto, Brad Copping in Aspley, Ontario, and the Vancouver 6 (the latter graduates of the famous Pilchuck school in Seattle). Copping, a graduate of Sheridan and former resident of the studio program at Harbourfront in Toronto, is a typical Canadian glassmaker, creating a signature line of hand-carved production vessels (goblets, candlesticks and vases in deep purples and acid greens) for distribution across Canada. Since 1993 he has designed a series of limited-edition bottles for Inniskillin's Icewine (FIG. 11.6). This production work supports his primary interest in mixed-media glass sculpture.

PREVIOUS PAGE: 11.6 Brad Copping's annual collector's series for Inniskillin Icewine began in 1993.

TOP RIGHT: 11.7 William Hughes decorated this creamer and sugar bowl with his signature Corn Flower pattern.

BOTTOM LEFT: 11.8 In the fifties, gift shops across Western Canada sold Altaglass candy dishes and bowls.

BOTTOM RIGHT: 11.9 Robert Held's art nouveau–engraved glasses, 1984, adorned the governor general's table in Ottawa.

11.10
Melmac

The American military developed melamine tableware for its navy in the late thirties. When fabricators launched Melmac for civilian consumption a decade later, they boasted that the unbreakable dishes would last forever. Retailers, concerned about loss of repeat sales, were at first reluctant to carry the product, forcing Rainbow Plastics in Montreal and Maple Leaf Plastics in Toronto to sell door-to-door.

Plastic tableware was initially listed at higher prices than china, but when it became apparent that it would scratch and fade, Melmac producers competed based on value. To save on material, they reduced the thick walls and rounded edges, resulting in thinner designs in the sixties. Earlier decorator pastels and mottled colours were replaced by floral and pictorial transfer patterns fused into the plastic, making Melmac appear more like china. In Toronto, Glenn S. Woolley and Company, working with E.S.&A. Robinson (Canada), introduced decal patterns.[9]

Higher-quality Melmac can be identified by the absence of compression-moulding seams and the removal of production-related spurs, which have been sanded down to a polished finish. However, most companies also continued to produce thick institutional ware. While Canadian firms often relied on American moulds, some original Canadian shapes emerged.[10]

MAPLE LEAF PLASTICS
CST bowl and cup, Moderne cup and saucer

Sid Bersudsky sat on the specification committee that commissioned the all-new tableware for the Department of Defence (TOP). The Canadian Standard Tableware was developed in co-operation with the British and the Americans. General Plastics, Glenn S. Woolley and Company and Maple Leaf Plastics produced the line under various names. Maple Leaf Plastics also offered Moderne, a rounded square design (BOTTOM LEFT) which resembles the American Brookpark Modern Design by Joan Luntz. Maple Leaf produced the first Canadian melamine cream and sugar set in 1950. It closed in the eighties, when the plastics industry redirected its market to automotive parts.

1950s
MANUFACTURER: Maple Leaf Plastics, Toronto
MATERIALS: melamine
DIMENSIONS: CST cup: h 12 x dia 8 cm (h 5 x dia 3 in.); Moderne cup: h 9 x dia 10 cm (h 3½ x dia 4½ in.)
MARKINGS: CST cup: GPL Melmac 330 Made in Canada; Moderne cup: Moderne by Maplex, Toronto

RAINBOW PLASTICS
Cup and side plate

Rainbow Plastics designed an original triangular snack plate (BOTTOM RIGHT) for its Stylized Dinnerware line for the residential market. The company also made the institutional Rainbowware (later known as Beaufort), modelled after Russel Wright's commercial line Meladur. The company, known as RPL, opened in the 1940s and was bought out by Cyanamid of Canada when it moved into product manufacturing. Maple Leaf Plastics purchased the company and its moulds in 1970.

1960s
MANUFACTURER: RPL, Buckingham, Quebec
MATERIALS: melamine
DIMENSIONS: irregularly shaped side plate: w 19 cm (w 7.5 in.)
MARKINGS: moulded on back: Genuine Melmac, Quality Molded Melamine, Stylized Dinnerware, RPL, Made in Buckingham Canada

11.11 **WILLA WONG**

Sonoma tableware

In the late nineties, Precidio helped to revive Melmac. Rather than relying on retro patterns and shapes, the designer Willa Wong introduced a fresh look, creating mix-and-match dotted, ribbed and grid-patterned embossing to add texture and design.

2000
MANUFACTURER: Precidio, Brampton, Ontario
MATERIALS: melamine
DIMENSIONS: dinner plate: dia 28 cm (11 in.)
MARKINGS: ribbed bowl: made in China exclusively for Precidio, Precisioncraft; dot plate without markings

11.12, 11.13 **JACK LUCK**

Drip coffee pot and cookware

Jack Luck's coffee pot represents one of the first new designs introduced after the war, when restrictions on aluminum for domestic use were eased. The aluminum tip of the Bakelite handle reveals Luck's careful attention to detail. It was exhibited and promoted by the NIDC in 1949, made the NIDC's 1954 *Design Index* of recommended Canadian products and until 1959 was distributed in the U.S. under the Wear-Ever brand. New manufacturing technology allowed Luck's designs to evolve from simple cylindrical shapes, like his drip coffee pot, (ABOVE RIGHT) to organic shapes, like a set of coloured cookware featuring a decorative turquoise pattern of concentric rings (ABOVE LEFT).

1949

MANUFACTURER: Aluminum Goods, Toronto
MATERIALS: spun aluminum, Bakelite
DIMENSIONS: coffee pot: h 24 x dia 14 cm (h9 ½ x dia 5 ½ in.)
MARKINGS: stamped on base: Wear-Ever Aluminum Made in Canada

11.14 **SID BERSUDSKY**

Model 250 Magnajector, magnifying projector

Sid Bersudsky's adaptation of an overhead projector into a child's toy sold for over thirty years. The design was also manufactured under licence in the U.S. Used to project and enlarge images, the "magic lantern" was moulded in two pieces from Bakelite, allowing fast assembly and good heat insulation.

1954
MANUFACTURER: Kelton Corporation, Peter-Austin Manufacturing, Toronto
MATERIALS: compression-moulded Bakelite
DIMENSIONS: h 19.5 x w 11 x length 26 cm (h 7 ¾ x w 4 ¼ x length 10 ¼ in.)
MARKINGS: Moulded Magnajector, marked Model 250
Peter-Austin Mfg. Co., Toronto, Ontario

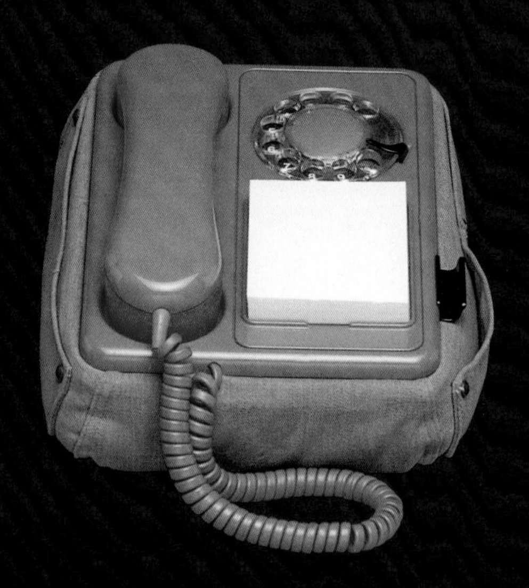

4 3 2 1
5
6
7 8 9 0

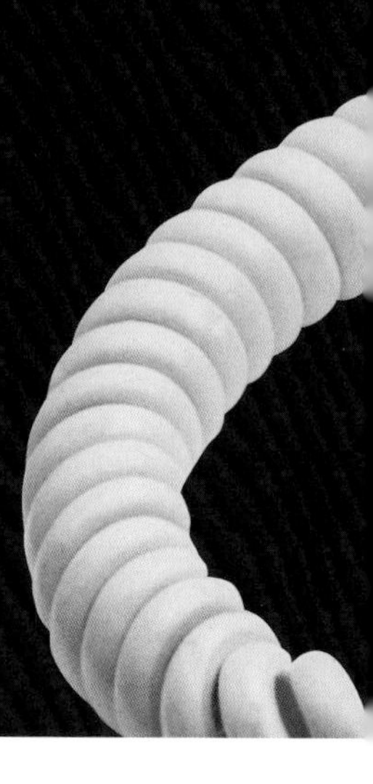

11.15, 11.16 **JOHN TYSON**

Contempra telephone and Imagination series

The Contempra's sleek and angular shape (BELOW) offered an alternative to the organic curves typical of sixties telephone design. John Tyson, fresh out of industrial design school, chose the name to reflect the phone's progressive style. This homegrown model marked the end of Western Electric's (the manufacturing arm of AT&T) hold on Northern Electric and the birth of Northern Telecom and the Bell Northern Research department (BNR). It was the first product developed after Northern Telecom's final ten-year licensing agreement with its former American parent expired in 1966.

Once Tyson won the corporation's approval of his wooden model, the project was fast-tracked and developed in secret in the Ottawa plant. Quebec City and London, Ontario, were the first cities to receive this high-style dial-in-handset telephone. Market research indicated the design would be successful. First offered in a rotary dial, then a push-button version, the phone's receiver sits along the length rather than the width of the set, reflecting the shape of its base. The product used more than seventy components and a multitude of plastics, including ABS, PVC and polycarbonate, to create a lighter handset that could be wall mounted. Northern Telecom manufactured approximately seven thousand sets a week for almost twenty years, selling to more than fifteen countries.

In 1977 Tyson worked on the Imagination series (LEFT), a group of novelty telephones for the residential market. Intended to keep the London plant busy as it made the transition from electrical mechanical to fully electronic, the fun phones were made by mixing and matching existing components. Dawn (defined by its circular shape), Kangaroo (with notepad and beanbag pouch) and Alexander Graham Plane (an airplane-shaped telephone with a propeller telephone dial) sold in the millions in Canada and in the U.S.[11]

1968

MANUFACTURER: Northern Electric Company, Ottawa

MATERIALS: Contempra ABS handset, PVC spring cord, polycarbonate dial finger wheel, polypropylene, acrylic dial number plate, styrene volume control knob

DIMENSIONS: h8 x w14 x length23 cm (h3¼ x w5½ x length9 in.)

MARKINGS: Northern Electric Company Limited Made in Canada, Contempra

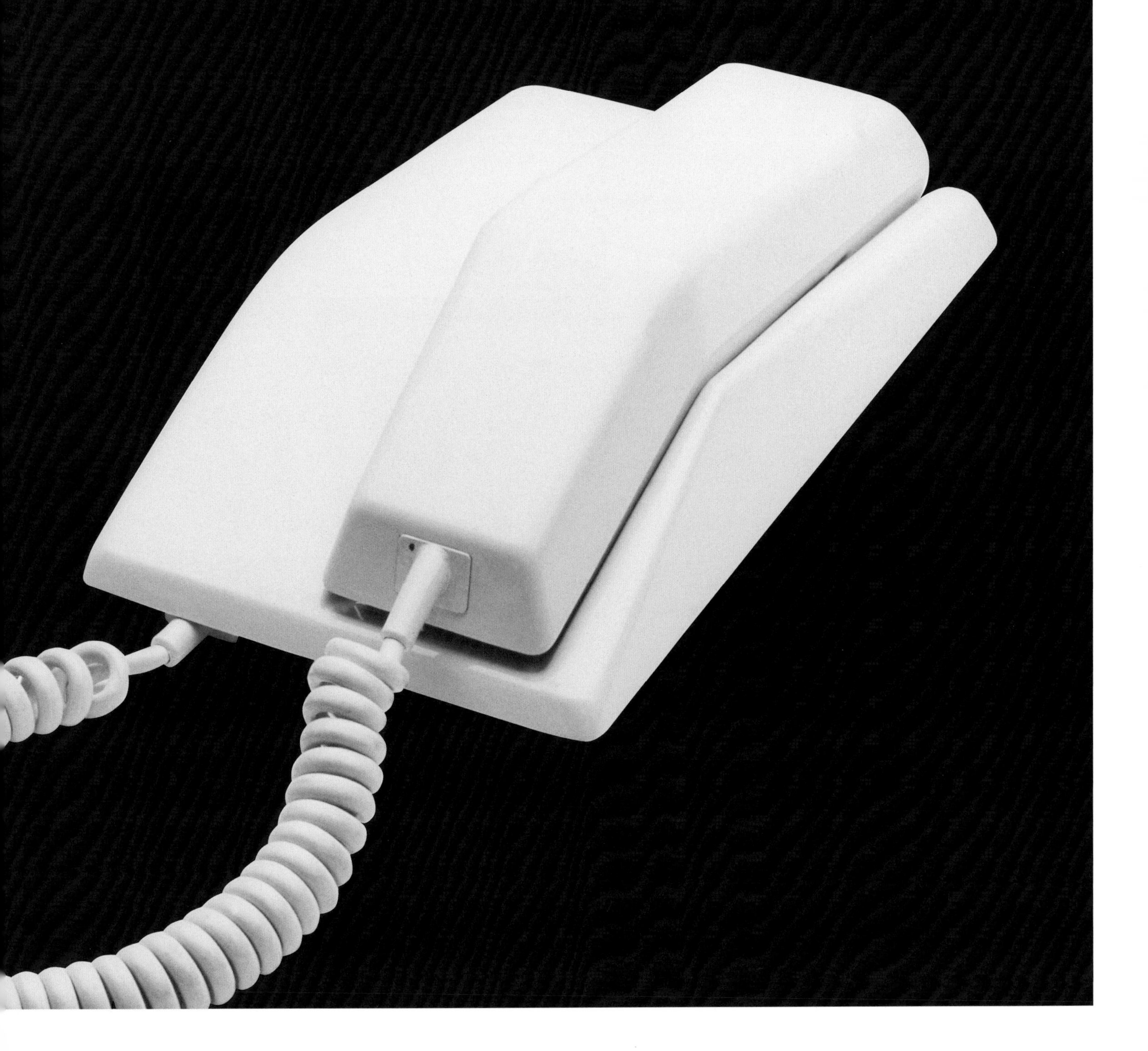

11.17 **THOMAS LAMB**

Lummus casserole prototypes

When the pre-eminent furniture designer Thomas Lamb explored the field of tableware, he chose the highly acclaimed studio potter Ruth Gowdy McKinley to make the prototypes. The stepped designs with simple wooden handles came with interchangeable tops and could be inverted into serving dishes. *Lummus* means "lamb" in Finnish, and the designer intended the products to be made in ceramic and/or metal. In the early eighties, Supreme Aluminum Industries produced a limited run.[12]

1979
MANUFACTURER: Supreme Aluminum Industries, Toronto
MATERIALS: anodized aluminum with wooden handle
DIMENSIONS: h12 x d33 cm (h5 x d13 in.)
MARKINGS: none

11.18 **MARCEL GIRARD AND IAN BRUCE**

Tukilik salt and pepper shakers

Marcel Girard and Ian Bruce, who went on to design the Laser sailboat with the Olympic champion Ian Kirby, created the simple, elegant Tukilik salt and pepper shakers for their firm Girard Bruce Garabedian in Montreal. Aimed at the federal government's program to promote Canadian design at Expo 67, the product was too late to appear at the World's Fair. The set of mix-and-match asymmetrical blocks consists of acrylic boxes encased in sliding aluminum frames. The design echoes the cubed composition of Moshe Safdie's showcase Habitat apartments. Innovative for the time, the combination of materials kept production costs high, due in part to the necessity to hand-polish the cast acrylic. Danesco produced approximately forty thousand sets until 1971. The line also included aluminum ashtrays, cigarette cases and jewellery boxes.

Girard consulted an anthropologist for the Inuit name Tukilik, which means "a well-conceived idea" and reflects the interest of designers in aboriginal culture during this period. For example, the Canadian pavilion was named Katimavik, meaning "meeting place." In 1970 the master potter Jean Cartier called his pottery line for Céramique de Beauce Skimo.[13] The Musée des Beaux-Arts in Montreal has a set in its collection.

1967
MANUFACTURER: Danesco, Montreal
MATERIALS: anodized black as well as silver aluminum, cast acrylic
DIMENSIONS: large: h6.4 x w3.8 x d3.8 cm (h2½ x w1½ x d1½ in.); small: h3.8 x w3.8 x d3.8 cm (h1½ x w1½ x d1½ in.)
MARKINGS: impressed on inside top of acrylic box: Patented-USA-DES-2114 Tukilik Rd. 1967

11.19 **MICHAEL SANTELLA**

Compackt CD holder

Ever since the CD player arrived in the marketplace in the late eighties, designers have created inventive CD holders. This elegantly minimalist one uses Michael Santella's favourite medium, bars of extruded aluminum. To allow for fifteen-disk storage, he needed a ten-inch (25 cm) aluminum extrusion—two inches (5 cm) larger than industry standard. Finding a professional mill to create the design was problematic, but the persistent Santella discovered an accommodating fabricator in Pickering, Ontario, and more than 100,000 units have been produced to date, in colours like black, gold, blue and red. The MOMA's design store sells the Compackt CD holder in its influential catalogue, and in 1992 it won Virtu 7, as well as Germany's Design Zentrum Award for product quality and design innovation.

Santella, a graduate of Carleton University's industrial design program, founded the studio Dibis in 1990. He produces a collection of colourful anodized aluminum products, finishing and polishing them himself.[14]

1991
MANUFACTURER: Dibis, Montreal
MATERIALS: solid extruded aluminum in anodized silver, with soft PVC anti-skid underpads
DIMENSIONS: h1.5 x w25 x d5 cm (h¾ x w10 x d2 in.)
MARKINGS: none

11.20 **MICHEL DALLAIRE**

L'Attaché Collection

More than four million copies of L'Attaché, a simple plastic briefcase, have been sold, initially for under $14 at bargain retailers. The briefcase languished until the sales boutique at New York's Museum of Modern Art ordered five hundred units in 1986, and it caught fire within the design community. The design evolved from the need to keep manufacturer Resentel's injection-moulding equipment constantly in production. Made from two identical polypropylene shells, the briefcase is assembled face-to-face. The line (still in production by Plasticase, Montreal, since 1997) now consists of satchels—ranging from a lunchbox to a suitcase—in a variety of sizes and colours.

1985
MANUFACTURER: Resentel, Montreal
MATERIALS: polypropylene copolymer
DIMENSIONS: large case: h31.1 x w8.9 x length41.3 cm (h12 ¼ x w3 ½ x length16 ⅛ in.)
MARKINGS: Design, Michel Dallaire Designers inc., Fabriqué au Canada, Made in Canada, La Companie Resentel limitée, Made in Canada, La Companie Resentel, Modele enregistré 1985, Registered Model 1985

11.21 **MARTHA STURDY**

Elephant end table and bowl

In the late eighties, Martha Sturdy took on the challenge of designing home products and revived the use of colourful plastic resins. The Elephant table demonstrates her interest in bold shapes. She includes the irregularities of the casting to maintain an element of handcrafting. The table is distinguished by four individually coloured thick legs in blue, yellow, pink and green.

Made in her Vancouver studio, the large oval bowl has become a signature Sturdy product.

1993
MANUFACTURER: Martha Sturdy, Vancouver
MATERIALS: cast resin
DIMENSIONS: table: h47 x w30 x d46 cm (h18 x w12 x d18 in.); bowl: h10 x dia30 cm (h4 x dia12 in.)
MARKINGS: metal insert label

11.22 **KARIM RASHID**

Garbo garbage can

Garbo's sensual curves and skinlike translucent plastic reflect Karim Rashid's fascination with the voluptuous female figure—a favourite motif. The raised paper-thin sculpted walls (which resemble blown glass) result from technical advances in polypropylene moulding. The luminous plastic has become popular in both product design and architecture.

The sculptural Garbo functions as a garbage can, champagne bucket, container or what have you. The design raised the international profiles of both its designer and its manufacturer, and over a million units have been sold. It has spawned a line of smaller Garbino and Garbini cans, as well as the larger Garbonzo. In 1997 it won a Good Design award from the Chicago Athenaeum Museum of Architecture and Design.

1996

MANUFACTURER: Umbra, Scarborough, Ontario
MATERIALS: high-impact polypropylene
DIMENSIONS: h 43 x dia 34 cm (h 17 x dia 13 ½ in.)
MARKINGS: back stamp; after 1998 manufacturer's logo added to front of canister

12.1 **FRED MOFFATT**

K42 electric kettle

The design originated when an engineer working at a steel-stamping factory at Canadian Motor Lamp Company, a subsidiary of CGE, turned the headlight of a McLaughlin Buick upside down and discovered that it was the perfect receptacle for an electric coil. To achieve this shape for the kettle's production, a powerful hydraulic press draws the dome from a flat circular sheet of brass. Fred Moffatt added a Bakelite handle, raised away from the hot spout. The first model was made of forty-five separate parts, and required no fewer than sixteen polishings.1 Moffatt's 1940 CGE version became the industry standard, with its two-quart (about two-litre) capacity and water-spout mouth. In an effort to capture CGE's sales, competi manufacturers introduced new shapes: beehive, flat top, streamline oval.

Despite its many competitors, the chrome dome remained in production as the "economy kettle" until the late sixties and sold in the millions.

1940

MANUFACTURER: Canadian General Electric Company, Barrie, Ontario
MATERIALS: chrome-and-nickel-plated brass shell, steel base, Bakelite handle
DIMENSIONS: h (including handle) 20 x dia 21.5 cm (h 8 x dia 8 ½ in.)
MARKINGS: stamped on bottom plate: K42 Canadian General Electric Company, Toronto, Made in Canada

12 Small Appliances

Independents Compete with Branch Plants

Small appliances (toasters, irons, kettles and so on) arrived in Canada in the early 1920s. Initially the high cost of electrical power made them accessible only to the wealthy, who bought them as status symbols. After the war, electricity was widely available, and utility prices dropped dramatically.[2] War-rationed families craved the gleaming new products, and Canadian industry willingly delivered their "labour-lightening" goods.

Small appliances entered most homes before large appliances because of their relatively low cost and portability. Equipped with war-related production capacity and technical know-how, manufacturers enjoyed a sellers' market.[3] Further boosting sales were quotas and tariffs imposed to deal with the foreign exchange crisis between 1948 and 1953. The protected market also helped members of the newly formed Association of Canadian Industrial Designers: industry turned to homegrown talent, and they introduced styling and inexpensive production techniques.[4]

Canadian General Electric Company dominated the appliance market. The result of a late-nineteenth-century merger between Edison Electric Light Company of Canada in Hamilton and Thomson-Houston Electric Light Company in Montreal, CGE began operating in Canada to circumvent tariffs and other laws. The company produced large metal stamped goods like refrigerators in Montreal and small appliances in Barrie. In the thirties, Fred Moffatt was the company's first freelance industrial design consultant, a role that continued for the next fifty years. Like other industrial designers of the era, Moffatt also "modified" American products for the Canadian market.[5]

Similarly, Canadian Westinghouse Company, a division of Westinghouse U.S., opened a factory in Hamilton to produce both small and large appliances and became a significant player. In contrast to CGE, it had an in-house designer, Thomas Penrose.

Subsidiaries like CGE and Canadian Westinghouse were encouraged to innovate. CGE offered its engineers a $25 cash bonus for every new patent.[6] Canadian-designed products, however, were rarely distributed in the U.S. by the parent companies as the differences between the Canadian Standards Association and its American equivalent inhibited export of Canadian products. Even today, variations as minor as a cord length complicate the export of appliances, so only 20 per cent of production leaves Canada. In addition, plain old-fashioned rivalry between the North American plants was a frequent obstacle.[7]

Independent manufacturers, operating primarily in Ontario and Quebec, offered an alternative to the branch plants. General Steel Wares (GSW) of London, owned by Ralph Barford, was one of the few independents to produce both small and large appliances.[8] Superior Electrics, Pembroke, Ontario, is one of the last independents still active in the industry. Toastess in Pointe-Claire,

Quebec (since shuttered), began after the war in the garage of Harry and Louis Solomon, who had been toolmakers for the aviation industry. These small companies survived by supplying regional markets and subcontracting private-label brands for the national retailers. Department stores, in turn, were happy to use the independents to prevent the American subsidiaries from controlling the market.[9]

Branding Expands the Market

In the late fifties, market saturation combined with a recession forced the industry to become more aggressive. CGE (following the model set by its American parent) joined forces with the Hydro-Electric Power Commission of Ontario to organize the 1958 "Live Better Electrically" advertising campaign.[10] That year the Canadian Manufacturers' Association launched its "Buy Canadian" crusade, dispensing maple leaf tags to Canadian companies and staging displays at Eaton's and other major retailers.

With competition growing fierce, branding emerged as a powerful competitive tool for the small appliance industry. Producing a comprehensive line ensured customer brand loyalty. The subsidiaries had the advantage, offering a multitude of models and a diverse line of products. CGE, for instance, produced American designs in its Barrie plant but imported hair dryers from overseas.

To maintain full product lines, Canadian manufacturers often hired one another to design and make products. Proctor-Silex (formerly called Proctor-Lewyt), in Picton, Ontario, bought finished kettles from nearby McGraw-Edison of Canada for its label. In turn, Samson-Dominion, a division of McGraw-Edison, bought parts from CGE.[11] While subcontracting and private labelling helped to keep the industry afloat, by the late sixties Canadian branch plants increasingly specified products from offshore.

Escalating competition from the Pacific Rim precipitated a realignment of the industry.[12] The General Agreement on Tariffs and Trade (GATT) pushed for the end of protection. By the mid-seventies, the federal government urged both the small and large appliance industries to unite to compete internationally. In 1977, to achieve economies of scale, CGE, GSW (the Canadian independent) and Moffat merged, then acquired Westinghouse Canada, to create the single major appliance manufacturer, Canadian Appliance Manufacturing Company (CAMCO). CGE sold its small appliance plant to Black & Decker, which closed the factory in 1984. Among the independents, Superior Electrics in 1989 bought Creative Appliance, which had previously acquired the assets of McGraw-Edison. Now that there are fewer players and fewer component manufacturers, existing companies must rely on imported parts.

Canadian Designers Love Kettles

Americans launched the first electric kettle, an electrical coil strapped to the underside of a metal drum, at the World's Fair in Chicago in 1893. In 1907 the German architect Peter Behrens modernized electric kettle manufacturing by standardizing production. In the twenties, new fully immersible heating elements were introduced, and by the fifties, automatic shut-off mechanisms replaced buzzers and whistles.[13] Since Americans, among industrialized countries, purchase fewer electric kettles, Canadian designers had a freer rein. The electric kettle became part of our vernacular tradition, and virtually every high-profile Canadian industrial designer has created one. Canadians seem to prefer the "kitchen workhorse" and have made the polished chrome dome a favoured design (FIG. 12.1).

TOP RIGHT: 12.2 Canadian industrial designers styled fans and heater. (LEFT) Canadian Beauty heater, Renfrew Electric, 1953; (MIDDLE) A.P. Whelan fan, 1953; (RIGHT) Fred Moffatt CGE heater, 1964.

BOTTOM RIGHT: 12.3 Lawrie McIntosh's Lady Torcan hair dryer for Rotor Electric sold more than half a million units in the sixties.

GENERAL ELECTRIC

HAIR DRYER

It's no trick
at all
to have
GLEAMING
FLOORS
all the time!
GENERAL ELECTRIC
FLOOR POLISHER
You Just Guide...IT DOES THE WORK

Plastic Influences Kettle Design

Plastic played an important role in the development of the electric kettle, first appearing in the handle as a heat insulator as well as styling device. Handles were jet-black Bakelite (fifties), orange polypropylene (sixties) and variously fat or thin, or attached to the spout or stand-alone. Metal parts were also replaced by plastics to save costs.[14]

The material initiated a race to create the first all-plastic electric kettle, though plastics were not yet strong enough to withstand high heat. In the early eighties, the industrial designer Ingo Glende, in Toronto, designed a plastic drum kettle, which still required a copper bottom plate. Herman Herbst, who had operated H&H Metal Stampings, a manufacturer of component kettle parts, founded Creative Appliance to pioneer Glende's innovation,[15] but the 1989 sale to Superior Electrics meant that in 1990 Superior produced Canada's first all-plastic kettle, only a year after Philips in Holland had made one.

With the all-plastic kettles came the injection-moulded tall coffee-pitcher shapes, but the design remained unpopular until the 1986 invention of a detachable electrical base, which allowed electric kettles to be cordless.

In colour, kettles followed the general trend of housewares, moving from avocado green (seventies) to white, almond and black (eighties), and translucency (nineties) (FIG. 12.13).

LEFT: 12.4 This fifties ad features Fred Moffatt's award-winning floor polisher from the forties.

ABOVE: 12.5 The DKR team, including Jan Kuypers, Julian Rowan and Frank Davies, review their model for an electric kettle in the sixties.

12.6 **SID BERSUDSKY**

51-2 electric kettle

A pioneering version of the "flat-top" kettle, it was launched in 1948 and was in production for over a decade. In addition to its flat top, the design is distinguished by speed bands that appear on the angled jet-black handle and encircle the base of the drum.

1948
MANUFACTURER: General Steel Wares, London, Ontario
MATERIALS: chrome-and-nickel-plated brass shell, steel base, Bakelite handle
DIMENSIONS: h (including handle) 21 x dia 21.5 cm (h 8½ x dia 8½ in.)
MARKINGS: stamped on bottom plate: Made in Canada
General Steel Wares, London, Canada

12.7 **THOMAS PENROSE**

K2 electric kettle

The Canadian Westinghouse staff designer Thomas Penrose created a beehive variation of the original CGE K42. The design strongly resembled a car headlight and was produced until the late sixties.

c. 1943
MANUFACTURER: Canadian Westinghouse Company, Hamilton
MATERIALS: chrome-and-nickel-plated brass shell, steel base, Bakelite handle
DIMENSIONS: h (including handle) 27 x dia 20 cm (h 10½ x dia 7¾ in.)
MARKINGS: stamped on bottom plate: Canadian Westinghouse Company, Hamilton Canada

12.8 **JULIAN ROWAN, DUDAS KUYPERS ROWAN (DKR)**

Supreme K69 electric kettle

This model was the first Canadian electric kettle designed with a separate plastic bottom plate, although the basic design still resembles the CGE K42 kettle.

1952–55
MANUFACTURER: Filtro Electric, Toronto
MATERIALS: stainless steel, Bakelite handle and thermoplastic base
DIMENSIONS: h (with handle) 20 x dia 21 cm (h 8 x dia 8¼ in.)
MARKINGS: stamped on base: Model K69 Filtro Electric

12.9 **LAWRIE MCINTOSH**

Electric kettle

Lawrie McIntosh responded to Fred Moffatt's dome shape but used an innovative and inexpensive one-piece plastic handle and spout. In addition to contributing bright colour, polypropylene eliminated the costly step of soldering a spout. It is still being produced by Superior Electrics from the original dies acquired from McGraw-Edison.

1968
MANUFACTURER: Superior Electrics, Pembroke, Ontario
MATERIALS: stainless steel, polypropylene handle
DIMENSIONS: h (with handle) 20 x dia 22 cm (h 8 x dia 8½ in.)
MARKINGS: on metal plate above spout: Superior

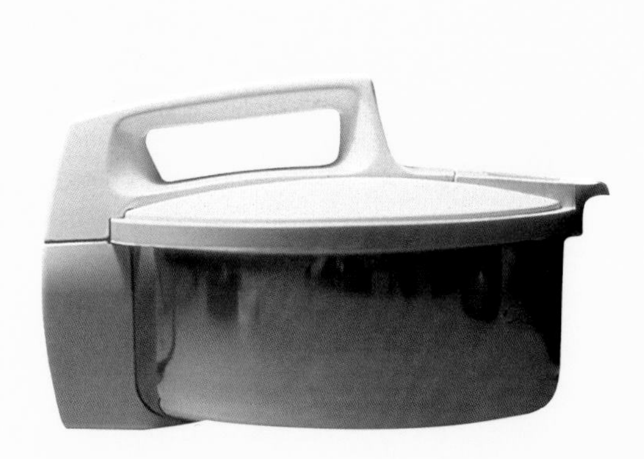

12.10 **JERRY ADAMSON, KAN INDUSTRIAL DESIGN**

Life Long model 89005 electric kettle

The manufacturer hoped for an all-plastic kettle but settled for a plastic lid and handle as suitable thermoplastics were not yet available. Sections could be disassembled and replaced as needed, but the design was not a commercial success.[16]

1970
MANUFACTURER: Proctor-Silex, Picton, Ontario
MATERIALS: stainless steel, polyethylene lid
DIMENSIONS: h (with handle) 16 x dia 21 cm (h 6½ x dia 8½ in.)
MARKINGS: on metal plate above handle: SCM Proctor Silex

12.11 **FRED MOFFATT**

K 840 electric kettle

In 1950, Moffatt introduced an elliptical design. The ovular form discouraged imitations because it required more complex tooling to bend the compound curves. Moffatt's final kettle has an integrated spout and handle, similar to Lawrie McIntosh's design. Its handle is "sculpted" and wider at the front to protect knuckles from the steam. Krups distributed it under licence in Europe.

1980
MANUFACTURER: Canadian General Electric, Barrie, Ontario
MATERIALS: chrome-and-nickel-plated shell, steel base, thermoplastic handle and spout
DIMENSIONS: h (with handle) 20 x dia 28 cm (h 8 x dia 11 in.)
MARKINGS: stamped on bottom: Canadian General Electric, Barrie, Ontario; GE logo above spout

12.12 **GLENN MOFFATT**

PK 502 electric kettle

Canada's first all-plastic kettle, including the base plate, was produced by Superior Electrics and designed by Fred Mofatt's son Glenn.

1990
MANUFACTURER: Superior Electrics, Pembroke, Ontario
MATERIALS: polypropylene
DIMENSIONS: h (with handle) 19 x dia 25 cm (h 7½ x dia 10¼ in.)
MARKINGS: stamped: PK 502 Superior Electrics

12.13 **TOASTESS**

Model 7304 electric kettle

Similar to Glenn Moffatt's PK 302 design, this kettle shares the circular lid and squared-off handle. However, Toastess introduced it in translucent candy-coloured plastic, giving it a high-tech sheen.

1997
MANUFACTURER: Toastess, Pointe-Claire, Quebec
MATERIALS: injection-moulded polypropylene
DIMENSIONS: h (with handle) 22 x dia 15 cm (h 8½ x dia 6 in.)
MARKINGS: labelled Toastess around kettle drum

12.14 **SID BERSUDSKY**

Lighto-Matic electric iron

The electric iron, invented in 1881, was a household favorite by the thirties. After the war, improvements in the thermostat allowed cast-iron plates to be replaced by lighter-weight aluminum. Most irons in Canada were made from imported American designs, but a few worthy products emerged, especially from the independent manufacturers. Sid Bersudsky added aerodynamic streamlining to his Lighto-Matic iron. He also moulded parallel lines into the handle.

1955
MANUFACTURER: Superior Electrics, Pembroke, Ontario
MATERIALS: steel, thermoset resin
DIMENSIONS: unknown
MARKINGS: Superior Lighto-Matic on handle

12.15 **THOMAS PENROSE**

IB22 electric iron

Handles on irons, as on electric kettles, evolved from metal to lathe-turned wood to plastic. Thomas Penrose developed the "open" black Bakelite handle for Canadian Westinghouse in 1951. Since it "permitted greater ironing depth," it was quickly copied. Produced until the sixties, the iron earned the distinction of being exported to the U.S.

1951
MANUFACTURER: Canadian Westinghouse Company, Hamilton
MATERIALS: steel-plated aluminum alloy, phenolic handle
DIMENSIONS: h11.5 x w12 x d22 cm (h4½ x w4¾ x d8¾ in.)
MARKINGS: stamped on metal plate: Canadian Westinghouse Company model IB22; Westinghouse label moulded on handle

12.16 **LAWRIE MCINTOSH**

Electric iron

Lawrie McIntosh contributed a modern handle made from thermoset resin and shaped the hot surface into two sharp points. His iron won an NIDC award in 1954 and a gold medal at the 1954 Milan Triennale. Steam Electric Products also produced an iron by John B. Parkin's short-lived industrial design department, headed by Charles Shepherd.

1953
MANUFACTURER: Steam Electric Products, Toronto
MATERIALS: steel, thermoset resin
DIMENSIONS: h 13 x w 11.5 x d 24 cm (h 5 x w 4 ½ x d 9 ½ in.)
MARKINGS: moulded on plastic: Steam Electric Products, Toronto Canada Patent 471417-1951

13.1 **HAROLD STACEY**

Candy dish with lid

In 1951 New York–based Steuben Glass (a division of Corning) hired Harold Stacey to create a line of silver and silver-and-glass objects to complement the company's crystal designs. Assisted by Solve Hallqvist (who later became a noted silversmith too), Stacey embraced the medium of glass and attempted to translate its fluidity into silver. He designed some fifty pieces, ranging from a traditional candlestick to an eccentric bonbon dish with a ram's horn handle. Some designs featured applied ornamentation, while others indulged in the luxury of ivory handles.

The candy dish is a successful amalgam of both media. When America went to war against Korea, material shortages caused the fledgling unit to be shuttered, and sadly, none of the designs were put into production. The prototypes eventually became part of the holdings of the Corning Museum of Glass, Corning, New York, and were exhibited as a group in 1988.[1] In his best work, Stacey, a meticulous craftsman, highlights the richness and quality of silver, expressed simply through shape, proportion and surface (usually well burnished).

1951–52

MANUFACTURER: Harold Stacey for Steuben Glass, New York
MATERIALS: sterling silver, glass
DIMENSIONS: h 19.2 x dia 14.7 cm (h 7½ x dia 5¾ in.)
MARKINGS: unsigned and unmarked

13 Metal Arts

Silver Represents Status

Making vessels from silver has been a part of Canadian life since settlers arrived more than 250 years ago.[2] Although the first objects were often either utilitarian or ecclesiastical, the difficulty in handcrafting metal, as well as its high cost, imposed limitations that influenced the design, shape and type of object—restrictions that inhibit design developments to this day.

Virtually from the moment that silver was employed to make household objects, it became an expression of wealth and status, associated with refined customs and European aristocracy. Since silverware is purchased for its heirloom qualities (beauty, durability and permanence), few companies dared to step outside a narrow range of conservative designs. Even established, well-capitalized firms like Henry Birks & Sons continue to produce traditional English tea sets or classic Grecian soup tureens in silver.

The introduction of Sheffield plate in the 1840s, followed by electroplating in 1859, made the luxury of silver accessible to the middle class. The new technology also facilitated mass production, as less expensive metals were rolled, stamped or turned on lathes to create the desired shape, then finished with a "coat" of silver. As many of these shapes were made from standard dies, additional hand decoration (embossing, engraving, gilding and enamelling) took on more significance. Increasingly, shapes became eccentric and ornamentation more elaborate.

In Canada, industrialization also shifted silversmithing from its eastern small-town roots to factories in Ontario, where silver mines were nearby and electrical power plentiful. One of the earliest manufacturing plants (as opposed to artisan workshops) was the Ontario Silver Corporation (later McGlashan Clarke Company). The company set up shop in the 1880s near the Niagara Falls generating station and remained in operation at least until the 1950s.

Silverware split into three streams: domestic, presentation and souvenir, and ecclesiastical. Domestic silver consists of hollowware (bowls, vases, pitchers, goblets and so on), flatware (cutlery and serving pieces) and occasional-use objects (tea sets, trays, cocktail shakers and candlesticks). Sports trophies and retirement plaques typically are classified as presentation silver, while objects for display rather than domestic use, such as spoons and other mementos, are considered souvenir silver. (Collecting spoons as a hobby is said to have begun in the 1890s, with the introduction of a Salem "witch" spoon.) Ecclesiastical silver has a strong tradition in Quebec, where firms such as Desmarais & Robitaille (established in the 1880s) supported the careers of artisans like Gilles Beaugrand with commissions for chalices, incense boats, processional crosses and altar candlesticks.

The revival in handcrafted silver began at the turn of the century with the Arts and Crafts movement. Its practitioners refocused on an object's function, eschewed excess decoration and

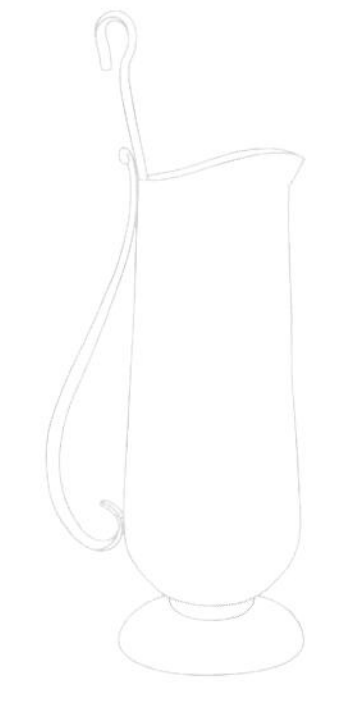

drew inspiration from the natural world. (In Canada, representing natural flora and fauna in silverware was already a well-established trend. Maple leaves appear as early as 1700, beavers even earlier.)[3] Although handcrafted in appearance, many Arts and Crafts objects had some machined elements, such as a cast or formed body, which was then welded to its base. Even some "hand-hammered" finishes were applied mechanically.

Ironically, the "machine age" of the twenties and thirties re-established the practice of raising silver forms entirely by hand. These hand-wrought works were considerably more expensive than machined products. War, and the continuing encroachment of mass production, depleted the ranks of artisan silversmiths, but handcrafters are the only silversmiths who have consistently paid attention to design. In the latter half of the twentieth century, a small coterie of metal workers attempted to create works of art in one of the most hidebound mediums.

Innovation Blossoms after the War

Following the war, the steady demand for ecclesiastical silver in Quebec helped to support a generation of silversmiths such as Beaugrand, Georges Delrue and Maurice Brault. Beaugrand, trained in Paris, produced more than eleven thousand chalices using a variety of techniques and methods until his retirement in 1995, yet he also created exhibition-quality hollowware and enamels. The avowed modernist Delrue accepted commissions from the Roman Catholic Church and was honoured for his biomorphic jewellery and forays into art deco tableware. Although their work is well regarded and collected by museums, little, if any, was destined for mass production.

Montreal also became home to Carl Poul Petersen. His firm, C. P. Petersen & Sons, became one of the few to bridge the divide between craft and industry. Unlike its much larger competitor, Birks, Petersen essentially created only original designs, and in the modernist style (FIGS. 13.7, 13.8).

From his shop in Victoria, the former engineer William Maurice Carmichael became British Columbia's top silversmith, producing everything from communion vessels to souvenir spoons that borrowed from Haida mythology. The bulk of his trade, however, during an active career between 1924 and 1953, was in reproduction Georgian hollowware.[4]

For a brief period, Toronto became the centre of activity for hollowware in Canada. After the Second World War, federal and provincial governments began to support crafts (via educational funding) to create jobs and rehabilitate workers. Many of the vocational-school courses in metalworking focused on less expensive metals like pewter, copper and brass. Nevertheless, the artisans preferred to work in silver, as attested by entries into competitions sponsored by local guilds.

Under the leadership of a flamboyant Swedish educator, Rudolph Renzius, a number of silversmiths (Harold Stacey, Andrew Fussell and Douglas Boyd) were launched into the craft. Stacey made attempts to design objects for mass production, but the projects never came to fruition (FIG. 13.1). Fussell, Boyd and Renzius all produced large bodies of work, but save for a few souvenirs, their hollowware consisted of custom, one-of-a-kind pieces.

By the mid-fifties, many of these artisans were making the transition from hollowware to jewellery. Toronto-based schools such as the Ontario College of Art, Ryerson Institute of Technology and Georgian College shifted their emphasis to jewellery design. To shore up interest, the Metal Arts Guild of Ontario introduced an award (the Steel Trophy) to recognize hollowware designs. In 1954 the NIDC (along with the Canadian Jewellers Institute) launched a design competition for a sterling silver tea service, complete with cash prize.

RIGHT: 13.2 The silversmith Harold Stacey, active from the thirties to the seventies, always worked wearing a tie.

There was little interest in silver tableware. Many post-war families were struggling to buy a house and purchase new appliances and had neither the income nor the fancy dining room and buffet to show off heirloom silver. Those who could afford it preferred the traditional lines stocked by national retailers, which also carried replacements and matching products so that buyers could build a "family" of specialty silver over time. By the end of the decade, only a smattering of silversmiths occasionally made domestic objects in silver.

By 1966 the decline in household silver was such that its manufacturers, refiners and retailers jointly launched a program to educate schoolgirls about silverware, but it was too late to stem the tide. For the past four decades, most Canadian silversmiths have designed custom jewellery, supplemented with an occasional piece of hollowware. For example, the work of Lois Etherington Betteridge (who studied at the renowned modernist American institution, the Cranbrook Academy of Art, Bloomfield Hills, Michigan) is exhibited in galleries and museums rather than sold in gift shops. Likewise, Donald Stuart, who creates both jewellery and hollowware, concentrates on exhibition-quality pieces rather than mass production. A rare exception is Anne Barros, Toronto, who has achieved some success marketing specialty flatware such as silver baby spoons (FIG. 13.9).

Pewter and Other Metals Reign

Most artisans, regardless of their personal preference, have worked in metals other than silver. For some, less expensive materials opened the door to producing in quantity. Harold Stacey designed and made several pewter coffee and tea services aimed at the mass market (FIG. 13.4). For others, new materials expanded the creative possibilities, allowing them to experiment with a lower-risk medium.

In Toronto Renzius was a key advocate for pewter. An apprentice, Douglas Shenstone, later wrote a book titled *For the Love of Pewter: Pewterers of Canada*, which showcases, among others, the work of Quebec-based pewter artists like Bernard Chaudron and Paul Simard (both of whom studied in France). While most of the featured pewterers have a catalogue of designs, their output is handcrafted individually.

In the seventies, a few high-volume pewter manufacturers appeared in Atlantic Canada. Aitkens Pewter in Fredericton, New Brunswick, continues to create popular lines of giftware, holiday pewter and jewellery. Although it has produced a goblet designed by Betteridge, it is best known for its more traditional offerings. Similarly, Seagull Pewter & Silversmiths, Pugwash, Nova Scotia, has had international success with pewter souvenirs and giftware. Working with sand-cast aluminum, Hoselton Studio, Colborne, Ontario, has produced sculptural objects (sold primarily through gift and crafts retailers) since 1970.

Many metal workers also executed custom architectural works like andirons, light housings and mirrors in various metals, often in collaboration with interior designers or architects. Sometimes a commission for a silver chalice would result in a contract to make chapel doors and organ screens. In more recent times, the metalsmith David Didur has fabricated everything from custom wrought-iron gates to the bizarre gynecological instruments in the David Cronenberg film *Dead Ringers*. Currently working as an industrial designer in New York, he has also designed metal lamps and vessels, although the biggest production run to date is one hundred vases.

Many metal artisans frequently experimented with enamel, but the works were often single objects rather than designs for production. Fay Rooke, Toronto, and her two daughters all work in enamel. Alan Perkins, also in Toronto, has been active since the late sixties. In addition to award-winning fine-art pieces, he has designed and produced lines of enamel ashtrays and desk sets and ceramic hardware.

Perhaps the best-known non-precious metal Canadian objects are the DC3 airplane ashtrays. These were originally made as souvenirs for the 1939 World's Fair in New York but were re-issued after the war with marble bases in either freestanding or tabletop deluxe models. Each features an interior-lit plane, complete with propellers.

In the eighties, artisans began to experiment with patinated and anodized metals, but most efforts were directed toward jewellery. Unusual metals like titanium and niobium also came into vogue, with Alison Wiggins and David McAleese, both in Toronto, handcrafting flatware from coloured titanium and sterling silver. Recently the respected industrial designer Helen Kerr has created a series of stainless steel flatware (FIG. 13.10), signalling a return to the utilitarian roots of metalworking.

BELOW: 13.3 Andrew Fussell added charm to silver salt and pepper shakers, 1947.

13.4 **HAROLD STACEY**

Coffee pot

Although the quality of Harold Stacey's hand-hammered work elevated him to the stature of an artist, he often explored ways to mass-produce his designs. Here, he uses inexpensive pewter and machinery to create a contemporary coffee pot with ebony handle. Sheets of pewter were "spun" by a lathe over a wooden form, creating the hollow bodies efficiently and uniformly. The Metal Arts Guild of Ontario awarded its design in 1948. Stacey would later make a graceful (but structurally more complex) coffee pot of silver that featured a rosewood handle.[5]

1947
MANUFACTURER: Harold Stacey, Toronto
MATERIALS: pewter, ebony
DIMENSIONS: h23 x dia12.5 cm (h9 x dia5 in.)
MARKINGS: probably scratched Stacey and date

13.5 **DOUGLAS BOYD**

Cocktail pitcher

Douglas Boyd worked in spurts, occasionally driving his patrons to distraction as they waited for wedding gifts to appear on time. Although his work was never as refined as Harold Stacey's, this silver cocktail pitcher won the Metal Arts Guild of Ontario's Steel Trophy for best in show in 1957.

1950s
MANUFACTURER: Douglas Boyd, Toronto
MATERIALS: sterling silver
DIMENSIONS: h22.6 x dia6.3 cm (h9 x dia2½ in.)
MARKINGS: typically, scratched on the bottom: Douglas Boyd Richmond Hill, plus the date

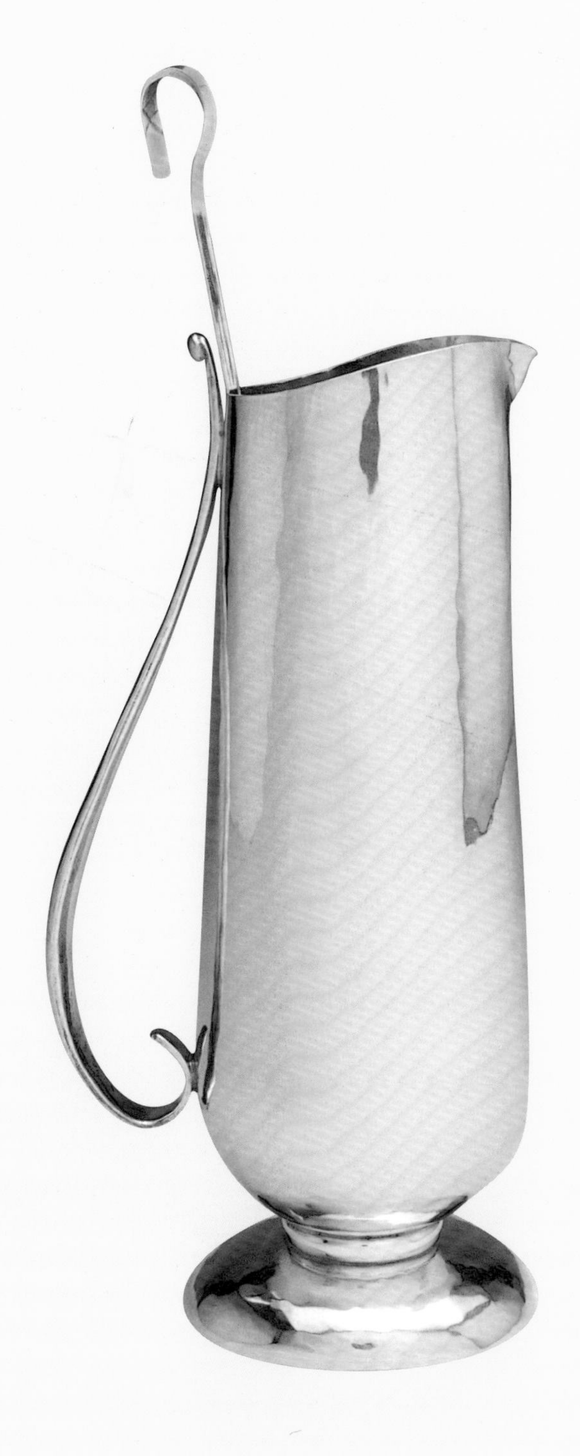

13.6 **ANDREW FUSSELL**

Bowl

In 1959 Andrew Fussell received the Metal Arts Guild of Ontario's Steel Trophy for this bronze bowl. Its finely hammered surface and undulating rim are characteristic of Fussell's higher-end work. Undecorated, it relies entirely on its shape and craftsmanship to create its effect.

1958
MANUFACTURER: Andrew Fussell, Toronto
MATERIALS: bronze
DIMENSIONS: h26 x w20.5 x d9 cm (h10 x w8 x d3 ½ in.)
MARKINGS: stamped with his registered trademark, Fussell, national mark, Handmade, identification of the metal, Toronto, and a four-digit date.[6]

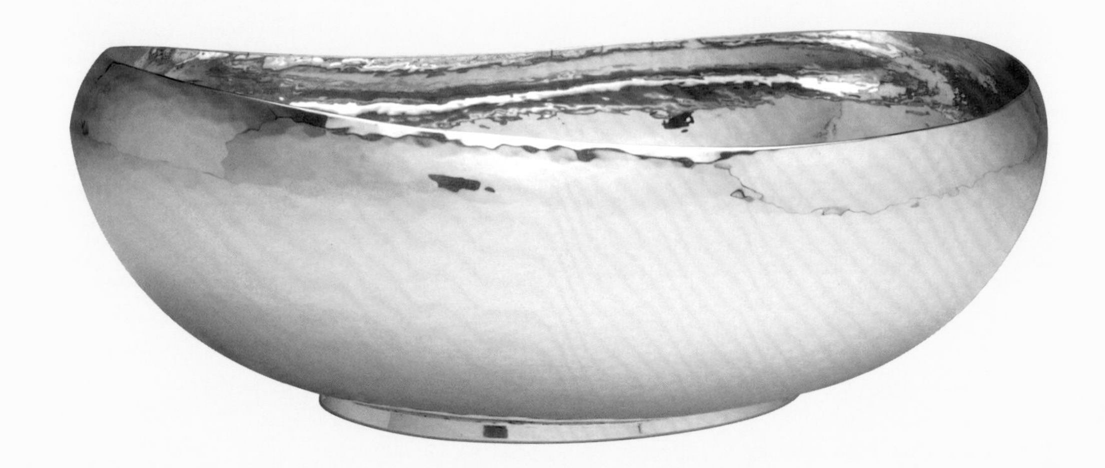

13.7 **CARL POUL PETERSEN**

Bread tray

Carl Poul Petersen created the designs for his firm and did not allow his assistants to vary from them.[7] This straightforward serving tray (FOREGROUND) reflects the balance and restraint that characterize the firm's best work. Organic, tactile and true to its material, it bears the mark of a mature artisan. The corn or pea-pod motifs have been successfully integrated into the overall design. Over time, the silver oxidation process emphasizes the delicate details. With a few deft strokes, Petersen's craftworkers hand-hammered a classic tray with appropriate mass and density to signal that it is of enduring quality and style.

Mid-twentieth century
MANUFACTURER: C. P. Petersen & Sons, Montreal
MATERIALS: 925 sterling silver
DIMENSIONS: 38 cm long (15 in.)
MARKINGS: hollowware stamped with one of two markings: PP Sterling or Petersen Handmade Sterling, followed by a lion's head hallmark

13.8 **CARL POUL PETERSEN**
Dolphin flatware

C. P. Petersen & Sons made eleven flatware patterns in traditional and modern styles, including Dolphin. Finishes included satin, high-polish or hammered. Carl Poul Petersen was influenced by the Arts and Crafts movement and often employed images from nature in his designs. Dolphin is somewhat oversized, giving it a masculine stature and heft. Evidence of hand-hammering was retained to emphasize that it was made by craftworkers.

Production of flatware ceased in 1969
MANUFACTURER: C. P. Petersen & Sons, Montreal
MATERIALS: sterling silver
DIMENSIONS: knife: 25 cm long (10 in.)
MARKINGS: impressed with the word Petersen

13.9 **ANNE BARROS**

Baby spoon

Anne Barros consulted a pediatrician to create the correct bowl shape and tine length for her baby cutlery.[8] The Avanti pattern (with bells) features three spoons, a fork, a teething ring and a bubble blower. Alexis (named after her son) has two applied horizontal bands at the base. More than two thousand pieces have been sold across North America. The silver is cut with a die and hand-hammered to create the delicate jewel pattern.

1982
MANUFACTURER: Anne Barros, Toronto
MATERIALS: sterling silver
DIMENSIONS: 12.7 cm long (5 in.)
MARKINGS: maker's mark A, and st for sterling

13.10 **HELEN KERR**

Ellipse flatware

Gourmet Settings (formerly Trupco) commissioned Helen Kerr to create a series of original flatware designs for a market accustomed to knock-offs of eighteenth-century designs and reissues of classic Georg Jensen patterns. To date, the gutsy gamble has paid off. Five designs (the Soshu Collection) have gone into production, with total runs (over time) estimated to reach ten thousand a line.[9] Ellipse has received the strongest reaction from wholesalers and retailers.

The designs evolved from extensive research into today's globally oriented mix-and-match food choices and styles of eating. A contemporary lifestyle product, Ellipse reflects both Asian and European habits and imagery. The swordlike knife blade provides an element of drama. The ergonomically balanced designs are die-cast and stamped or forged from stainless steel (depending on the line) to keep the cost of a setting low.

1998

MANUFACTURER: Gourmet Settings, Richmond Hill, Ontario
MATERIALS: die-cast stainless steel
DIMENSIONS: knife 25.8 cm long (10¼ in.)
MARKINGS: Gourmet Settings logo, Korea

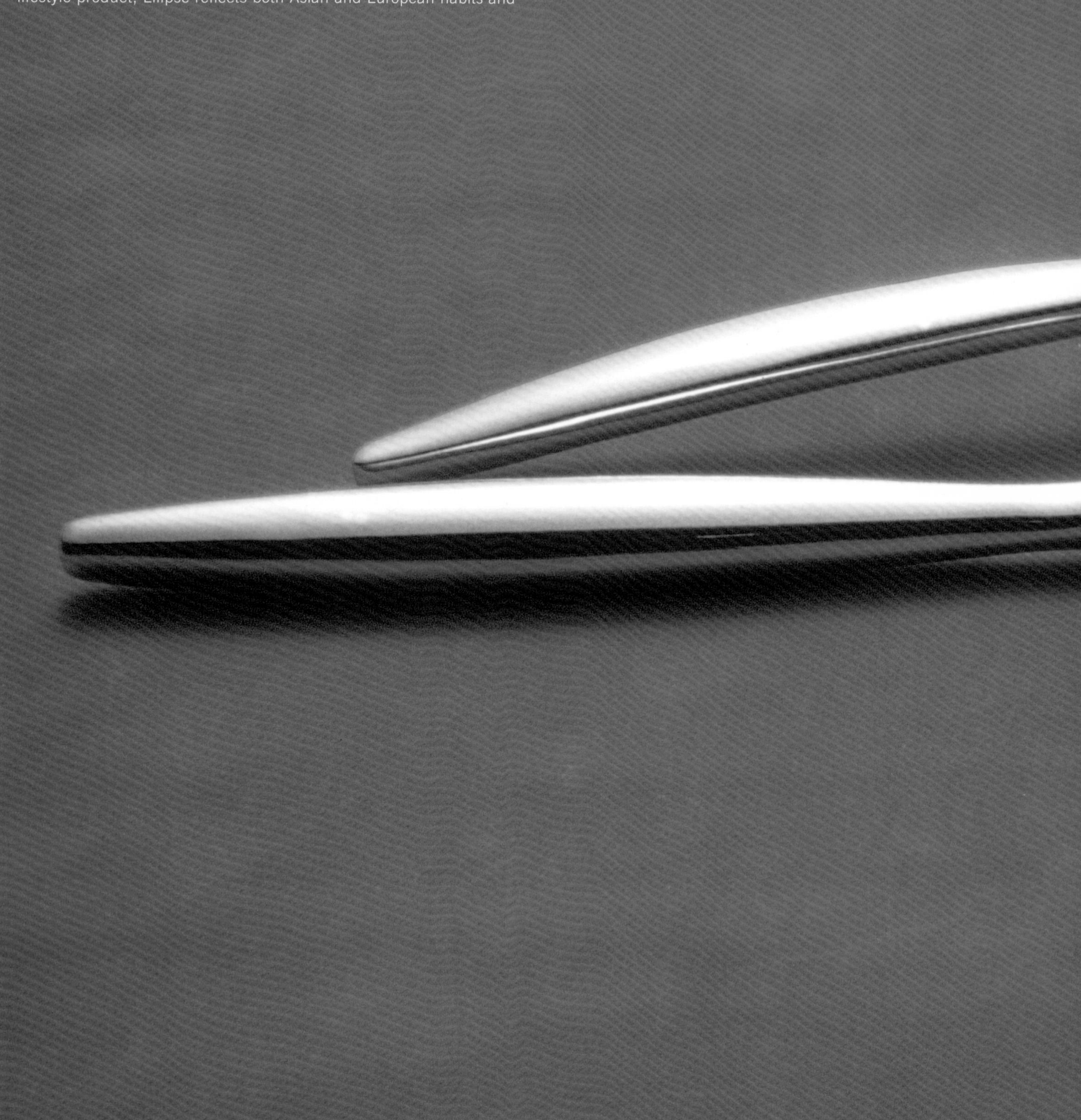

Biographies and Corporate Histories

Biographies and Corporate Histories

ADDISON INDUSTRIES
Late 1930s–1955, Toronto

Launched in the late thirties to distribute General Electric's line of Norge appliances in Canada, Addison Industries eventually manufactured electric appliances (refrigerators, stoves, washers and water heaters), small appliances (toasters, irons, vacuum cleaners and heating pads) and consumer electronics (console and tabletop radios, televisions and electric pianos). The company is best remembered, however, for its colourful plastic radios, which have become highly collectible in the international market.

Harry Addison, born in 1905 and lamed in a childhood accident, met the hockey entrepreneur Conn Smythe, who helped him obtain the franchise for Norge appliances. The distribution arrangement was successful, and he soon brought his hard-living, promotion-oriented brother, Jack (born in 1902), into the business. When the Second World War broke out, they purchased a neighbouring Toronto firm, A. Cross & Company Limited (ACCL), which made communications equipment. A meeting with the prime minister, Mackenzie King, helped the company obtain a military contract for walkie-talkies. A portable model's effectiveness increased sales, and the company's fortunes grew. In 1939 it began making Crosley radios for the armed forces (advertising the models as all-Canadian), a contract it held until 1942. It also manufactured Webster-Mohawk intercom systems, licensed by CGE and Northern Electric.

Addison Industries returned to civilian production by 1944. Products were marketed under Norge, Addison or combined brand names. ACCL became the national sales and service operation, eventually establishing additional offices in Moncton, Montreal, Winnipeg, Calgary and Vancouver. In 1945 the first two, and now most collectible, Addison radios (Model 2 and Model 5) were promoted to the trade.

In the ensuing five years, more radios were offered, although most were wooden console or tabletop models (retailing for as much as $239) in more conservative styles. The company also introduced record players and electric pianos, as well as a complete line of small appliances. Purportedly

BALL, DOUGLAS
Born 1935, Peterborough, Ontario

Douglas Ball is Canada's most important and successful systems furniture designer. For most of his career, he worked with a single client and earned his living primarily from royalties. He designed office furniture, including the RACE system, which anticipated the impact of technology on offices. Ball's innovations allowed Sunar Industries to successfully compete against the well-established American manufacturers Herman Miller and Knoll International.

Ball graduated from the industrial design program at the Ontario College of Art in 1958. After a brief stint in Europe, he joined Sunshine Office Equipment in Waterloo, Ontario, and was associated with the firm in various professional capacities for nearly thirty years. Sunshine (a division of Massey-Ferguson Industries) specialized in metal stamping and made everything from garage doors to baby carriages. As chief designer, Ball focused the firm's output on office furniture.

By the late sixties, the company (now named Sunar Industries) embraced open-concept, modular systems furniture. In 1967 the first line, System F, was made from wood. Two years later, System S (for steel) was launched, and in 1972 Ball followed up with PAS, an acoustic panelled system. With RACE in 1978, he introduced the market to a groundbreaking design that banished unsightly cabling. RACE continues to sell well in North America, Europe and Asia, and in 1999 *Contract* magazine named it one of the most significant products of the past twenty-five years.

Sunar was sold to U.S.-based Hauserman in 1978. Ball's designs, including RACE, PAS and Uniwall, the storage system, as well as wooden furniture (under the brand names Douglas and Cameron) sold alongside products by international superstars such as Michael Graves, Niels Diffrient and Frank Gehry. In 1990 certain assets of the firm were purchased by the Michigan firm Haworth, which now produces the lines RACE and Cameron (Ball's middle name).

More recently, Ball's designs have included lounge seating for Atelier International (1991); Ballet folding tables for Texas-based Vecta, a division of Steelcase (1991); and the Activity Products line for Steelcase (1993). Ball's current passion is the Clipper CS-1, an enclosed capsule that functions like a virtual office. He was made a fellow of the Association of Canadian Industrial Designers in 1993, and at the Chicago NeoCon World's Fair in 2000, he won two golds for the Logistics chair (Arconas Corporation, Toronto) and the Lucy chair (Vecta).[3]

BARROS, ANNE
Born 1939, Watertown, New York

Anne Barros has sold thousands of pieces of complementary cutlery such as baby silver, demitasse spoons and serving pieces in stores ranging from Bloomingdale's and Barneys in New York to William Ashley in Toronto.

during this period, the company's chief designer was Carl Nielsen; Alex Pringle was in charge of making television cabinets.[1] By 1950 Addison was offering everything from vacuum cleaners to televisions and reminding buyers that they emanated from "the laboratory that built the famous walkie-talkie." The TV, however, had problems with its horizontal oscillator, and sales, which were rarely hampered by Addison radio's poor-quality sound reproduction, plummeted.

A large factory and broad product line strained the company's resources, and Harry and Jack split the company into ACCL and Addison Industries. Some branches were shuttered by 1952, and within three years Addison closed. Jack died in 1958, and Harry reshaped ACCL into Addison Electric and founded Addison on Bay, a General Motors dealership in Toronto. Harry died in 1985, but his sons continue to operate the car dealership.

AMBIANT SYSTEMS
1968–1989, Toronto

Ambiant Systems was an innovative manufacturer of public furniture. Simple blond or stained woods, exposed construction and quick and easy on-site assembly were the signature features of Ambiant's twenty-piece collection. The firm received the prestigious Chairman's Award for Design Management from Design Canada in 1982.

Keith Muller (b. 1938) and Michael Stewart (b. 1940), both graduates of the Ontario College of Art in 1963, formed the industrial design partnership Muller & Stewart in 1967. Stewart is the son of the graphic designer Clair Stewart, who ran the prominent graphic design firm Stewart & Morrison, and the son-in-law of the Finnish architect Viljo Revell, who designed Toronto's New City Hall, 1958. Muller worked at Stewart & Morrison in 1964 and at Dudas Kuypers Rowan (later known as KAN Industrial Design) in 1965. After two critically acclaimed designs, the MS stacking chair and the Image Series furniture, they founded Ambiant Systems to distribute their contract furniture, and it eventually became a manufacturer. Muller left Ambiant in 1975 and formed his own design studio in Toronto, specializing in environmental graphics.

Ambiant's most lucrative product was the Series 700, introduced in 1970. This modular precast concrete "street system" featured stacking bases for seating, tables and display cases. (Early in his career, Stewart had spent ten months in Helsinki studying concrete.) In 1980 the company expanded its concrete line into a wall and ground system, known as Groundscape.

Paul Epp joined in 1983 as design director, notably designing the Nexus chair. The company briefly produced Thomas Lamb's Steamer chair, bought a wood factory in Pickering and founded an upholstery division. Showrooms were opened in Chicago and New York City. The expansion proved untimely, and Ambiant Systems closed in 1989.[2]

In the eighties, her product line swelled to fifty items, including makeup brushes, bookmarks, buttons and tea-for-one pots. The Canadian Museum of Civilization, Hull, Quebec, the Canadian Craft Museum, Vancouver, and the Design Exchange have collected her work.

Barros earned a B.A. from New York's College New Rochelle in 1961. As a second career, she studied metal arts, gold and silversmithing at Toronto's Humber College, graduating in 1978. That same year, she attended the Sir John Cass School of Silversmithing in London, England. While studying, she designed for Burkhardt Jewellers (1977–78), and later for Rembrandt Jewellers (1982–83), both of Toronto.

When Humber's metal shop closed in 1985, Barros set up shop at home, forcing a switch from hand-hammering vessels (too noisy) to die-cutting flatware blanks. She makes her own dies using the R.T. Blanking System and finishes the pieces by hand. She occasionally contracts other artisans to complete her designs and has experimented with materials such as Corian plastic handles. In 1996 Barros wrote a reference book, *Ornament and Object: Canadian Jewellery and Metal Art 1946–1996*.[4]

BAZZ
Founded 1978, Montreal

In the eighties, Bazz was part of Quebec's flourishing design scene, catering to its loft living and dynamic night life. Simon Ben Ghozi, a graduate of McGill University's School of Architecture, founded the company in 1978 and made it a full-time commitment three years later. Bazz's line progressed from colourfully painted spun aluminum lamps to high-style halogen lighting. In 1989 Bazz moved production offshore to compete with inexpensive imports from Southeast Asia. Economic circumstances returned manufacturing to Canada. It now produces high-volume lighting for retailers like Home Depot.

B.C. CERAMICS
c. 1952–1967, Vancouver
Herta Gerz born 1918, Hanau, Germany

B.C. Ceramics, founded by Herta Gerz and her husband, Walter Gerz, was one of the first art potteries in Canada to specialize in modern design. Their best work favoured freehand and abstract patterns and clean, simple shapes. Herta studied at the Staedel Art Institute in Frankfurt in 1930 and at the Keramische Fachshule (school of ceramic art) near Koblenz until 1933. She ran a faience factory in Gelnhausen, Germany, from 1946 to 1951. The couple immigrated to Vancouver in the fifties and immediately established B.C. Ceramics.

Herta handled design and modelling, while Walter, a ceramics engineer, developed the glazes and ran the business. They published a catalogue and produced a standard line of more than fifty patterns, including vases, pitchers, bowls, ashtrays and serving dishes (no dinnerware). Hand-decorated abstract patterns included Bamboo, Mardi Gras and Crackshell.[5] Amoeba-like shapes and crackle-shell glazes represent the Gerzes' strongest designs.

BEAUDIN, GAÉTAN

Born 1924, Toronto

Gaétan Beaudin is one of the founders of the pottery craft revival in Quebec. A master craftsman, he came from the studio pottery movement to launch a manufacturing facility to mass-produce ceramic dinnerware.

Beaudin graduated from the École des Beaux-Arts in Montreal and moved to Rimouski to teach war veterans at the local technical school. There he founded Décor Pottery, which operated between 1945 and 1953. He also taught at the Penland School of Crafts in North Carolina for three years.

When the school in Rimouski closed, Beaudin opened a summer school in North Hatley, Quebec, in 1953, which he ran until the sixties. After a year's stay in Japan, he teamed up with Pierre Legault and Bertrand Vanasse in 1965 and created Sial, which fabricated and distributed glazes and clay materials. Sial II opened in the seventies to produce tableware designed by Beaudin.

BERSUDSKY, SID

Born 1915, Odessa, Ukraine; died 1993

Trained as an illustrator, Sid Bersudsky became one of Canada's first industrial designers, modelling his career after American pioneers like Raymond Loewy and Walter Dorwin Teague. He embraced plastic, then a new material, and earned twenty patents in Canada and the U.S. for products like a dust remover and fibreglass bowling ball.

Bersudsky, who was raised in New Brunswick, studied illustration at the New York Art Students League (1937–38), where he was exposed to the design profession and its leaders. On his return to Saint John, he worked as a cartoonist, graphic artist and illustrator. By 1946 he found the call of industrial design irresistible. He established Sid Bersudsky and Associates in the small city of Sydney, Nova Scotia, and created a prototype chair formed from a single sheet of acrylic.

Bersudsky relocated to Toronto in 1948. In addition to plastics, he gained some repute designing housewares and small appliances for clients such as General Steelwares and Superior Electrics. He won several NIDC awards (including for an automatic iron and a gas-fired furnace) and became the council's representative to the Society of Plastics Industries. A staunch supporter of professional accreditation, he was a longtime member of the Association of Canadian Industrial Designers and served as its president in 1963. Eleven years earlier, he had been the first Canadian accepted as a member of the American Society of Industrial Designers. Bersudsky was awarded a Centennial medal in 1967.

In 1947 they won a contract from Eaton's to supply one hundred yards in their Flight pattern for drapery for the new airport at Malton, Ontario. Their initial client base came from John's brother, Philip, later a partner in Brisbin Brook Beynon architects.

By 1952 John Brook (who had studied engineering at Queen's University) had joined his wife's silkscreening business full-time, and they opened a studio to produce a small line of twenty-five hand-blocked printed textiles.

The success of the collection led the Brooks to venture into interior and furniture design through a subsidiary, Contemporary Distribution, Toronto. Elizabeth Wilkes Hoey took over the production of the printed fabrics in Bronte, Ontario, until the mid-fifties, when they began to be produced by Jim Farquhar in Toronto. The new company sold this output as well as fabrics by other Canadian designers, notably Micheline Knaff and John Gallop. It also carried modern furniture designed by John Brook and built by Gary Sonnenberg's company, Precision Craftwood. Sonnenberg also collaborated with Contemporary Distribution on the Sonos System of metal library shelving, which became one of their best-sellers. The industrial designer William Sloan created a popular plastic-and-metal stacking chair for the company.

Significant J & J Brook interior design projects include Union Carbide headquarters, Toronto (Shore and Moffat architects, 1959), Gander International Airport in Newfoundland and Great Western Life in Winnipeg. John and Joanne ran Pax, a design boutique in Toronto, from 1966 to 1972, when they also closed J & J Brook.[8]

BULOW, KAREN/KAREN BULOW LTD.

Born 1899, Skannerborg, Denmark; died 1982

Karen Bulow was one of Canada's leading weavers in the post-war era. Her company, Canadian Homespuns, introduced Scandinavian-style fabrics to corporate and residential interiors.

Bulow arrived in Montreal in 1929 with only a hand loom and a few dollars. To survive, she sold handwoven neckties, sashes, belts, skirt lengths and scarves. By the mid-thirties, interior decorators like Gert Lamartine of Montreal's Artists' Workshop and Toronto's Freda James helped Bulow secure important commissions for drapery and upholstery fabric. Significant projects were tapestries for Holy Blossom Temple, Toronto, drapery for the Bank of Nova Scotia headquarters in Montreal and blinds for Trans-Canada Airline's jets (now Air Canada).

Bulow was also an important teacher, establishing a weaving school in her studio between 1933 and 1949. Her best students became employees, hand-looming Bulow's designs from their homes. She was a strict teacher with a formidable eye, so pupils rarely strayed from her authoritative samples. By the fifties Canadian Homespuns employed weavers in-house, hand-looming upholstery fabric, wall hangings, window treatments and rugs. Still, her signature ties remained the company's bread and butter, with fifty-six thousand selling in one year.

BOSTLUND INDUSTRIES

Founded 1954, Oak Ridges, Ontario

For a local version of Danish modern lighting, Canadians turned to Gunnar and Lotte Bostlund, who made their handsome designs in a makeshift "factory" at their farm in Oak Ridges. With their six children, the Danish immigrants became the lighting industry's equivalent of the musical Von Trappe family, participating in international trade fairs and a National Film Board documentary. Lotte designed, and Gunnar handled production. Lotte also created a line of colourful tiles and tableware.

Gunnar Bostlund, an electrical engineer specializing in porcelain insulators, and his wife, who graduated from the Royal Danish Academy of Fine Arts in Copenhagen, founded the company in 1954, two years after their arrival from Denmark. Their lamp line included fifteen shapes, all made in slip-cast stoneware. Annual production averaged three thousand units.

In 1963 the company introduced fibreglass shades and expanded their line to include wall sconces and pendant lighting featuring elegant combinations of fibreglass, string and teak. Unable to expand the factory owing to zoning restrictions, Bostlund Industries relocated in 1975 to its subsidiary outside Buffalo, opened ten years earlier. Members of the family sold the company in 1997 and returned to Oak Ridges, where they currently offer a limited production of the original lighting line.

BOYD, DOUGLAS

Born 1901, Toronto; died 1972

Like Andrew Fussell, Douglas Boyd approached metal arts with more enthusiasm than formal training, had a long and varied career and supplemented his income making spoons, cufflinks and dresser sets. Unlike most artisans in the forties and fifties, he had an international outlook and completed commissions destined for China, Czechoslovakia and Germany, perhaps as a result of training under Rudolph Renzius. His down-to-the-wire approach came in handy in 1951 when he completed a silver cigarette box for HRH Princess Elizabeth and the Duke of Edinburgh in two and a half days.[6] Boyd began as a hobbyist in the 1930s, and won his first award at the Canadian National Exhibition in 1937. He was a member of the Metal Arts Guild of Toronto, was its president in 1949–50 and was awarded its Steel Trophy in 1957.[7]

J & J BROOK

1948–72, Toronto Joanne Brook born 1917, Cleveland, Ohio; John Brook born 1914, Salmon Arm, British Columbia; died 1997

In the late fifties and early sixties, J & J Brook was the modernist corporate interior design firm. It was founded in 1948 by the husband and wife John and Joanne Brook, who initially designed textiles.

Joanne, an art school graduate, began silkscreening textiles at home. In 1960 Bulow sold Canadian Homespuns to her assistant, Margareta Steeves, and her husband, Edward Steeves, although she remained involved until 1963. Capitalizing on the Bulow name, they called the company Karen Bulow Ltd. The couple closed the Montreal factory in 1978 and moved to Toronto, where they merged with a sister company, Loomloft Designs, and introduced machine-powered looms. The firm operated until 1987, employing as many as eighty weavers. Notable clients included Jacques Guillon & Associates in Montreal, Alison Bain in Toronto and Arthur Erickson in Vancouver.[9] Swedish-born Margareta added her own designs while continuing to produce Bulow signature items.

Donald Stuart, as a representative of Karen Bulow Ltd., organized in 1969 the successful government initiative to introduce weaving as a means of employment for the Inuit hamlet of Pangnirtung. The Arctic weaving cottage industry continues today.

BÜLOW-HÜBE, SIGRUN

Born 1913, Linkoping, Sweden; died 1994

Sigrun Bülow-Hübe brought Scandinavian modern design to Canada. She designed high-quality furniture for AKA Works of Montreal, became the first woman to join the Association of Canadian Industrial Designers and wrote extensively on design policy.

Bülow-Hübe (whose much younger sister is the noted jewellery designer Vivianna Torun Bülow-Hübe) graduated in 1935 from Copenhagen's Royal Danish Academy of Fine Arts. Under the tutelage of the influential Danish modernist Kaare Klint, she specialized in furniture and interior design. She apprenticed as a cabinetmaker, worked in Stockholm as an interior architect (1936–42) and became the chief designer for Malmo City Theatre (1942–43). In 1947 she helped to organize the Swedish pavilion at the Milan Triennale and was awarded in 1949 a grant to research and publish a book about housing and mass-production furniture.

Bülow-Hübe moved to Montreal in 1950 to work for Eaton's as an interior design consultant. Three years later, she joined Arnold Kasak's new custom furniture company, AKA Works, became chief designer, and until 1968 worked with Reinhold Koller, designing Scandinavian-style furniture that won twelve NIDC awards. Designs included commercial case goods, custom built-in cabinetry and wood panelling, as well as beds, chairs and sofas, usually upholstered with handwoven textiles by Karen Bulow. The firm supplied furniture to Montreal's McGill University and Place des Arts, Ottawa City Hall and Air Canada's offices.

Her designs were exhibited at the 1958 Brussels World's Fair and appeared in Habitat. Between 1957 and 1974, Bülow-Hübe was a shareholder in AKA Furniture Company, a retailer that, in addition to showcasing her designs, sold imported products. In the late sixties, Bülow-Hübe completed a study on residential kitchen design for the federal government's

Central Mortgage and Housing Corporation. As senior design consultant to the National Design Council/Design Canada (1971–77), she oversaw its Scholarship Grant Programs and its controversial Record of Designers, a list of industrial designers eligible for government grants.[10]

BUSH, ROBIN

Born 1921, Vancouver; died 1982

Robin Bush embodies the meteoric rise of post-war industrial designers. Like his idols Charles Eames and George Nelson, Bush capitalized on the attractive image of designers of the day and made his way into corporate boardrooms as well as marketing departments. As his reputation grew, he was able to impose an emphatic design aesthetic on everything he touched. He also persuaded the iconic American firm Herman Miller to subcontract manufacturing in Canada.

Bush attended the Vancouver School of Art in the forties, studied architecture by correspondence and served in the Royal Canadian Navy during the Second World War. He and Earle Morrison took over the Standard Furniture Plant in Victoria, B.C., in 1950 and began manufacturing their own wooden and wrought-iron domestic furniture. The lines sold in Eaton's department stores across Canada and the pair won several NIDC awards.

Constrained by the limits of the island market, Bush departed for Vancouver in 1953, formed Robin Bush Associates and won a major contract to supply metal frame furniture, which he had previously designed for the hotel/motel market, to Alcan's new company town in Kitimat, B.C. Contracts with the Vancouver Public Library and the B.C. Electric building permitted him to specify Herman Miller furniture, thus cultivating an important relationship. In 1957 his Prismasteel line went into production at Canadian Office and School Furniture, the 120-year-old firm in Preston, Ontario. Bush anointed COSF the Herman Miller of the north and himself its independent designer. Within two years, Max DePree, president of Herman Miller, was marketing Prismasteel under the Herman Miller imprimatur, and COSF and Snyder Bros. Upholstery Company, Waterloo, became subcontractors until Herman Miller withdrew from manufacturing in Canada in 1965.[11]

The Bush–COSF partnership resulted in other office furniture lines, including Sheerline (1959), Dynaform (1960) and the Preston Executive Group (1961). The COSF team scored with Lollipop seating, when it was selected in 1960 for Toronto's new international airport. Bush negotiated for generous royalties and all-encompassing input into furniture design, marketing and manufacturing. At times, the designer–client relationship was fractious. Prototyping the futuristic Radial Office System Desk, a circular workstation that was advanced for its time, further tested the partnership, and in 1966 Bush and COSF parted company.[12]

Following Eames's lead, he turned to exhibition design, set up a new division, InterDesign, and won major contracts at Expo 67, the Stratford Festival, the Canadian National Exhibition and the National Arts Centre.

models produced by Rogers-Majestic and other manufacturers.

Throughout the forties and fifties, the engineer Thomas Penrose oversaw the production of small appliances, including electric toasters, kettles and irons. In 1971 the company received an IDAP (Industrial Design Assistance Program) grant from the National Design Council to co-develop the partially plastic Galaxy electric kettle with the Toronto-based industrial design firm Savage Sloan. Westinghouse Canada (as it was now named) merged with the major appliance company CAMCO in 1977.

CANADIAN WOODEN AIRCRAFT

1947–1949, Stratford, Ontario

The short-lived Canadian Wooden Aircraft was founded by the Polish émigré Hilary Stykolt. Waclaw Czerwinski served as chief engineer, directing the design and manufacture of plywood parts for de Havilland Aircraft of Canada and later producing prototypes for a moulded plywood glider plane. In 1947 the company parlayed its skill with plywood and bent laminated wood into the residential furniture market and produced modern dining room and bedroom suites as well as radio cabinets. A few years later, Imperial Rattan (later Imperial Furniture Manufacturing Company), also of Stratford, acquired the company.

Stykolt (1894–1974) studied chemistry at the Sorbonne in Paris under Madame Curie. He switched to law and became a prominent judge outside Lodz, Poland. Escaping the Nazi invasion, he arrived in Canada in 1941, secured a contract from de Havilland and sponsored a group of Polish engineers (including Czerwinski) to join his new company.

Czerwinski (1900–1989) was born in Poland and graduated from the prestigious Polytechnic University of Lwow, Poland. He taught aviation design at the college in Lwow from 1927 to 1930 and throughout the thirties was an engineer in glider and airplane factories in Lwow and Krakow, including a stint as chief designer at PWS, a government-owned airplane manufacturer. During his career, he designed and constructed eighteen types of gliders and three types of airplanes. In the early fifties, he joined the aircraft division of A. V. Roe Canada in Malton, Ontario, which developed the fabled Avro Arrow supersonic interceptor jet.[13]

CÉRAMIQUE DE BEAUCE

1940–1989, St-Joseph-de-Beauce, Quebec

Over a period of more than forty years, Céramique de Beauce (named the Syndicat des Céramistes Paysans de la Beauce until 1965) produced several million vessels that were sold across Quebec as well as in Ontario and the U.S. Beauceware reflects a range of stylistic influences, but its robust country pottery was its most popular.

The Collège de Beauceville founded its ceramics school in 1940 with funding from the provincial department of agriculture. Three years later, a manufacturing collective was formed in nearby St-Joseph-de-Beauce.

Between 1972 and 1975, Bush was director of the Sheridan College School of Crafts and Design. He drowned in Vancouver while taking photographs.

CANADIAN GENERAL ELECTRIC COMPANY
1892–1976, Montreal

Canadian General Electric Company was Canada's premier manufacturer of large and small appliances. Its brand name was so well known among consumers that it rarely gave in to national retailers' demands for less expensive private-label versions of its popular designs. The company first produced electric lamps, generators, transformers, motors and cables. In its heyday, during the fifties, it operated plants across Ontario and Quebec.

The small-appliance factory in Barrie, Ontario, opened in the post-war era. It produced clocks, small appliances, heaters and outdoor equipment. The freelancer Fred Moffatt of Toronto was its principal designer. After the Second World War, the Toronto factory manufactured consumer electronics, although most radios, for example, were based on U.S.-designed moulds. In the late sixties, Robin Bush designed wooden cabinetry for stereos, including an adventurous spherical design that never went into production. CGE ran a large plastic moulding plant in Cobourg, Ontario, which sometimes subcontracted its services to other companies.

Industry consolidation resulted in Black & Decker taking over the small-appliance division in 1984, and it was closed the following year. CGE's major-appliance factory, based in Montreal, merged with Canadian Appliance Manufacturing Company (CAMCO) in 1977. GE Canada, a subsidiary of General Electric Company, opened in Toronto in 1987.

CANADIAN WESTINGHOUSE COMPANY
Founded 1896, Hamilton

Westinghouse Manufacturing Company was founded in 1896 in Hamilton to sell and manufacture air brakes for steam railways. Seven years later, Canadian Westinghouse Company was incorporated to manufacture electrical equipment and large and small household appliances such as fridges and toasters. In 1922 the American and Canadian branches collaborated to produce the first radiotron (vacuum-tube-powered radio) manufactured in Canada, although it was held back from the market for five years because of technical unreliability. The most collectible Canadian designs to emerge from this century-old Canadian branch plant are its Personality radios.

After the Second World War, Westinghouse produced a range of long- and short-wave wooden console and tabletop radios as well as radio/phonograph combinations in traditional furniture designs. The branch plant's first foray into industrial design appears to have been in 1947, when it launched a Bakelite radio, Model 578A. Its shape was largely defined by the parameters of compression moulding, resulting in a similarity to

The group helped former students enter the commercial market by permitting them to use the school's equipment. Despite its humble origins, Beauce produced seven thousand designs.

Raymond Lewis, a graduate of Montreal's École des Beaux-Arts, was its first influential designer. As deputy director and artistic director, he ran the modelling and glaze department between 1942 and 1964. His unsigned rustic designs in local red clays were listed in the NIDC *Design Index*, and he was also responsible for "reinterpreting" urbane contemporary English and American patterns (including Nelson McCoy and Hull Pottery). In 1946 the studio switched from local to imported white clay, which allowed for more sophisticated moulding and higher firing temperatures, improving the quality of glazes. Largely during this period (1947–65), goods produced for English-speaking markets were stamped Beauceware.

Jacques Garnier (1934–1998), an important studio potter running L'atelier L'Argile Vivante in Beloeil, Quebec, collaborated with the ceramics collective between 1963 and 1970. A line of his own designs, Argile Vivante Beauce, introduced Scandinavian design. In 1965 the company changed its name to Céramique de Beauce and introduced the trademark (cb).

Jean Cartier (1924–1996), a graduate of the École du Meuble and a renowned studio potter, became design director in 1971, holding the title until 1974. He introduced a colourful, primitive style, typically incorporating exposed buff clay in his designs. After a fire in 1974, Céramique de Beauce was rebuilt and commissioned designs from the art potters Goyer-Bonneau until it closed in 1989.

CLAIRTONE SOUND CORPORATION
1958–early 1970s, Toronto/Stellarton, Nova Scotia

Clairtone Sound Corporation was launched in 1958 with $6,000. The intent was to become an export market leader in top-of-the-line stereos and televisions. During its heyday, the company contracted a Who's Who of Canadian designers: Carl Dair, Frank Davies, Al Faux, Burton Kramer and Hugh Spencer. To achieve its goal, Clairtone employed innovative design, technology and marketing—factors some commentators claim led to its very public downfall.

Initially, the company imported Braun radios, imprinting them with the Clairtone logo. The founders, Peter Munk and David Gilmour, then launched a plan to capture the high-end stereo market in North America and Europe by housing top-notch electronics in modern-style teak and rosewood cabinetry. Its first breakthrough was with Princess, a Scandinavian-style stereo cabinet designed by Gilmour, which won an NIDC award. Encouraged, the partners hired Hugh Spencer to create the ground-breaking Project G stereo, which became an icon of the swinging sixties.

The stereo was launched with an equally innovative campaign spearheaded by the former political strategist Dalton Camp. Foreshadowing present-day marketing campaigns, the program involved product endorsement by contemporary celebrities like Frank Sinatra, Tuesday Weld and Dizzy Gillespie, and the stereo was displayed in advertisements in *Life*, *Vogue* and *The New Yorker*—all firsts for a consumer electronics company. The renowned photographer Irving Penn photographed the stereo, and brochures featured the acclaimed jazz pianist Oscar Peterson. It was sold in department stores such as Bloomingdale's and Macy's in New York and Selfridges in London, England.

Despite the high visibility of the G lineup—which included a tabletop design (G3)—Clairtone relied on traditional stereo cabinets (with names like Countess and Duchess) to achieve its sales volume. In 1965 Clairtone accepted provincial tax incentives and subsidies to relocate to Stellarton, Nova Scotia. A year later, it launched five models of colour television, one being the GTV, with traditional as well as Pop styling. Consumers weren't ready to upgrade to colour, and the company foundered. This circumstance, combined with a disastrous foray into auto parts assembly, forced Munk and Gilmour out of the firm, and the plant was closed in the early seventies. By 1974 the company's assets were liquidated and a competitor, Electrohome, briefly took over the plant. Munk now heads the multinational real estate and mining conglomerates TrizecHahn Corporation and American Barrick Resources.[14]

COTTON, PETER

Born 1918, Merritt, British Columbia; died 1978

Peter Cotton's underappreciated designs represent modernism at its best: compact and flexible and made from low-maintenance materials. While he was still an architecture student, Cotton's steel-rod furniture and lighting won numerous awards, were distributed nationally and were featured in international publications.

Cotton's schooling at the University of British Columbia was interrupted by the Second World War. He served overseas with the Canadian forces for five years, then returned to study architecture intermittently at UBC from 1947 until he graduated in 1955, during which time he helped to establish its School of Architecture. Between 1951 and 1954, Cotton, along with the interior architect Alfred Staples, also manufactured and marketed their own designs, initially under the name Cotton Lamp Studio (1951) and then as Perpetua Furniture (1952–54) in Vancouver's South Granville district. Morgan's department stores sold the line in Eastern Canada. Cotton and Staples also accepted custom commissions and provided interior design services. The pair shared NIDC awards for a settee and matching armchair in 1953, and Cotton won an individual award for his high-back armchair. By this time, about a dozen of his well-known designs were registered in the NIDC *Design Index*.

With the closing of Perpetua in 1954, Cotton moved to Victoria, B.C., and later supervised the reconstruction program at Victoria's Government

Dallaire has designed electrical kitchen appliances for CGE, the official torch for the 1976 Summer Olympics in Montreal and, along with André Jarry, modular furniture for Montreal's Olympic Village (1972), of which forty-seven thousand pieces were made. He won the Canada Award for Business Excellence three times: bronze awards for an ASPRI acoustical guitar reverb system in 1989, and for a barbecue tool set in 1988; and gold for bicycle hand-brake levers for Resentel, Montreal, in 1986. He was given the Paul-Émile Borduas prize for visual arts in 1991 and made a member of the Order of Quebec in 1994.

Dallaire has been a visiting professor in industrial design at the Département Supérieur des Designs Lausanne, Switzerland, since 1988 and is associate professor of industrial design at the Université de Montréal and at the École Nationale Supérieur de Création Industrielle, Montreal. He also teaches graduate courses at the Faculty of Environmental Design at the University of Calgary. His 1991 solo exhibition, *Beauty in Utilitarian Forms* (*La Beauté des formes utiles*), was featured in the Centre de Design at the Montreal campus of the Université du Québec.[16]

DANESCO

Founded 1963, Montreal

Danesco produced contemporary housewares and lighting from the late sixties until the mid-eighties. Knud Petersen, the former commercial secretary at the Danish consulate, founded Danesco (short for Danish company) as an importer of Scandinavian candles and ceramics, and he eventually expanded into Danish furniture and fabrics. With the impetus of Expo 67, Danesco began manufacturing using local Canadian designers. Products included the Tukilik salt and pepper shakers by Girard Bruce Garabedian and Associates (1967), and the Glo-Up lighting collection by Ball-Berezowsky Associates (1969).

In 1979 Danesco invited Koen de Winter, a Belgian designer working in Sweden, to direct its design department. Some of its notable products were the Aurora Borealis lighting collection (1982) and Porcelaine de Chine tableware (1985). Since the nineties, the company has focused on distributing imported housewares and limited its production. Petersen retired as president in 1999.

DE WINTER, KOEN

Born 1943, Antwerp, Belgium

Koen de Winter, former design director at Danesco, is best known for his ability to manipulate ceramics and metal into housewares and lighting.

In 1962 de Winter earned a diploma in ceramic technology at Belgium's École des Métiers d'Art de Maredsous. Seven years later, he added a diploma in product design from the Akademie voor Industriele Vormgeving in neighbouring Eindhoven, Holland. In 1970 and 1971 he worked for Volvo in Sweden, then returned to Holland as a product designer for

House. He was a founding member of the Northwest chapter of the Society of Architectural Historians and the Crystal Gardens Preservation Society Foundation and also a director of the Hallmark Heritage Society.

CRINION, JONATHAN

Born 1953, Liverpool, England

In the 1990s, Jonathan Crinion staked his claim as a leading member of the next generation of Canadian industrial designers. Based in Toronto, he became a sought-after systems furniture designer, with the talent to work with a variety of global manufacturers. He was the first Canadian to be included in *ID* magazine's International Design 40, an annual list of the top forty designers in the world, in 1994. In 2000 Crinion received the Toronto Arts Award for Architecture and Design.

Crinion graduated from the Ontario College of Art in 1980 and joined the local multi-faceted design firm KAN Industrial Design. Three years later, he set out on his own and by 1986 designed the playful Gazelle chair. While on vacation in England in 1988, Crinion visited the architectural firm of Sir Norman Foster and was hired to help refine the Nomos furniture system for Tecno, the Italian manufacturer. That led to assignments for panel-based office systems for Teknion, Toronto, and Knoll International, allowing Crinion to build a freelance practice that balances his artistic sensibility with a rational, problem-solving approach. In the mid-nineties, he designed a café chair named Bebop for Kiosk Design in Toronto. It echoes some of the themes he explored with the Gazelle chair.[15]

DALLAIRE, MICHEL

Born 1942, Paris, France

Michel Dallaire is Montreal's leading industrial designer of consumer products. Son of the noted French modern painter Jean Dallaire, he was educated at Montreal's Institut des Arts Appliqués. He studied interior and industrial design under Julien Hébert (and later worked with him) and graduated in 1963. Awarded a grant from the NIDC, he continued his studies at Stockholm's School of Arts, Crafts and Design (Konstfackskolan) until 1965.

On graduation, he joined the Montreal firm Jacques Guillon & Associates, where he worked on Concept B furniture for Habitat, produced by the manufacturer Paul Arno. In 1967 he became an associate at Bosse, Coutu, Dallaire Associés, where he stayed until forming his own firm, Dallaire Morin Designers Associés, two years later. The company spun off a subsidiary, Dallaire, Morin, DeVito, and Dallaire became its president. Since 1974 he has operated his own firm, Michel Dallaire Design Industriel, in Montreal, and has earned countless prestigious design awards, including two from Geneva's Salon International des Inventions (Jardibac gardening system, 1998, and Angelcare Sound and Breathing Monitor, 1999).

Mepalservice. He headed the company's design and development department until 1979.

Knud Petersen, president of Danesco, which imported Mepalservice products, persuaded de Winter to immigrate to Montreal to become the firm's design director. During his early years in Canada (1982–84), de Winter also taught design at the Université du Québec à Montréal. By 1985 he was promoted to vice-president of design at Danesco, a position he held until 1990, when he launched his own firm, HippoDesign. He continued to teach part-time until 1992 and still serves as Danesco's independent design consultant.

DEACON, TOM

Born 1956, Toronto

Tom Deacon emerged as a pre-eminent designer of contract seating in the 1990s, introducing warmth and style to corporate office furniture.

Deacon graduated from the University of Toronto School of Architecture in 1982. After practising for one year, he chose furniture design as a career. Deacon, with Lee Jacobson, founded the Toronto furniture manufacturer AREA in 1984, producing Le Corbusier's 1928 Grand Confort seating. With this hands-on manufacturing experience, the company expanded to produce his own designs and others by new Canadian talent like Jonathan Crinion. Deacon's first design to receive critical acclaim was the York chair (1987).

Deacon withdrew from AREA in 1989 to focus on design. Among his freelance clients were the American Bernhardt Furniture Company (the Studio chair, 1993) and Nienkämper (the Cirrus table, 1992). He also designed the Vector compact disc shelf for CSL, Toronto (1992). He continues to design for AREA on a royalty basis, including the Promenade chair (1993) and has a gentlemen's agreement to design office seating exclusively for Keilhauer. The Deacon (1989) is Tom Deacon's first chair for the company, and it helped to launch both designer and manufacturer into the international arena. The Tom chair (1997), also for Keilhauer, is his best-selling chair.

DEICHMANN, ERICA AND KJELD

Erica born 1913, Wisconsin;
Kjeld born 1900, Copenhagen, Denmark; died 1963

Kjeld and Erica Deichmann were innovative potters. Inspired by their friend Bernard Leach, the founder of the global studio pottery movement, the couple embraced pottery making as a way of life.[17]

Kjeld studied philosophy at the University of Copenhagen and sculpture at the Royal Danish Academy of Fine Arts. He immigrated to Alberta in 1927, where he married Erica, the daughter of a Danish clergyman. In 1933 Kjeld apprenticed in Denmark with the potter Axel Bruhl. The Deichmanns returned to Canada and in 1935 opened Dykelands pottery in Moss Glenn, New Brunswick. They moved the studio to nearby Sussex in 1956. Celebrated in their time, they demonstrated their skills at the Rockefeller Center in New York City in 1952 and exhibited at the Brussels World's Fair in 1958. The National Film Board made documentaries about the Deichmanns; one was the popular *Story of Peter and the Potter* (1953).

In addition to domestic ware, Erica also hand-modelled animal figurines, like Goofus. They created a thirty-one-piece stoneware coffee service for Princess Elizabeth and the Duke of Edinburgh on the occasion of their visit to New Brunswick in 1951. Their diverse designs evolved from stoneware to porcelain, using local as well as imported clay, and they favoured simple functional shapes with narrow necks and natural glazes. Kjeld threw the pots on the foot-powered kick wheel, and Erica created the glazes, producing more than five thousand of them in her career and recording the ingredients by hand in recipe books. Her signature colours are the Kennebecasis blue (named after the nearby river) and purple patch. Erica closed the pottery after Kjeld's death in 1963. The noted writer Elisabeth Harvor is their daughter.

DISMO INTERNATIONAL
Founded 1989, Montreal

Founded by Sylvain Faucher, DISMO (Distribution Mobilier) International emerged in the early nineties as a catalyst for new furniture design. As the only furniture *éditeur* in the country (modelling itself after European promotional agencies), it united designers with manufacturers and served as a marketing arm for these collaborations.

DISMO specialized in hospitality furniture and promoted the careers of Jean-François Jacques, Michel Morelli and Alain Degasné. It made its reputation furnishing the Casino de Montréal and other important commissions like the Musée du Québec, and the Toronto Dominion Centre and CN Tower in Toronto. At the end of the decade, DISMO withdrew from production to operate the retail store Latitude Nord in Montreal.

DODDS, HUGH
Born 1913, Toronto; died 1993

Hugh Dodds was an inventor specializing in post-war moulded plywood and fibreglass products. His Dodds stacking chair (1952) won critical acclaim and furnished countless schools and churches in Ontario.

Dodds founded Aero Marine Industries in Toronto in 1944, moving to Oakville in 1954. Self-taught, he acquired skills in moulded plywood when he worked for Massey-Harris aircraft. Dodds's company purchased the latest glue presses for compression moulding, and his state-of-the-art workshop was regularly visited by industrial design students.[18] Aero Marine's moulded wood products included motorboats in three standard sizes and school furniture. Like its contemporary, Canadian Wooden Aircraft in Stratford, the University of Manitoba and helped to establish the Affiliation of Canadian Industrial Designers. During his twelve-year tenure at the U. of M., he maintained an architectural practice, designing noteworthy buildings such as Niakwa Country Club (1952), the university's School of Architecture (1958) and Monarch Life's Winnipeg headquarters (1960). Between 1963 and 1981, Donahue taught architectural design at the Technical University of Nova Scotia (TUNS), Halifax, again designing the university's School of Architecture (1976), as well as the Nova Scotia Archives (1977) and the Halifax Police Station (1975). He worked as a consultant for TUNS Design Centre (1985–89), and completed the HMCS Sackville Theatre in 1991.[20]

DUCHARME, MAX
Active 1960s and 1970s

Max Ducharme insisted that designers be valued for their market knowledge rather than their aesthetic sensibilities. A 1958 graduate of the Ontario College of Art, Ducharme joined the industrial design division of Philips a year after exploring Europe on an Eaton's design scholarship. The company was phasing out the Rogers-Majestic brand (purchased in 1954), although it continued to operate out of its former plant. Within a few years, Ducharme became chief designer in the consumer electronics division, a position he held until 1980, when Philips closed it as a result of competition from Asia.

The division, which reported to marketing rather than engineering, designed televisions, stereos, tuner/amplifiers and cassette recorders, as well as a few radios and small appliances. About 90 per cent of its designs were original rather than adaptations from its Dutch parent, although most reflected the styles of the day, like French Provincial. The department Ducharme led was respected internally but often ignored externally, as the bulk of its design output was in traditional styles. He also designed some furniture for the wholly owned subsidiary Strathroy Furniture in Strathroy, Ontario. Ducharme was a director of the Association of Canadian Industrial Designers. After 1980, he designed exhibits for seventeen years and now works with a company that makes products for theatres.

ECANADA ART POTTERY
1939–53, Hamilton

George Emery born 1891, Newcastle-under-Lyme, England; died 1959

Ecanada Art Pottery, founded by George Emery in 1939, blended the Wedgwood pottery tradition with Canadian iconography such as settlers and fur traders or local flowers like the trillium and lily of the valley.

Emery was one of Canada's leading figures in the ceramics industry and served as president of the Canadian Ceramic Society in 1930. Trained at the prestigious Wedgwood factories in England, Emery immigrated to Hamilton in 1912 to work for Campbell Pottery Company, a manufacturer

the company unsuccessfully attempted to make plywood glider planes.

By 1960 Aero Marine began to specialize in FRP (fibreglass reinforced plastic) boating equipment. Dodds patented many FRP products, including life rafts, diving boards, catamarans and tourist boats. He moved to Picton, Ontario, when he retired in 1975 and continued to develop new fibreglass products, such as tiles and food containers.

DOMINION GLASS COMPANY/DOMGLAS

1913–1990 Montreal

The Dominion Glass Company was Canada's largest and longest-running manufacturer of glass tableware. The Diamond Flint Glass Company of Montreal (founded in the nineteenth century) reincorporated in 1913 as the Dominion Glass Company after merging with Sydenham Glass in Wallaceburg, Ontario, and the Jefferson Glass Company in Toronto, both also originally founded in the nineteenth century.

Dominion Glass became fully automated by the twenties and produced more than five thousand products, including tableware, lamps, bottles, birdcages, doorknobs and insulators. By the thirties, it phased out most of its household products, although the Wallaceburg plant continued to make some kitchen glass and tableware. The Wallaceburg glassworks became known as the St. Clair division of Dominion Glass in 1974 and concentrated on beverage ware (goblets and glasses).

The large manufacturer Libbey Glass of Toledo, Ohio, merged with the division in 1978 and the newly incorporated Libbey St. Clair Glass produced some beverage ware and candleholders until it ceased production in 1999 (last known as Libbey Glass of Canada). In 1989 Consumers Glass (the Montreal glass container company founded in 1917) bought Domglas (so named after 1976). Now known as Consumers Packaging, it is the largest manufacturer of glass containers in Canada.[19]

DONAHUE, JAMES

Born 1918, Regina; died 1997

James Donahue was an influential teacher and architect with a passion for furniture design. His moulded plywood lounger, often referred to as the Canadian Coconut or Winnipeg chair, is well known in the collectibles market.

Donahue obtained his Bachelor of Architecture from the University of Minnesota in 1941 and his master's from Harvard University in 1942, where he studied under the celebrated International Style architects Marcel Breuer and Walter Gropius. He returned to Canada to work at the National Housing Authority in Ottawa (1943–44) and with Donald Buchanan on a touring exhibition titled *Wood in Canada* (1945). While in Ottawa, he designed a plywood-framed webbed chair and, with A.G. Medwin, a moulded plywood radio cabinet.

In 1946 Donahue was appointed professor of architectural design at

of porcelain insulators. The following year, he joined the Canadian Porcelain Company to serve as its assistant superintendent for the next thirty years.[21] In its heyday, Ecanada employed more than twenty people, and many of its artisans, like the British model maker Bertram Watkin, came from Sovereign Potteries. Its line included over seventy slip-cast designs, including jardinieres, candle holders, vases and lamp bases, often imitating the Wedgwood Jasperware colours of pale blues, pinks and greens. It sold at Henry Birks & Sons and Eaton's, and Emery himself distributed the line to gift stores in the Niagara Peninsula.

Emery's son, George Jr., took over the company in 1952 and introduced a short-lived ovenware line and restaurant teapots. He closed Ecanada in 1953, six years before his father died of silicone poisoning. In 1976 Jack Orme of St. Catharines, Ontario, reproduced Ecanada designs using the original moulds under the name EMRO-CANADA.

ELECTROHOME

Founded 1907, Kitchener, Ontario

During its near century in business, Electrohome has manufactured consumer electronics, small appliances and furniture, has owned and operated radio and television stations and has made large-screen projection televisions for the commercial and staging markets.

During its heyday producing consumer goods (mid-fifties to mid-seventies), the company made private-label products for Eaton's, Simpson's and the Hudson's Bay Company, as well as for other department stores such as Woodward's and MacLeod's in Canada and Wanamaker's in the U.S. It also absorbed about a dozen smaller manufacturers, so that virtually every Canadian home, however unwittingly, contained an Electrohome product—televisions, heaters, humidifiers, lamps—under names such as Viking, Elmira, Flexsteel, Serenader and countless more. In 1967 a conveyer belt in the company's finishing department was three-quarters of a mile (1.2 km) long and turned out a TV or stereo cabinet every thirty seconds.[22]

Arthur Pollock founded the company to design, assemble and market wind-up phonographs. A decade later, it expanded into manufacturing the cabinets to house the radios it was producing. By the mid-twenties, a furniture division, Deilcraft (Dominion Electrohome Industries) was created. It was also licensed to produce some designs by Imperial Furniture of Grand Rapids, Michigan, under the name Deilcraft Imperial. In the mid-thirties, the company hired the designer George Eitel to create a unified image for its in-house furniture, which was sold across the country in branded boutiques within department as well as independent stores.

Arthur's son, Carl Pollock, who served on the board of the NIDC for ten years, heralded a new era of design innovation, particularly in the area of consumer electronics. When Eitel died in 1961, Gordon Duern, a graduate of New York's Parsons Design School, was promoted to chief designer. Duern also conceptualized the Telesphere, a forerunner of the company's move into projection television. The department included Keith McQuarrie, Michael Baldwin and Fred Bent. Freelance designers like John Murray and Luigi Tiengo also designed furniture, lighting and consumer electronics for Electrohome.

During the sixties, Carl Pollock and his wife, Helen, established Albon Reproductions, which briefly imported furniture from Charak in Boston and produced original designs from the former Murawsky Furniture Company in Kitchener. Pollock formed Knoll International Canada, and along with the firm Leif Jacobsen, the consortium manufactured Knoll-designed furnishings for Toronto City Hall.

Electrohome also produced wooden contract furniture on behalf of Sunar Industries of Waterloo. The company's own Deilcraft contract furniture division won large commissions from hotels like the Four Seasons chain, the Plaza in New York, and the Pan Pacific in Vancouver. This business continues under the ownership of Art Craft (The Art Shoppe).

FAUX, AL

Born 1931, Toronto; died 1978

Al Faux inspired a generation of students, and in 1964 he designed a flexible drafting table that transformed the Norman Wade Company in Toronto from a small local manufacturer to an internationally successful firm. He avoided commercial styling, preferring to design entire systems, and later became preoccupied with ecological housing.

Faux began his career as a journeyman machinist, taking apart steam engines for the Algoma Railway in Sault Ste. Marie, Ontario. Encouraged by an aptitude test, he studied furniture and interior design at Ryerson Institute of Technology, graduated in 1957 and worked for both Robin Bush Associates and McIntosh Design Associates, Toronto. By 1965 he formed Design Collaborative in Toronto with the graphic designers Rolf Harder, Ernst Roche and Anthony Mann, and their clients included Clairtone Sound Corporation.

Design Collaborative won a $1.3 million contract to supply furniture to the University of Guelph. Assisted by the former Ryerson students Earl Helland, Thomas Lamb and Ian Norton, Faux created a system of flexible units for student residences and designed moulded plywood furniture, special lighting and a flexible floor grid for its library. This led to commissions like furnishing the Ontario Pavilion at Expo 67.

Although Faux continued to work on the University of Guelph with Gerald Beekenkamp, he left Design Collaborative and established Al Faux Associates in 1968. Along with the architect Roderick Robbie (who later designed Toronto's SkyDome sports stadium), he designed furniture for his alma mater. L'Enfant Company in Toronto manufactured many of his institutional designs, like plywood beds, chairs and spun steel accessories. His lighting designs were produced by Galaxi Lighting, Toronto. By the end of Parliament Hill. He taught metal arts from his home-based studio for decades and conducted classes at Central Tech for about ten years.[25]

GOODMAN, JEFF

Born 1961, Vancouver

Jeff Goodman is a creative, versatile designer specializing in glass. Using the techniques of both blowing and casting, he makes furniture, vessels and sculpture as well as custom architectural glass for commissions from clients like the Princess of Wales Theatre, Toronto. He incorporates materials like stone, concrete and metal in his furniture, a mixed-media approach that developed from the studio glass movement in the late eighties.

Goodman graduated in fine art from the University of Illinois in 1986. He also attended Sheridan College School of Crafts and Design (1981 to 1983) and Alfred University in New York State (1983 to 1984). Since 1994 he has operated his own glass studio in Toronto.[26]

GOYER-BONNEAU

Founded 1970, Carignan, Quebec

Denise Goyer born 1947, Montreal; Alain Bonneau born 1946, Montreal

Denise Goyer and Alain Bonneau operate one of the few art potteries in Canada. They collaborate on all their work, Goyer taking responsibility for design, Bonneau focusing on construction. After opening their studio in 1970, they broke away from the predominant fashion for earthy Far Eastern ceramics. Instead, they made porcelain castings in strong colours and irregular forms, creating triangular teapots, asymmetrical bowls and plates in non-traditional shapes.

Goyer graduated from the École des Beaux-Arts, Montreal, in 1966 and the Institut des Arts Appliqués de Montréal in 1970. Bonneau graduated in 1968 from the Académie des Arts du Canada, Montreal, in graphic art. Together they developed their porcelain skills at the prestigious Sèvres school in France in 1986 and at Wedgwood in 1995.

In 1985 Goyer-Bonneau created a 317-piece service for the Quebec consulate in New York City. They also sold some designs to Céramique de Beauce. Goyer-Bonneau exhibited work in Virtu 3 and Virtu 8.

JACQUES GUILLON & ASSOCIATES/GSM DESIGN

Founded 1954, Montreal

Jacques Guillon & Associates is one of the largest and longest-running design firms in Canada. Now known as GSM Design, the company creates everything from office furniture to subway car interiors, exhibition pavilions, signage and even tractors. An astute businessman who never fully explored his early potential as a furniture designer, its founder, Jacques Guillon, mastered an ever-changing market. Like KAN Industrial Design in Toronto, GSM is Quebec design's teacher, lobbyist and advocate.

Guillon was born in Paris in 1922, the son of a Canadian architect,

the 1960s, Faux was teaching at the Ontario College of Art. At his farm in Tottenham, Ontario, he and his students designed and built ecologically sensitive shelters.[23]

FINKEL, HENRY

Born 1910, London, England; died 1996

As the world embraced plastics, Henry Finkel became one of the Canadian specialists designing new moulds for mass production. Known more for his diverse body of work than for any individual design, Finkel, during his prolific career, produced everything from intercoms to ballpoint pens. He was an early supporter of the Association of Canadian Industrial Designers and helped found ADIQ (the Quebec chapter), where he was made a fellow in 1975.

Finkel graduated in architecture from McGill University in 1934. The dearth of architectural jobs in the thirties led him to find work in mechanical engineering. During the Second World War, he headed the drafting department of an aircraft machinery factory, where he gained hands-on experience in mass production. In 1945 he co-founded the plastic injection-moulding plant Die-Plast Company, Montreal, and designed its first product line. Two years later, he established his pioneering consulting practice in industrial design. In addition to creating designs that used compression and extrusion-moulded plastic, Finkel designed products for the aluminum industry, including tube garden furniture for Featherweight Aluminum Products and aluminum-edged Formica tables for Whitehouse in Quebec.[24]

FUSSELL, ANDREW

Born 1904, Leipzig, Germany; died 1983

Largely self-taught, Andrew Fussell attracted important patrons and as a result was able to forge a long career and create a substantial body of work despite the vagaries of the market. Fussell was known for working in a range of metals (silver, pewter, copper, aluminum and even gold). The objects he produced were equally varied: hollowware, jewellery and commissions for everything from ecclesiastical silver to architectural metalwork.

Fussell immigrated to Toronto from England in 1926 and spent the next six years struggling to fund his night courses in sculpture at Central Technical School by working as a construction labourer and architectural draftsman. He learned to work in pewter under Rudolph Renzius at Northern Vocational and Technical School and in 1932 opened his own shop selling pewter hollowware and jewellery.

In 1937 he participated in the Paris Exposition. After the war, he became a founding member of the Metal Arts Guild of Ontario. He produced hollowware via the traditional method of hand-raising (rather than machining) the forms. His work was known for its simple, clean lines, which attracted commissions from architects and designers as well as the clergy. One of his projects, for example, was the clock face in the library on who brought his family to Canada in 1940. He returned to Europe as a Royal Canadian Air Force pilot. After the war, he studied architecture at McGill University, graduated in 1952 and practised with the Montreal architect Max Roth until 1954. While in university, he designed an acclaimed nylon cord chair.

With his wife, Pego McNaughton, Guillon owned and operated Pego's, two retail stores that imported Scandinavian furniture to Montreal (1954–62) and Quebec City (1956–61). In the mid-fifties, he launched a production arm, Ebena Manufacturing, to build furniture. The designer Christen Sorensen managed the division until 1962, when Art Woodwork (which in turn was acquired by Sunar Industries) bought it. During the fifties, Guillon also designed metal rod furniture for Park Manufacturing Company, Montreal, and in 1954 won an NIDC award for a fireplace rack.

Guillon established his own design firm in 1954. Four years later, it became Jacques Guillon & Associates when Sorensen joined the firm to design seating for the new airport at Dorval. The company specialized in commercial office planning and interiors, and by the early sixties clients included BP Canada and the Montreal showplace headquarters for the Aluminum Company of Canada. As an offshoot, the firm designed hardware and furniture for these large-scale projects. It also benefited from the Expo 67 boom by working on three pavilions (Man and His Life, Belgium and Algeria) and designed furniture for the Montreal manufacturer Paul Arno. During that heady year, Guillon was president of the Association of Canadian Industrial Designers.

Early in the sixties, the firm planted the seeds for a new direction in transportation design. Awarded the contract to co-ordinate the design of the Montreal Métro subway cars (including body, frame and interiors), it hired Morley Smith, a graduate in industrial design from Syracuse University. Under the direction of the Swiss designer Laurence Marquart, the firm added visual communications to its repertoire and by 1978 was renamed Guillon, Smith, Marquart & Associates. Five years later, it was simply GSM Design. Guillon retired in 1987.[27]

HÉBERT, JULIEN

Born 1917, Rigaud, Quebec; died 1994

The sculptor Julien Hébert represents one of the best examples of the successful confluence of industrial design and art. He designed graceful folding furniture, became an influential teacher who trained a new generation of designers and used his knowledge of aluminum manufacturing techniques to enhance his work as a sculptor.

Hébert studied fine art at the École des Beaux-Arts in Montreal, graduating in 1941. Three years later, he earned a master's degree in philosophy from the Université de Montréal. Between 1947 and 1948, he studied sculpture in the Paris atelier of Ossip Zadkine, then returned to Montreal to teach fine art at his alma mater.

In the early fifties, Hébert designed lightweight metal furniture for Siegmund-Werner (under the Sun-Lite Outdoor Furniture brand name) that won five NIDC awards and was produced well into the sixties. Hébert claimed he knew a hundred ways to fold a chair.

Between 1951 and 1961, Hébert and a partner, Yves Groulx, produced wood and steel office desks, tables and commodes under the name Grébert. Hébert received a grant to study industrial design at the Massachusetts Institute of Technology in 1953. Three years later he organized an exhibit, *Good Design in Aluminum*, at the National Gallery of Canada. In 1963 Hébert designed trapezoid school desks for the Montreal manufacturer Paul Dumont.

Between 1956 and 1966, he taught design at the École du Meuble (later the Institut des Arts Appliqués), where he led the drive to bring professional design techniques to Quebec's craft-oriented manufacturing industry. A decade later, he founded the faculty of industrial design at the Université de Montréal and trained such well-known designers as Michel Dallaire, Albert Leclerc and Marcel Girard.

In 1967 Hébert and his former pupil Girard designed "La Ronde," the widely known logo for Expo 67. This led to a commission to manage the design of both the Canadian and Quebec pavilions at the Osaka, Japan, world's fair in 1970. Hébert was also a noted sculptor whose works in aluminum grace the foyer of the Place des Arts in Montreal (1963) and the Opera Hall ceiling in the National Arts Centre in Ottawa (1966).[28]

HELD, ROBERT

Born 1943, Santa Ana, California

Robert Held pioneered the studio glass tradition in Canada, first as an influential teacher in the seventies, then as a commercial artware glassmaker in the eighties. He is known for reviving art nouveau–style lustre glass.

Held studied ceramics at San Fernando Valley State College in 1965, then earned an M.F.A. in ceramics at the University of Southern California. In 1969 he joined Sheridan College, where he founded the first hot-glass program in Canada. Six years later, he moved to Calgary and became the manager of the hand-rolled stained-glass manufacturer Canadian Art Glass. In 1979 he founded the studio Skookum Art Glass in Calgary and later taught at the Alberta College of Art (1983–86).

Since 1987 he has operated Robert Held Art Glass in Vancouver, the largest hot-glass studio in Canada. The company's 150-piece product line includes giftware like paperweights, goblets and perfume bottles, which sell at William Ashley in Canada and Gump's and Bloomingdale's in the U.S. In 1984 his Tiffany-style goblets won the Governor General's Perfect Setting competition. His work is in the collection of the Musée des Beaux-Arts in Montreal.

fifties. Kuypers's subsequent designs for Imperial, contracted after he left the firm's employ, were less adventurous, perhaps because Imperial heavily promoted its designs (by its president, Donald Strudley) as "down-to-earth, sound and sensible."

In the sixties, and as an independent designer, Kuypers created a series of furniture groupings for Imperial with names like Rideau Related and Sampler that humorously exaggerated their historical references.

JACQUES, JEAN-FRANÇOIS

Born 1957, Quebec City

Jean-François Jacques has excelled at whimsical furniture since he entered the design scene in the early eighties. After completing a Bachelor of Environmental Design at the Université du Québec à Montréal in 1982, he became one of many Quebec designers to vault from interior design into product design. He opened Météore Design in 1983 in Montreal and began making playful objects in small quantities (Zenith lamp, Blender bowl).

In 1994 the City of Montreal named Jacques Designer of the Year. Four years later, he won (with Claude Maufette) Alcan's AluDesign competition for AL 27, a prototype for an all-aluminum chair. His designs include Baby Face chair and the Bowling chair (both 1993), for DISMO International, and kitchen products for Trudeau Corporation.

KAISER, ROBERT

Born 1925, Detroit, Michigan

Robert Kaiser's experience in art, illustration and product and store design were all germane to his success during a long career designing furniture for homes and offices, as well as teaching design. His best work is individualist, born from his imagination rather than commissioned to fill a market niche.

After serving in the Second World War, Kaiser studied at the Meinzinger Art School, Detroit, and at the Layton School of Art, Milwaukee, Wisconsin, before settling on design at the Institute of Design, Illinois Institute of Technology, Chicago. There he was influenced by Buckminster Fuller, the engineer and architect best known as the designer of the geodesic dome. Kaiser immigrated to Canada in 1950 and became a technical illustrator for the airplane manufacturers A.V. Roe Canada and de Havilland Aircraft of Canada, designed retail stores for Toronto's Cameron McIndoo and Kinsella Design and freelanced as a furniture designer for nearly a decade before joining the Primavera Design Group. His most noteworthy designs are an armchair and an occasional chair.

In the mid-sixties, Kaiser created soft seating and systems furniture for Concept Furniture International, Toronto, which sold to clients across North America like the American Bar Association, Canada's Department of National Defence and the Levi Strauss Company. The hexagon-shaped

HUBEL, VELLO

Born 1927, Tallinn, Estonia; died 1996

Vello Hubel is most remembered as an outstanding teacher. He taught at the Ontario College of Art between 1965 and 1993 and won the A. J. Casson Teaching Award in 1988. His roster of former students who are now leading designers includes Miles Keller, Helen Kerr and Scot Laughton. With Diedre Lussow, Hubel authored *Focus on Designing*, 1984.

Hubel immigrated to Canada in 1947. He put himself through school working as a mechanical draftsman and graduated from the Ontario College of Art in industrial design in 1953. Five years later, he founded a design office, which had many blue-chip clients. He produced television and stereo cabinets for Sparton of Canada (1959–67) and CGE (1967–71), household lighting for William J. Campbell Company (1960–61) and residential furniture for Knechtel Furniture Company (1968–76).

In the sixties, Hubel entered an experimental phase and created several critically acclaimed prototypes, including the Playsphere playground (1966), children's cubed furniture (1966) and Infinitum outdoor seating (1969). Between 1974 and 1979, he partnered with Robert Kaiser. His best work was for Baronet Corporation, the family-run residential furniture manufacturer in Quebec. Designing for the company for more than thirty years, Hubel created contemporary updates of period designs that helped to make the company a leading manufacturer of residential furniture. Toward the end of his career, Hubel dabbled in the postmodern style, creating the overly cute Wink and Flirt chairs for Vogel. His more successful Clover Leaf table for CSL (1992) was widely acclaimed.[29]

IMPERIAL FURNITURE MANUFACTURING COMPANY

1910–1983, Stratford, Ontario

Named Imperial Rattan Company until 1949, this firm popularized modern furniture design. In 1941 it manufactured a whole-house grouping, sold exclusively through Simpson's, that was designed by the Finnish architect Eliel Saarinen. Bitten by the design bug, the founder's son, Donald Strudley, ventured into modernism with his own designs, then brought the Dutch designer Jan Kuypers to Canada.

In the early years, the firm's bread and butter was traditional solid wood furniture with names like Loyalist. By 1939 the company was manufacturing furniture in the Swedish modern style and had updated some of its rattan line. After Kuypers's arrival in 1952, it embraced Scandinavian modernism with zeal, producing dining, living and bedroom groupings called Stockholm, Copenhagen, Oslo and Helsinki. Kuypers also designed a number of well-regarded signature objects, and eventually the company trumpeted his national and international successes in advertisements—a rare occurrence of designer "branding" in the mid-Module 360 allowed more workstations in a smaller space. Kaiser also designed mid-priced seating like Sir Winnie and L'Esprit de Corbu, which were in production for years and included line extensions such as tables, ottomans and sofas. During this period, he designed a line of knock-down plywood residential case goods that appeared in the Suite of the Future in Montreal's Habitat at Expo 67. He also taught at the Ontario College of Art (1960–65 and 1970–89) and Ryerson Polytechnical Institute (1984–95).[30]

KAN INDUSTRIAL DESIGN (PREVIOUSLY DKR)

1963–1996, Toronto

Toronto-based KAN Industrial Design, as it was last known, was one of the longest-running design studios in Canada. It shaped Canada's products and environments, creating furniture, world's fair exhibits, stoves, lighting and playground equipment. Its diversity, scope and volume of commercially viable work are remarkable.

While the soul of the firm was the designer-philosopher Jan Kuypers, many of Canada's brightest young designers (including Karim Rashid and Jonathan Crinion) apprenticed at what became an institution. In 1960 the original partners, Kuypers, Julian Rowan and Frank Dudas, were the Product Design Unit within Stewart & Morrison, the Toronto-based design firm. After three years, they broke away and founded the "seed" company that by 1967 was known as Dudas Kuypers Rowan (DKR), had a staff of sixteen and boasted additional offices in Ottawa and Montreal. The firm worked on the Expo 67 pavilion Man the Producer and developed such diverse products as a cement mixer for Monarch Machinery, teakettles for Proctor-Silex and an X-ray machine for Picker International.

Julian Rowan was born in Edmonton, Alberta, in 1925 and studied science at the University of Alberta and plastics engineering at the Plastics Industries Technical Institute in Los Angeles. On his return to Canada, he created toys and housewares as well as fertilizers and explosives. As DKR's plastics specialist, he designed hockey helmets for CCM (Canada Cycle and Motor Company) and the first two-colour plastic vacuum bottle for Canadian Thermos Products.

Jerry Adamson was born in Cambridge, Ontario, in 1937, obtained a degree in industrial design from the Ontario College of Art and worked with the well-known designer Robin Bush. He joined DKR in 1963 and became a partner five years later. He designed many notable products, including the widely acclaimed Habitat chair, and was influential within the firm.[31]

In addition to Dudas, other partners in the firm were Ian Norton, who specialized in small appliance design, and Gerald Beekenkamp, whose designs helped King Products, Toronto (phone booths) and Paris Playground Equipment (Duraglide slide) dominate their respective markets.

KEILHAUER

Founded 1981, Scarborough, Ontario

This family-owned and -run company has, through the use of design and marketing, become a leading manufacturer of contract seating. Its success with mid-management task chairs dovetailed with the growth of personal computers. It is one of the largest privately owned contract furniture companies in North America, leading the new wave of export-savvy Canadian manufacturers.

Mike Keilhauer founded the company with his three brothers Ron, Steve and Rick. Their father, Ed, contributes to design and has deep roots in manufacturing. Trained as an upholsterer-saddler, Ed established Fine Art Upholstery in Scarborough in 1956 to supply custom leather fabrics to Precision Craftwood and Nienkämper. In 1965 he launched Cambridge Furniture to serve the home furnishing market. The two companies were integrated as Fine Art International in 1976, but licensing problems forced it into bankruptcy five years later.

When the sons opened Keilhauer (originally called Keilhauer Industries), they produced foreign designs under licence. Ed's Elite chair (1983) was the company's first in-house design, and it remains in production. In the early eighties, the Montreal designer Christen Sorensen created the Respons chair (1989), the company's first multi-task chair and its meat and potatoes for many years. It captured the silver at the NeoCon World's Trade Fair in Chicago.

President Mike Keilhauer moved the company into the global market by hiring the branding specialist Michael Vanderbyl of San Francisco. The firm now operates three factories in Toronto, and 80 per cent of its market is in the U.S. Since 1989 Tom Deacon has designed one-third of the company's line and introduced wood with chairs like the Deacon (1989) and the Danforth (1993). Andrew Jones, Scot Laughton, Gord Peteran and Jonathan Crinion have also designed for Keilhauer.[32] Kerr Keller Design created the Chit Chat chair (1996), the company's first entry into plastic components. Deacon's Tom chair (1997) solidified the move into plastics and has become the company's best-selling chair.

KERR KELLER DESIGN/KERR AND COMPANY

Helen Kerr born 1959, Montreal; Miles Keller born 1959, Calgary

In the early nineties, Kerr Keller Design in Toronto established itself as a leader in designing sophisticated and stylish kitchen products. The company follows a playful approach but tempers this with practical problem solving.

Before attending the Ontario College of Art, Miles Keller already had a B.A. from the University of Calgary and a Bachelor of Architecture from the University of British Columbia. Helen Kerr was awarded a Bachelor of Environmental Science from the University of Waterloo and graduated from OCA in 1988, where she currently teaches. Umbra became their first significant client, producing one of Miles Keller's student projects, a sleek beds in Canada, as well as a best-selling anthro-ergonomic task chair for Harter, Middleburg, Ohio.

In retirement, Kuypers worked with a young designer, Hiroshi Okano, to create a revolutionary high-performance chair that would perfectly express beauty and comfort following the logic of nature. Kuypers wrote, "Designs that are art are distinguished by creating emotion in the user."[34]

LAMB, THOMAS

Born 1938, Orillia, Ontario; died 1997

Thomas Lamb, a prolific designer with wide-ranging interests, is best known for his furniture produced in the seventies and eighties. His Steamer chaise longue was selected for the Study Collection at New York's Museum of Modern Art in 1979, one of the few Canadian furniture designs to hold that honour.

Lamb studied furniture design at Ryerson Institute of Technology, graduated in 1964 and went to work for his mentor and former teacher, Al Faux. At Design Collaborative, he helped to develop Faux's award-winning drafting table. Three years later, the pair designed furniture ranging from chairs to library storage units for the University of Guelph, then one of the largest furniture commissions in Canada. During this period, Lamb also worked with the noted designer Robin Bush on furniture for the Ontario Pavilion at Expo 67 in Montreal.

When Lamb established his own studio in 1968, his first important client was the Bunting Furniture Company in Philadelphia, an outdoor furniture manufacturer. His prototype for a perforated stamped metal chair gained critical acclaim, but prior to production its moulds were destroyed in a fire. After this loss, he designed the renowned Steamer, which is still in production.

In 1982 Lamb joined Toronto-based Nienkämper, an upscale office furniture manufacturer. During his eight-year tenure, he helped to transform the company from a subcontractor for Knoll International to a leading manufacturer of Canadian design. He designed the architectonic Embassy table in 1987 for the Arthur Erickson–designed chancery of the Canadian embassy in Washington, D.C.[35]

LAUGHTON, SCOT

Born 1963, Toronto

Scot Laughton has a distinctive style, putting him at the forefront of design since the nineties.

After graduating from the Ontario College of Art in 1986, Laughton ran Portico, the Toronto studio manufacturer, with fellow graduates James Bruer and Scott Lyons. The studio, along with subcontractors, produced and assembled limited runs of residential and restaurant furniture. Several designs earned awards, most notably the Strala lamp. Laughton left the partnership in 1990, and the studio disbanded three years later.

plastic dustpan with a snap-in brush (1990). Between 1990 and 1998, Kerr Keller designed more than forty products, and their work contributed to Umbra's pre-eminence in the housewares industry. Kerr Keller also designed furniture, including the Talking side table for Nienkämper (1995) and the Chit Chat chair for Keilhauer (1996), which won the Industrial Design Society of America (IDSA) bronze award (1997).

In 1997 Miles Keller left the firm after his marriage to Helen Kerr ended. He now works for Allseating Corporation, Toronto, where he designed the notable Os^5 chair (1999). Helen Kerr continues to run the firm, now known as Kerr and Company. Under her direction it has expanded into metal product design for Browne & Company and the Soshu flatware collection for Gourmet Settings (1999). In a male-dominated profession, Kerr has played an important role in making industrial design a viable career for women.[33]

KUYPERS, JAN

Born 1925, Nymegen, The Netherlands; died 1997

Jan Kuypers, an important designer of task chairs and an influential mentor, counselled young designers to keep things fluid, to never define something before it had a chance to develop. He maintained that flexibility with one firm (last known as KAN Industrial Design) that flourished for thirty-five years, and he had as partners noted industrial designers like Julian Rowan, Frank Dudas, Jerry Adamson, Ian Norton and Gerald Beekenkamp.

Kuypers studied at the Academy of Arts and Architecture in The Hague, was exposed to Bauhaus and De Stijl principles under the tutelage of Cor Alons and wrote a thesis titled, "Mass Production for Masses of Individuals." He graduated in 1947 and joined his sister in Manchester, England, where he found work with Grenfell Baines Architects. A year later, he joined H. Morris & Company in the competitive market of Glasgow, Scotland, where companies were designing and manufacturing wooden products ranging from luxury furniture to ships.

After four years in Glasgow, Kuypers immigrated to Canada to become the chief designer for the Imperial Furniture Manufacturing Company in Stratford, Ontario. During his tenure, the company was awarded four NIDC awards. He worked extensively with Eaton's design group to develop retail furniture as well as residential, hotel, motel and hospital furniture.

Tired of a "solid diet of furniture," Kuypers left Imperial in 1955 to study industrial design at the Massachusetts Institute of Technology, where he met his future business partner Rowan. Following a stint as a partner at Orr Associates, Toronto, Kuypers joined forces with Rowan and Dudas in 1960 at Stewart & Morrison. The trio later incorporated as Dudas Kuypers Rowan (DKR). Kuypers designed office furniture for Krug, Kitchener, and Office Specialty, Toronto, and developed the first made-in-Canada steel hospital furniture for Dominion Metalware Industries, Port Credit, Ontario. He also created one of the first automatic electric hospital

Currently he designs for Nienkämper (Tufold table [1996]), Pure Design (Jim stool [1997]) and Umbra (Juxta [1999]).[36]

LEIF JACOBSEN

Founded 1952, Toronto

In the heady 1960s, the firm Leif Jacobsen became known as the source for high-end corporate office furnishings and millwork. The plant often worked out design details with customers on the telephone and shipped the finished product to them sight unseen. Such was the design community's faith in the legendary Leif Jacobsen quality.

Leif Jacobsen, the firm's owner, came to Canada from Denmark as an infant and later worked with his father making ammunition boxes for the war effort. Unschooled in design or cabinetmaking, he founded his custom millwork firm in 1952. At that time, there was such a shortage of woodworking shops in Canada that Knoll International subcontracted Leif Jacobsen to produce a huge boardroom table for the B.C. Electric building in Vancouver. The company continued to work with Knoll, as well as with its competitor, Herman Miller—which profoundly influenced the style of the company's in-house designs and its commitment to meticulous craftsmanship.

The informal arrangement with Knoll persisted until the mid-sixties and remained the nucleus of the firm. Svend Nielsen, a Danish-trained cabinetmaker, joined the company in 1954, and it produced furniture for virtually everyone in the business: Metalsmiths, Nienkämper, Walter Nugent Designs and others.

One of the company's innovations in the late sixties was a semicircular desk with an attendant credenza, table and communications console to house a telephone, calculator and other "technology," which never caught on; however, a communications console to match an existing desk became a popular and expensive "must-have" for every executive. A typical Leif Jacobsen custom office suite would approach $20,000 and special orders twice as much. There was considerable local lore about the "unique" Leif Jacobsen furniture finishes with names like architectural bronze and swamp oak (which was merely a well-known European technique of curing furniture with ammonia fumes to emphasize the grain in the wood). Few of the designs were produced in quantities over a hundred units, which precluded the company growing beyond a well-regarded boutique manufacturer. Nielsen left in 1979 to form his own firm, Svend Nielsen, which continues to operate in Toronto. The Leif Jacobsen brand is still available through its new owner, Global Group.[37]

L'IMAGE DESIGN

1969–1988, Toronto

Bob Forrest, a former funeral director who became interested in design by working in a furniture store owned by the mortuary, founded L'image Design in 1969 and presided over its development into a popular retailer and wholesaler of contemporary residential furniture. The company's line was marketed as "furniture for the see-through era."

Nancy Bernecker, who studied at Parsons School of Design in New York and later became Forrest's wife, introduced him to the possibilities of acrylic. He launched L'image with a few of his own designs. In the early seventies, the interior designer Bryon Patton joined the firm to design condominium show suites and to establish an interior design division. L'image eventually had showrooms in Vancouver, Edmonton, Toronto and Montreal.

In the eighties, L'image focused on selling to the trade (interior designers and architects), operated as both a retailer and wholesaler and produced custom furniture, largely designed by Forrest. It also sold established furniture lines from Italy and manufactured knock-offs of classic designs, all in a modernist vein. Forrest's most popular designs included a bicycle bar cart (which used real bicycle wheels), the Habi cocktail table and free-form wall-mounted shelving. His design approach was that of a manufacturer intent on technical and engineering success.

Forrest bought Patton's share of L'image in 1986 and sold the furniture manufacturing operation to Robin Lauer of Brunswick Contract Furniture, Toronto, in 1988. Forrest now develops condominium and town home complexes.[38]

LISHMAN, BILL
Born 1939, Pickering, Ontario

Best known as a sculptor and as the environmentalist "Father Goose," Bill Lishman occasionally turned his hand to furniture design. His most successful object is a curvaceous rocker that reflects his aversion to straight lines.

A committed iconoclast, Lishman made sculptures that have appeared at city halls, in movies and at Canada's Wonderland and Expo 86. He also became a filmmaker, author and inventor, partly as a result of working with the naturalist William Carrick. Lishman, along with the photographer and pilot Joseph Duff, uses ultralight aircraft to train birds to follow a new migratory route, a technique documented by the ABC-TV news magazine show *20/20* and later in the film *Fly Away Home*. He is also involved with the Paula Lishman firm, an international fashion company.[39]

LUCK, JACK
Born 1912, London, England; died 1963

Many North Americans have used Jack Luck's simple, functional designs. He designed award-winning pots and pans, kettles and coffee makers, even door pulls under the brand names Wear-Ever Aluminum and Mayfair. His expertise in aluminum was acquired at Alcan, where he spent a

in 1913, it was renamed Medalta Stoneware in 1916 and then reincorporated as Medalta Potteries in 1924. It witnessed its best years between 1929 and 1953, when it produced utility ware, artware and later hospitality ware.

Much of the company's success was due to the Briton Thomas Hulme, who established and ran the art department between 1929 and 1953. His emphasis on surface ornament allowed Medalta to have a large inventory (over two thousand artwares) without the constant expense of mould preparation. He supervised a staff of mostly women (twenty-eight at its peak) who executed the transfer printing, stencils and rubber stamping. Artware included figurines, vases, jardinieres, plaques and lamps. Hulme often copied American designs, but his best work experimented with glaze decoration. Medalta also produced more profitable lines of crockery, mixing bowls and premiums. In 1937 Ed Phillipson joined as plant manager and industrialized the company, introducing assembly-line techniques and replacing the beehive kilns with more efficient tunnel kilns. He also developed a patented machine that formed a cup and handle from a single piece of clay.

Medalta sold hospitality ware in open stock (as opposed to sets) through Eaton's catalogues or from its own catalogue, the final one published in 1947. With Canadian Pacific as a major client, Medalta thrived. However, a change in ownership and a costly renovation closed the company. It reopened several times under different names and ownership, including as Medalta Potteries, operated in Redcliffe. Between 1966 and 1986, it produced crocks and flowerpots, sometimes using the stamps from the original Medalta pottery, which creates considerable confusion for collectors.[40]

MEDICINE HAT POTTERIES/HYCROFT CHINA
1938–1992, Medicine Hat, Alberta

Although less widely known than Medalta Potteries, Medicine Hat Potteries (called Hycroft China after 1957) produced residential dinnerware for over a half century. Hop Yuill of the tile and brick firm Alberta Clay Products Company founded the pottery in 1938 to compete with Medalta and attracted many of its employees. During the war, exempted from producing army utility ware because its china was too thin, it captured Medalta's dinnerware and artware market, thereby contributing to its rival's demise.

In 1955 the Marwell Construction Company bought Medicine Hat Potteries and later changed its name to Hycroft China, after its new apartment complex in Vancouver. It dropped the trademark stamp of a little native chief. A year later, Harry Veiner, the colourful self-made millionaire rancher and city mayor, acquired the company. Veiner continued the company's tradition of importing stencils from the U.S. in modern and traditional patterns. Its signature Hatina breakfast set with encircled ridges (renamed Matina after 1945) was offered as an alternative to the American Fiesta line.

large part of his career and in turn introduced the company to the process of creating and documenting sketches, working drawings and models.

Luck immigrated to Canada in 1930 and joined the Alcan subsidiary Aluminum Goods in Toronto. He later moved to its Montreal office to become a draftsman in the engineering department. When the firm founded Aluminum Laboratories in 1936, Luck joined the new division but continued to work concurrently as an artist and cartographer. When the lab moved to Kingston, Ontario, in 1949, he uprooted once again and helped to reduce wartime aluminum smelting overcapacity by using the metal to create products for the home. Over the years, Luck won four NIDC awards for aluminum cookware and by 1958 was lecturing on industrial design at the University of Toronto. He also served as president of the Association of Canadian Industrial Designers.

MCINTOSH, LAWRIE
Born 1924, Clinton, Ontario

Lawrie McIntosh earned a degree in mechanical engineering but was soon swept up into the emerging design profession. Throughout his long career, he gravitated toward assignments that could benefit from his engineering training.

McIntosh graduated from the University of Toronto in 1946. The NIDC awarded him a scholarship to the Illinois Institute of Technology, where he earned a master's degree in product design in 1951. Buckminster Fuller became a mentor. McIntosh returned to Toronto, established McIntosh Design Associates and in 1952 won an NIDC award for a stacking and folding chair. Two years later, he won another NIDC award, as well as a gold medal at the Milan Triennale for an automatic steam and dry iron, designed for Steam Electric Products, Toronto. During the fifties and sixties, McIntosh designed the Lady Torcan hair dryer, as well as appliance plugs, trouble lights and electric kettles.

By 1960, when McIntosh was elected president of the Association of Canadian Industrial Designers, he had already returned to his engineering roots by designing the Cobalt 60 therapy machine for Atomic Energy Canada Limited (AECL). Over the next two decades, he saw AECL's Theratron cancer treatment machine through numerous generations of design innovation. In 1981 McIntosh was given citations for outstanding achievement by Design Canada and ACID. Between 1983 and 1989, he taught design at the Ontario College of Art.

MEDALTA POTTERIES
1924–1954, Medicine Hat, Alberta

Despite its frequent changes in ownership and lack of a strong design strategy, Medalta Potteries was one of the most important commercial potters in Western Canada. Founded as the Medicine Hat Pottery Company

Hycroft China sold pottery at both Woolworth's and Eaton's. Souvenir ware with Canadian scenes, wagon wheels and Stetsons enhanced the company's revenues. In 1960 the company introduced toilets to supplement its dinnerware production. Hycroft China shut its factory in 1989 and the company was dissolved in 1992.[41]

METALSMITHS COMPANY
Founded 1926, Toronto

Metalsmiths Company is a mid-size contract furniture manufacturer that in its heyday in the sixties and seventies produced high-end contract seating and case goods for clients like the University of Ottawa, Winnipeg's Centennial Hall and Bentall's Vancouver headquarters. Its early principals were architects, a rarity for Canada.

The architect Kenneth Noxon established the Toronto-based artisans' foundry in 1926 to make hand-forged hardware, lighting fixtures and architectural ironwork. By the forties, the firm was making custom furniture and fireplace fittings, much of it reflecting the spirit of the Bauhaus. The increasing demand for metal rod furniture resulted in designs for tables with hairpin legs, glass-topped nesting tables and reproduction butterfly sling chairs.

Between 1956 and 1959, the company won five NIDC awards for designs by the founder's son, Court Noxon. He had graduated from the University of Toronto's School of Architecture in 1953. In the late fifties, the younger Noxon joined Metalsmiths and by 1963 became its president and owner. He expanded the firm into steel, wooden and upholstered furniture. Under licence, the firm produced component pieces for Knoll International and the designers Laverne International of New York. Production was automated, and the firm became known for its precision metalwork, such as articulated joints, executed in polished chrome.

When interior designers began to specify furniture for their corporate clients, Metalsmiths' lines were sold in co-operation with Leif Jacobsen, then later marketed via its own sales force. Rarely groundbreaking, Noxon's better-known designs include a wall-mounted coat rack (1955) and model 1148, a three-legged tub chair (1963). The firm was sold to Jim and David Thomson in 1982; Court Noxon returned to his architectural practice. The new owners moved the plant to Unionville, Ontario, and continue to produce some of Noxon's original designs.[42]

MOFFATT, FRED
Born 1912, Toronto

Fred Moffatt's design for an electric kettle became so pervasive that its chrome dome is considered both a fifties icon and a remarkable Canadian success.

Moffatt attended Central Technical School in Toronto, where he assisted the noted war memorial sculptor Alfred Howell. He cut woodblocks at Southam Press, then apprenticed as an illustrator at Rapid, Grip & Batten alongside well-known Canadian painters like Jack Bush and Charles Comfort. In the late 1920s, he worked at McLaren Advertising, where CGE was a major client, and by night took classes at the Ontario College of Art. He opened his own firm in 1931 and began making the transition into industrial design.

For the next fifty years, his principal client was CGE, for which he designed everything from kettles to electric lawn mowers. He won numerous awards, including two NIDC honours for a floor polisher and a food mixer, and a silver medal at the 1964 Milan Triennale for CGE's teardrop-shape electric space heater.

His son, Glenn Moffatt, joined the firm in 1967 and kept up the family tradition by designing kettles (mostly in plastic) for Black & Decker and Superior Electrics, as well as other consumer and industrial products.

MORIN, ANDRÉ
Born 1941, Montreal

André Morin, based in Quebec, has designed a variety of products over the past thirty-five years—consumer electronics, housewares, medical equipment and garden furniture. He won five Canadian Awards of Excellence in Industrial Design, sponsored by the government of Canada. Early in his career, he developed the Forma Collection for RCA Victor Company, Montreal. A decade later, he was best known for his plastic kitchenware collection for IPL, Montreal.

Morin graduated from the Université de Montréal and the Institut des Arts Appliqués, Montreal. His second job out of school (1965) was at Canadian Marconi Company in Montreal, where he designed a series of portable radios, record players and portable TVs. When Marconi abandoned consumer electronics in 1967, Morin joined RCA's larger industrial design division. He was responsible for all audio products (radios, cassette players and record players) as well as portable televisions. His Forma Collection became the company's best-selling line.[43]

Morin became a freelance designer in 1977, winning awards for the IPL kitchenware line (1980 and 1982); bathroom accessories for Fabri-Metal (1984); and the Match I Collection, a kitchen storage system (1985). For IPL, he also designed Oasis garden furniture. Morin's designs for international companies range from toys to computers. He frequently addresses design conferences in North America and Europe and served as president of the Quebec chapter of the Association of Canadian Industrial Designers between 1976 and 1980.

MORRISON, EARLE
Born 1923, Vancouver

Earle Morrison is best known for the furniture he and Robin Bush designed

NIENKÄMPER
Founded 1968, Toronto

Minimalist modernism and refined postmodernism define the Nienkämper style. For more than thirty years, the company has fashioned sophisticated office furniture.

The firm's ebullient founder, Klaus Nienkämper, rejected his family's antiques business and served a three-year apprenticeship with Knoll International in Düsseldorf, Germany. At twenty-two, he immigrated to Canada and worked with Gary Sonnenberg at Toronto's Precision Craftwood (now Craftwood). Two years later (1962) he joined forces with David Bain, who had licences to manufacture furniture by European modernist designers like Robert Haussmann, Verner Panton and Eero Aarnio. Fuelled by the baby boom and Expo 67 building bonanzas, the firm added local content to its repertoire, most notably a leather Safari chair and a simulated pony-skin seating cube, designed by Klaus. In 1968 Swiss Designs of Canada became Klaus Nienkämper Design, minus Bain.

By the mid-seventies Nienkämper, as it was henceforth known, began producing furniture under licence for Knoll International as well as De Sede of Switzerland, a world renowned leather furniture manufacturer. Notably, it produced custom furniture for Prime Minister Pierre Trudeau's office, created by the architect Arthur Erickson and the designer Francisco Kripacz. It furthered the commitment to Canadian design by hiring Thomas Lamb as director of design (1982–90). Significant Lamb designs of the period reflect a postmodern, curvilinear aesthetic, including the Sculpted armchair. The firm also produced custom furniture for Toronto's Roy Thomson Hall by Erickson and Kripacz (1982) and the Ambassador and Elle chairs by Christen Sorensen (both in 1985).

By 1987, when annual sales approached $10 million (about half with Knoll), the licence to produce Knoll product expired. Five years later, the company slipped into receivership but was quickly revived by private backers (including the package designer Don Watt and the branding guru Dave Nichol) until it formed a partnership with New York–based ICF Group. In the nineties, Canadian design leaders such as Tom Deacon, Scot Laughton and Yabu Pushelberg have created office furniture for Nienkämper. The industrial designer Mark Müller, a graduate of Barrie's Georgian College, joined the company as Lamb's assistant in 1988 and was appointed design director in 1995.[45]

NORTHERN ELECTRIC COMPANY/NORTEL NETWORKS
Founded 1882, Montreal

Founded in Montreal as the Mechanical Department of Bell Telephone Company of Canada, Northern Electric and Manufacturing Company was incorporated in 1895. Initially it made telephones and switchboards as well as gramophones and a radio receiver (1922). During the Second

for their own firm, Standard Furniture Plant, located in Morrison's hometown of Victoria, B.C.

In the early 1940s, Morrison studied aeronautical engineering at the California Institute of Technology. He worked for the American furniture manufacturers California Manor and Brown Saltman. During the Second World War, he stayed in Southern California and worked in the design and engineering department at Plywood Moulded Aircraft, a division of the Howard Hughes Aircraft Corporation.

When the war ended, Morrison returned to Victoria and established a custom-furniture manufacturing firm, Earle A. Morrison. In 1950 he and Bush took over the Standard Furniture Plant. The company favoured modernist designs, production runs typically approached five hundred units and three of their designs won NIDC awards. Their designs were exhibited at the 1954 Milan Triennale, in a Trend House in Toronto and at the Design Centre in Ottawa.

After Bush moved to Vancouver in 1953, Morrison continued to manufacture some of the designs. In the mid-fifties, Morrison worked in Knoll International's New York office on the furniture and interior design of Vancouver's landmark B.C. Electric building. A wrought-iron chair designed for Small & Boyes won an NIDC award in 1956.

MURRAY, JAMES
Born 1923, Ayr, Ontario

James Murray, a farmer turned furniture designer, is best known for his creations for his own firm, James Murray Furniture, over a period of twenty-two years. His work appeared in the *Design Index* in the late fifties and in *Canadian Design 67*.

After graduating from the Ontario College of Art at age twenty-nine, Murray designed furniture for a variety of manufacturers. In 1955 he built a factory in his hometown to manufacture his own designs. Three years later, his first submissions to the NIDC (two armchairs) won awards. The following year (1959), he won another NIDC award for a table.

Most of Murray's early designs were made from oiled woods such as walnut, but by the mid-sixties, he was working with more sophisticated materials. The GT series of chairs with chrome-plated steel frames signalled a move into office and institutional furniture.

Before closing the plant in 1978, he created his final design series, lamina 2, consisting of chairs, sofas and occasional tables made from only four hardwood plywood components. The line's unique interlocking machined joints created structurally sound furniture, even though its platform bases, arms and backs were held together without glue or metal fasteners. In the eighties, Murray retailed early Canadian furniture and folk art and is now committed to environmental and human rights issues.[44]

World War, it manufactured a wide range of military and commercial communications equipment, including radio navigational aids, broadcasting and motion picture sound reproduction systems and telephone systems. It began to promote its re-entry into consumer products in 1943 and by 1949 had two large manufacturing plants in Belleville, Ontario, which made amplifiers, electric organs, and emergency and alarm equipment, among other products. Its Bakelite plastic moulding plant was one of the largest in North America.

The launch into locally designed radios probably began in 1939. William Doig, the supervisor of the radio receiver engineering department in Belleville, was denied membership in the Association of Canadian Industrial Designers, despite his claim to overseeing all design and drafting, chassis engineering and cabinet design for plastic and wooden models.[46] The company began selling its plastic radios in 1949 (Baby Champ and Midge) and introduced the Panda in 1951.

In the late 1940s, the company introduced a combination hand-wound phonograph and battery-operated radio, as well as a miniature television with sixty-one square inches (393 cm^2) of viewing area. The TV was likely designed elsewhere, as it would be at least 1955 before broadcast signals were widely available in Canada. The company continued to advertise wooden console-model radios like the Laurentian, originally designed by the American Egmont Arens. Available through Eaton's, it appeared in the American Society of Industrial Designers' yearbook *U.S. Industrial Design, 1951*, but by that time buyers were gravitating toward the company's less expensive tabletop combination radio/record players.

In 1956 the U.S. government forced the breakup of foreign subsidiaries of AT&T/Western Electric, which since 1911 had owned up to 50 per cent of Northern Electric. A year later, Northern Electric Research and Development Laboratories was created to design proprietary products. That same year, Sylvania took over the company's radio, television and household appliance distribution business.

In the sixties, from a new plant in Ottawa, Northern Electric directed its R&D toward sophisticated telephone systems and satellite equipment (which the company sold to SPAR Aerospace in 1977). The renowned Bell Northern Research department (BNR) opened in 1971 and became a subsidiary of the corporation, renamed Northern Telecom five years later.

It launched its revolutionary telephone-exchange switch in 1981. Now known as Nortel Networks, the firm has been an international leader in telephony, data, wireless and wireline solutions for the Internet.

NUGENT, WALTER
Born 1913, Vancouver

The self-taught designer Walter Nugent shook up the Canadian furniture industry in the sixties with his patented design for a one-piece chair seat and back that is supported internally with tempered steel rod. He launched a company to produce various models of the chair, won two NIDC awards and earned major contracts across Canada.

Nugent attended Oakwood Collegiate in Toronto and served as an air force navigator during the Second World War. He became interested in design while working at the advertising agency McConnell Eastman, prompting a visit to Denmark. In 1957 he developed a prototype for a sprung steel chair. Two years later, Walter and his brother Doug, along with a private investor, founded Walter Nugent Designs in Oakville, Ontario, and Eaton's placed a significant order for the chair.

The chair design, in various guises, eventually won three awards, two from the NIDC. An evolutionary high-back model (#68), patented for hospital use, won an award in 1968 from the Ontario chapter of the Association of Canadian Industrial Designers.

Doug took over managing the business in 1968 to allow Walter to focus exclusively on design. In 1971 Walter's final design for the firm (#99) was for a plywood birch lathe-turned table and matching chair (without its customary steel core). Following disputes with investors, he left the company in 1973, although it continued to produce chairs until it closed in 1976. George Powers, who worked for the company between 1972 and 1974, founded Avenger Design, Oakville, Ontario, in 1976. His 1973 design for the Avenger stacking chair, which exposed its sprung steel interior, won a Trillium award.[47]

ORIGINA CANADA

Founded 1974, Toronto

For over twenty years, Gustavo Martinez has designed contemporary residential lighting. Initially the company made stylish Pop lights before moving into more mainstream contemporary design.

In 1956, as part of the Mexican Youth program, Martinez and his brother cycled across the continent and settled in Toronto. Martinez worked as a draftsman for Canadian National Telegraph, before graduating in mechanical engineering at the University of Toronto in 1967.

By the early seventies, Martinez founded Origina Canada. Its first product line featured large bulbs and highly polished metal bases. In the mid-eighties, the company manufactured swing-arm brass lamps in the classic style popularized by the American designer George Hansen. With the arrival of halogen, Origina switched to task lighting featuring stamped aluminum shades. All manufacturing, approximately ten thousand units a year, takes place in its small factory. Most of its output is sold to Canadian independent lighting stores.[48]

C.P. PETERSEN & SONS

1944–1979, Montreal

Fusing craft with industry, C. P. Petersen & Sons worked in both traditional table and chairs, ironically were used in malls across North America.

The brothers studied architecture and design at the University of Rome, graduating in the late fifties. They opened their first studio in the sixties in Beirut (where their father was stationed as an Italian diplomat) and helped to introduce the city to European modernism. In 1968, on the design strength of Expo 67, they immigrated to Canada and established a small practice in Toronto. Their interiors in the fashion, advertising and restaurant industries have been widely published internationally. The brothers have also designed low-cost housing projects.[50]

PRECISIONCRAFT/PRECIDIO

Founded 1963, Brampton, Ontario

Originally known as Precisioncraft, the company specialized in tricolour acrylic drinkware. Owned and operated by Hank Sawatsky, a chemical engineer, Precisioncraft was one of the first companies to use sonic welding to permit greater variety in the shape and colour of products. Welding different plastic components into a single finished design is now industry standard.

In 1995 the company went bankrupt and re-established itself as Precidio. The design director, Willa Wong, a graduate of the School of Architecture, University of Waterloo, joined in 1997. Her designs include Sonoma melamine tableware, produced offshore. Precidio also licenses designs from the Toronto textile designer Paula MacMillan and etches her Op patterns on acrylic and glassware blanks (supplied by Libbey Glass in the U.S.). The holding company U.S. Housewares bought Precidio in 1999. The firm sells 70 per cent of its $20 million annual production to the U.S. Merchandise is available at Bowring in Canada and Bloomingdale's and Bed, Bath & Beyond in the U.S.[51]

PURE DESIGN

Founded 1994, Edmonton

Launched as a studio manufacturer, Pure Design has grown into an adventurous company producing contemporary Canadian and international design for the residential market.

Daniel Hlus, Randy McCoy and Geoffrey Lilge, the founders of the company, graduated from the University of Alberta's industrial design program in 1991. With six other schoolmates, they formed Hothouse Design Studio, a co-operative. In 1992 the three original members left and formed Pure Design to take over McCoy's family welding factory and to fabricate their designs for metal coat racks and shelving. Their first success was with the Mantis CD rack.

The company commissioned Scot Laughton to design the UBU storage rack in 1995. Pure hired more young freelance designers, including Johnny Lim, 3rd Uncle Design in Toronto and the Generation X writer Douglas Coupland, and ventured into furniture.

and Danish modern styles, creating everything from reproduction tea services to assembly-line liturgical silver, as well as the Stanley Cup hockey trophy. At its height, the company imported four tons of silver annually and employed twenty silversmiths. It also produced works in gold and copper, alone or in combination with silver.[49]

Its founder, Carl Poul Petersen, was born in 1895 in Copenhagen, Denmark. At age thirteen, he apprenticed with the renowned silversmiths Georg Jensen and Kastor Hansen at the Georg Jensen Silversmithy. He immigrated to Montreal in 1929 and became a master goldsmith at Henry Birks & Sons, leaving only for a brief period (1937–39) to operate his own studio. During the Second World War, he made aluminum and brass wool filters for aircraft. Buoyed by commissions from Samuel and Saidye Bronfman of the Seagram distillers' dynasty, Petersen reopened his firm in 1944. He trained his sons, Arno, John Paul and Ole, and many of the other silversmiths, like Eigil Pedersen, who worked for the firm.

Petersen was steeped in the Arts and Crafts tradition and favoured motifs such as Canadian wildflowers and beavers in jewellery, tableware and gifts. The company made eleven flatware patterns with such fanciful names as Dolphin, Wild Berrie [*sic*] and Pearl. Two patterns, Viking and Corn, appear to be modelled after Georg Jensen's 1908 Continental flatware and Johan Rohde's 1907 Acorn respectively. To complement the flatware, the company made hollowware (covered entrée dishes, wine coolers and candelabra) in standard shapes from flat sheets of silver. The plain bases were raised by machine, then edged by hand with beading and bayleaf garlands or decorated with floral or key-fret borders. Floral and vegetal motifs such as grape, sweet pea, lily of the valley and cherry were stamped or soldered onto the vessels. It also marketed numerous smaller objects such as salt and pepper shakers and napkin rings and had some success selling to the American market.

The designs for jewellery for men and women were less successful, although about two hundred versions of brooches were produced. When the multi-millionaire Hunt brothers of Texas attempted to corner the market on silver, the price of the metal was temporarily driven up to $50 an ounce, forcing C. P. Petersen & Sons out of business in 1979, a mere two years after the death of its founder.

PICCALUGA, ALDO AND FRANCESCO
Born 1936 and 1938 in Genoa, Italy

Known for their fastidious detailing and sculptors' sensitivity to form, the brothers Aldo and Francesco Piccaluga are architects and interior designers who occasionally create custom furniture and lighting for their innovative projects. Their only mass-produced furniture, the System Sigma Series

Pure expanded in 1999, opening its own computer-aided wood manufacturing facility. It also took over production of the Aura stool by the Vancouver designer Niels Bendtsen. Eighty per cent of the company's $4 million annual sales is to the U.S.[52]

RASHID, KARIM
Born 1960, Cairo, Egypt

At the end of the twentieth century, Karim Rashid emerged as one of the most sought after designers. Art et Industrie in SoHo and Totem in TriBeCa (both in New York City) have organized exhibitions of his work. In 1998 the Brooklyn Museum named Rashid the Young Designer of the Year, and the following year he won the prestigious DaimlerChrysler design award.

Rashid emigrated from England with his family to Montreal in 1966, then moved to Toronto. Graduating from Ottawa's Carleton University School of Industrial Design in 1982, he was awarded one of the last Design Canada Scholarships and studied in Naples. The following year, he apprenticed at the studio of Rodolpho Bonetto in Milan. On returning to Canada in the mid-eighties, Rashid, Pauline Landriault and Scott Cressman created Babel, Toronto, a mens- and womenswear line that won considerable acclaim.

He gained industrial design experience at KAN. As the company's senior designer for seven years, he had corporate clients that included Brita, Semi-Tech Microelectronics and Imax. Simultaneously, he created art furniture that was featured in Virtu.

Rashid taught at the Rhode Island School of Design in 1992 and set up a studio in New York City the following year. Nambé, a manufacturer of metal-alloy tableware based in Santa Fe, New Mexico, was the first company to make Rashid's designs in high volume, producing over forty patterns of tableware accessories. For Pure Design in Edmonton, he created the Pura chair and Arp stool series. Umbra is a major client, and the Garbo garbage can (1997) has been a notable success for designer and company.

ROGERS-MAJESTIC/PHILIPS ELECTRONICS CANADA
Founded 1924, Toronto

In 1924 Edward Rogers successfully engineered improvements to the alternating current (AC) radio tube to make it reliable. It revolutionized the industry and turned a specialist hobby into a popular consumer product.

The first Rogers Batteryless radios (a play on "horseless" carriages) went into production in Canada in August 1925. Rogers undertook a dizzying array of mergers and acquisitions. In 1928 the company merged with Chicago's Grigsby-Grunow Company, makers of Majestic Radios. A year later, the combined firm, Rogers-Majestic, built the largest radio manufacturing plant in Canada and soon made Viking brand radios for Eaton's. In 1927 Rogers founded the world's first all-electric radio station, CFRB (Canada's First Rogers Batteryless), and later experimented with the fledgling television industry.

By 1934 the consumer electronics division manufactured car radios for both Ford and General Motors and acquired Consolidated Industries, makers of DeForest-Crosley radios, Norge electric refrigerators and Hammond clocks. Following Edward's untimely death in 1939, the division was sold to a British company and by 1941 operated as Rediffusion Canada, which made both Rogers-Majestic and DeForest-Crosley brand radios until Philips Electronics Canada acquired it.[53] Philips kept the brand alive until 1959.

While most of the company's wooden console radios (shaped like laundry hampers or styled like a jukebox and tuned to CFRB) were likely designed in Canada, the design for most of the plastic models probably emanated from American partners such as DeForest-Crosley and Majestic. The media component of Rogers (broadcasting, cable, telecommunications and publishing) continues to expand under the leadership of Edward's son, Ted Rogers, as Rogers Communications.

SÉRI +

Founded 1974, Montreal
Monique Beauregard born 1945, Montreal;
Robert Lamarre born 1950, Montreal

Monique Beauregard and Robert Lamarre, calling themselves creator-producers, made a commitment to textile design and education in Canada.

Beauregard attended the interior design program at the Institut des Arts Appliqués in Montreal between 1962 and 1966. Lamarre graduated in exhibit design at the same institution in 1970. In 1974 they opened SÉRI + so that each could independently design small production lines. To compete with Scandinavian imports like Marimekko, SÉRI + printed approximately twenty playful patterns in small quantities. After a fire in 1980, Lamarre and Beauregard redirected the company away from household items to higher-end designs for the interior and architectural markets.

In 1985 they founded the Centre de recherche et de design en impression textile in Montreal, which serves as a training ground and production house for freelance designers. The following year, they received a scholarship to study computer-aided textile design at Goldsmiths College, University of London. This led to explorations in unorthodox materials and computer-generated imagery. SÉRI + has presented its collections at the annual Salon international du design d'intérieur de Montréal (Sidim). The Victoria and Albert Museum in London holds examples of their work in its collection.[54]

with both respect and sadness that design "was a calling, a terrible discipline. Design is [Siwinski's] whole life."[55]

SNYDER BROS. UPHOLSTERY COMPANY/JAMES C. SNYDER

Founded 1885, Waterloo, Ontario

From the late forties to the early fifties, Snyder's was the largest furniture manufacturer in the British Commonwealth, with two hundred employees in four plants: two in Waterloo, Ontario, and one each in Montreal and Vancouver (Small & Boyes). During that productive decade, the company's entire output was dedicated to modern design, including three thousand to thirty-five hundred units a month of its popular Vista wooden furniture line. The company turned down an opportunity to franchise the La-Z-Boy line in Canada, as it was too busy marketing its own Sandman recliner.

Founded by a pharmacist with ambitions to consolidate the Ontario furniture manufacturing market, the firm over a century later remains in the hands of a fourth-generation Snyder (Jamie), albeit as a niche producer. In the 1930s, the company manufactured designs by the well-known American modernist Russel Wright. By the end of the war, it was making radio cabinets.

In the late forties, the firm introduced sectional sofas to the Canadian market and turned the concept into a marketing success. A variety of furniture lines were sold through Eaton's. As competition increased, Snyder's imported upholstery fabrics—ranging from ribbed mohair to Harris tweeds—and gained a reputation for quality. In the fifties, Snyder's added commercial case goods to its lineup and became a licensed Herman Miller manufacturer, producing designs by George Nelson as well as Robin Bush.

While a Snyder's family member always oversaw its design committee, the firm employed a succession of industrial designers: L. C. Ruby, George Soulis (who later taught design at the University of Waterloo) and Lorne Winkler. Occasionally, it contracted independent designers like John Murray to develop specific lines. More recently, the company has focused on mass-market upholstered furniture. Its name was charged to James C. Snyder in 1966. Notably, in 1999 the Toronto interior designer Richard Eppstadt designed the Lux Sofa, which won the Best of Canada Award from the magazine *Canadian Interiors*.[56]

SORENSEN, CHRISTEN

Born 1921, Esbjerg, Denmark

Christen Sorensen was schooled by Hans Wegner and was a contemporary of Poul Kjaerholm, both noted Danish designers, but his work transcends its Scandinavian roots and employs new materials to "add romance" to industrial products. Even so, it remains true to the social democratic principles of his youth.

Sorensen was born into a close-knit community of furniture designers,

SIAL II
1974–1981, Laval, Quebec

Like Céramique de Beauce, Sial II worked with noted studio potters, and its designs by Gaétan Beaudin and the ceramic sculptor Maurice Savoie reflect the firm's best work.

The company originated in 1965 as Sial (meaning "the earth's crust"), when three well-known professional potters, Beaudin, Pierre Legault and Bertrand Vanasse, formed an association to sell clay and glazes to local potters. Inspired by a report from the federal Department of Industry, Trade and Commerce identifying stoneware dinnerware as a growing market, they founded Sial II to fill that niche. Awarded a provincial grant, they built a large plant with the latest machinery and clays and opened a warehouse in Albany, New York. Sial exported to Europe under the trademark Serval. Design Canada undertook a product design case study promoting the company's successful synthesis of craft and industry.

Beaudin served as the company's master craftsman, developing dozens of moulds to find the right design that looked handcrafted but could withstand the requirements of modern living. Beaudin's work for Sial was controversial among the craft community, and he later claimed that he felt "like a defrocked priest." Sial I continues to operate as a distributor of clay materials.

SIWINSKI, STEFAN
Born 1918, Wturek, Poland

A passionate perfectionist, Stefan Siwinski mastered many materials—wood, metal and plastics—in his quest to design and manufacture high-quality modern furniture. His monochromatic sculptural designs of the fifties and sixties are remarkable for fusing industrial materials with accomplished hand labour. His endless drive to improve his designs rather than standardize production kept his company in a constant state of financial instability but simultaneously earned it media attention and the respect of designers.

Siwinski immigrated to Canada in 1952. Two years later, his workshop in Toronto, sometimes called Korina Designs, employed ten craftsmen. The small carpentry studio specialized in commercial furniture and also accepted commissions, such as making a table for the Canadian modern architect John C. Parkin (1960). That same year, Siwinski's lounge chair 100-1 was specified for the Toronto International Airport departure lobby. The contract led to commissions for chair designs for York University, the Toronto Dominion Bank and the Art Gallery of Ontario.

In the seventies and eighties, Siwinski designed furniture for a Toronto subway station and for the international airports in Calgary and Toronto. Harold Town, the late Canadian painter and Siwinski admirer, once said

retailers and manufacturers. In 1934, at age thirteen, he was already "working on the bench" in his father's cabinetmaking shop. He attended the famed School of Arts and Crafts in Copenhagen and bested Kjaerholm in an upholstered-furniture competition. Throughout the early fifties, his work was regularly exhibited in Denmark, and by 1954 it appeared in the Milan Triennale, where Sorensen was introduced to the work of the Canadian designer Robin Bush.

In 1956 he immigrated to Montreal, surviving on a series of odd jobs and custom furniture commissions. Briefly, he represented the Herman Miller, Knoll International and Paul McCobb product lines for J & J Brook. Two years later, Sorensen became a partner in Jacques Guillon & Associates. He worked on the interior design of Montreal's Dorval airport, where a notable two-seater sofa of his own design appeared, and on Alcan's headquarters in Place Ville-Marie. He also managed the subsidiary, Ebena Manufacturing, which produced designs by the American Ward Bennett and DUX (the Swedish company best known for popularizing Bruno Mathsson designs). He became Ebena's president and merged it with LaSalle, an office furniture manufacture, but left in 1962 to refocus on design.

Sorensen and Bush formed InterDesign and created the main Expo 67 theme building, Man and His Community, and the Atlantic Provinces pavilion. Sorensen also designed a Habitat suite. After business declined in Montreal, Sorensen rejoined Bush, who was director of the Sheridan College School of Crafts and Design, to teach furniture design. In the eighties, he designed the Elle chair for Nienkämper, which has been re-issued, and the Respons series for Keilhauer, his most successful commercial design.[57]

SOVEREIGN POTTERS
1933–1974, Hamilton

After the war, Sovereign Potters led the design and production of informal modern dinnerware, marketed as "everyday indoor and outdoor living." At its peak, the pottery employed a staff of four hundred.

Founded by the local businessmen William Pulkingham (who later bought Medalta), Alfred Etherington and James McMaster (formerly of McMaster Art Pottery in Dundas), Sovereign Potters imported equipment from Syracuse, New York, to produce whiteware china for the hospitality market, which would remain its mainstay. The ROM holds a collection of the first dinnerware to come off the Sovereign assembly line (inscribed February 1934).

Family connections with the influential American designer Russel Wright and his wife, Mary Wright, enabled Sovereign Potters to produce some of their patterns in 1954 under licence.[58] The company "borrowed" the Wrights' design elements for its own line. The Carnival line was created by Etherington, with assistance from his children. Lois Etherington Betteridge went on to become one of Canada's leading metalsmith artists, and Bruce Etherington became a respected architect.

In addition to the modern dinnerware, Sovereign Potters produced over twenty traditional British-style china patterns, as well as butter dishes for Canadian Westinghouse Company refrigerators. In 1947 Sovereign became affiliated with H & R Johnson Brothers, the largest tableware manufacturer in England. Ten years later, Sovereign Potters manufactured only industrial tiles. Tableware blanks, imported from the parent company, were decorated locally. In 1974 the company changed its name to H & R Johnson (Canada). It closed in 1993.

SPANNER, RUSSELL
Born 1915, Toronto; died 1974

The former wrestler and largely self-taught furniture designer Russell Spanner created three lines of residential furniture (Ruspan Originals, Catalina and Pasadena) that reflect the Canadian furniture industry of the fifties: solid and earnest but with enough personality to transcend its manufacturing compromises. He made it a virtue to reuse (and eventually, continuously improve) standard components, thus creating a recognizable style.

Spanner Battery Separator Company was founded in the early twenties by Russell's father, grandfather and uncle, initially to make wooden battery parts and, later, battery boxes. As the demand for battery boxes declined, the company phased in furniture manufacturing. Russell, who had studied architectural drafting at Toronto's Northern Vocational and Technical School, joined the company as night foreman in 1941, three years after the firm's name was changed to Spanner Products. His designs had an impact only after his father, Albion Spanner, retired in 1948 and Russell and his brothers, Doug and Oliver (Herb), took over the firm.

Russell's first line, Ruspan Originals, aimed at young post-war families, also found favour with the chic hostesses of the day. A 1953 *Canadian Homes and Gardens* article shows well-known women using Originals dining chairs. It became Spanner's most extensive line, ultimately featuring twenty-eight modular pieces that could be combined in a myriad of ways.

Two years later, Spanner followed up with the aesthetically more refined Catalina line. Distributed through Eaton's, Simpson's and independent retailers, it appears most frequently in the collectors' market. Spanner reached the apex of his design career by 1953, with the Pasadena line, the centrepiece of which is an elegant cork-topped table, No. 534.

By the late fifties, the business's rapid expansion and changing tastes conspired against the family-run firm. Spanner reinterpreted the prevailing Scandinavian style and was awarded by the NIDC but not by buyers. The firm produced contract furniture for hotels and motels and laboratories that often exhibited more verve than was customary or, perhaps, necessary or practical. By 1961 Spanner had parlayed his factory-floor skills into a position as plant manager at Ontario Store Fixtures in Toronto. Two years later, the assets of Spanner Products were auctioned off.[59]

skills (such as enamelling and lost-wax techniques) to his repertoire.

Stacey attended Toronto's Central Technical School and later Northern Vocational and Technical School, where the Swedish designer Rudolph Renzius influenced him. He opened a studio in 1931 (working mainly in pewter) and between 1936 and 1940 took over Renzius's classes at Northern. During the Second World War, Stacey built radar equipment for Research Enterprises in Toronto. When the war ended and metals became more widely available, he specialized in silver liturgical and architectural works. During the same period (1944–50) he taught night classes for war veterans at the Ontario College of Art.

In 1948 the federal Department of External Affairs commissioned Stacey to design flatware for Canadian embassies and consulates. He completed several prototypes featuring a stylized maple leaf or wheat sheaf on the handles but was unable to raise the money required to finance production. The following year, he attended a silversmithing workshop at the Rhode Island School of Design under the tutelage of Baron Erik Fleming, silversmith to the Swedish royal family.

Stacey joined the New York office of Steuben Glass (Corning) in 1950 to supervise its newly formed silver department. He returned to Toronto in 1952 and worked as an industrial designer for Massey-Harris and other companies. In 1953 he won an NIDC award for a stainless steel sink designed for Kitchen Installations in Toronto. Stacey operated a metalwork studio from home and continued to teach, including a stint at Humber College (1968–75).[61]

STENE, JOHN
Born 1914, Stavanger, Norway

The bomber pilot, air force instructor and war hero John Stene began a "second" career importing Scandinavian furniture to Canada after the war. In 1957, at age forty-three, he formed a company to produce his own designs and one year later won two NIDC awards for a dining table and a chair, the former made from Scandinavian-style oiled walnut, the latter with a traditional Norwegian rope seat.

Stene graduated as a civil engineer in 1938 in his native Norway. When the Germans invaded during the Second World War, he escaped through Russia and took a circuitous route around the globe before arriving in Canada. Shelagh Vansittart, who became Stene's wife, introduced him to contemporary furniture, and together they imported and sold Scandinavian furniture through Shelagh's of Canada, the influential Toronto retailer, launched in 1954. After their divorce, Stene founded Brunswick Manufacturing Company in Toronto. His creative partner, Rudolph Rataj, a master woodworker from Germany, often influenced Stene's initial designs.

The company profited from the buoyant school and institutional market

SPENCER, HUGH

Born 1928, England; died 1982

Outspoken and audacious, Hugh Spencer created some of the most avant-garde designs to come out of this country, causing *Time* magazine to proclaim in a 1965 issue, "The unit [Clairtone's Project G stereo] comes not from Mars but from Canada." Spencer, however, stated that a designer's role was merely to "observe, comprehend, select, reason and re-assemble."

Spencer studied painting at the Slade School in London, England. On graduation, he became head of design for Granada Television in Manchester, then moved to the British Broadcasting Corporation to design sets. He immigrated to Canada in 1956 at the behest of the Canadian Broadcasting Corporation, but worked for Robert Lawrence Productions.

Spencer founded the Toronto design firm Opus International in 1960. Six years later, he launched a retail division, Form/Factor, which marketed commercial and residential furniture designed by the Opus team as well as other Canadian designers. Its mandate to promote Canadian furniture and decorative arts was a rarity.

Spencer's philosophy of total design attracted the image-conscious Clairtone Sound Corporation. Between 1962 and 1967, he designed the company's logo, factory, exhibitions, showrooms and a series of stereos, including the Project G. For Expo 67 he designed the Western Canadian Pavilion, furnished a Habitat suite with existing and new designs and created the Club chair U30 for a private lounge for visiting dignitaries.

Spencer also designed packaging and promotional materials for clients such as Canadian Westinghouse Company, Philips Electronics Canada and Kodak Canada. In these endeavours, he partnered with the celebrated graphic designers Paul Arthur (Arthur + Spencer) and Don Watt. In 1969 Spencer left Canada for Dallas, Texas, where he designed in-flight food-service systems and hospital directional graphics.[60]

STACEY, HAROLD

Born 1911, Montreal; died 1979

Harold Stacey was active in Toronto for nearly fifty years, as both an important silversmith and an influential teacher. The quality of his craftsmanship was much admired, and many consider him the most skilled silversmith of his generation. His work was shown at the 1937 Paris Exhibition, in the Art Gallery of Toronto and in the National Gallery of Canada. He was the founding president of the Metal Arts Guild of Ontario in 1946 and later an original member of the Canadian Craft Foundation.

Stacey preferred silver but worked in virtually every metal, employing both modern and traditional styles and producing in excess of five hundred objects, including flatware, hollowware, ecclesiastical items, jewellery and architectural metalworks. A lifelong learner, he gained knowledge through the sheer variety of his collaborations and continuously added of the 1960s, producing a line of oak chairs with interchangeable components. These blocky, sleigh-based chairs (reminiscent of the designs of Walter Nugent) were virtually indestructible, and many remain on duty at York University, Seneca College and the University of Toronto.

Stene sold the firm in 1986 to Robin Lauer, a graduate of the Ontario College of Art's design program. Brunswick Contract Furniture continues to make custom office furniture such as boardroom tables.[62]

STURDY, MARTHA

Born 1942, Vancouver

Vancouver's Martha Sturdy is a pre-eminent designer whose work with plastic resins has earned her an international reputation.

Sturdy was a thirty-two-year-old mother of three when she returned to art school and studied sculpture at the Emily Carr College of Art and Design in Vancouver. She created a jewellery line in 1974 (to pay for her children's daycare), and her bold oversize designs caught the eye of top fashion magazines like *Vogue* and *Harper's Bazaar*.

Sturdy moved into houseware design, and since 1987 she has operated a retail store as well as selling her semi-custom products at Barneys in New York City and Holt Renfrew in Canada.[63] The small studio manufacturer pursues artistic concerns, primarily in plastics and most recently in metal. Production staff comprises six artisans.

SUPERIOR ELECTRICS

Founded 1917, Pembroke, Ontario

Superior Electrics is one of Canada's last independent manufacturers of small appliances. Over the past fifty years, it has made products by many stellar Canadian designers including Sid Bersudsky, Lawrie McIntosh and currently Glenn Moffatt. It is the last company to produce the chrome dome kettle in North America.

Founded in 1917 by three Pembroke businessmen (within a semi-finished bankrupt hotel), the company originally specialized in electric heaters for barns. After the war, Superior Electrics produced an array of items such as electric irons, fans and compact stoves. In 1972 it was briefly owned by the Japanese consumer electronics company Magnasonic, which then sold it to its present owner, Harold Shifman.

Superior Electrics entered the electric kettle market in the late seventies, when it acquired the tools and dies from McGraw-Edison Canada, Toronto. Superior moved into plastic manufacturing in 1989, after it purchased the assets of Creative Appliance. Superior subcontracts for the American companies Sunbeam and Toastmaster, which enables it to compete with the multinationals. Currently the electric kettle is its most popular appliance and its largest production run, selling tens of thousands of units annually.

SVERIGE

1983–1992, Ste-Thérèse, Quebec

The lighting company Sverige (Swedish for "Sweden") was part of a coterie of Quebec manufacturers producing contemporary design in the eighties. Like Bazz, its line evolved from Scandinavian modernism to Italian halogen. Michel Dallaire designed its Piccolo standing lamp, which won a Design Canada award.

At its height, Sverige produced more than seventy designs, including torchère, track, floor, desk and pendant lamps. The industrial designer Michel Morelli, who had previously worked with Dallaire, served as in-house designer between 1985 and 1991 (Tom-2 torchère and Rapolla pendant lamps). The company also produced designs by André Desrosiers.

In 1986 the multinational Luxo, specializing in commercial and medical lighting, bought the company and distributed its line in the U.S. Shortly thereafter, Luxo merged with another company, closing Luxo-Sverige to avoid duplicating lines.

TEKNION

Founded 1982, Toronto

In the nineties, Teknion replaced Sunar Industries as one of Canada's largest manufacturers of office systems furniture. The company is known for its quick response to market trends and its use of good design.

Teknion, a division of the privately held Global Group, was founded to make high-end panel furniture. It went public in 1989, although Global remains a major shareholder.

Teknion plays an important role in cultivating a strong Canadian design climate. The vice-president of design, John Hellwig, a graduate of the Carleton University School of Industrial Design, Ottawa, runs the large department and commissions designs from Jonathan Crinion and Andrew Jones. Like Keilhauer, the company consults corporate branding specialist Michael Vanderbyl of San Francisco. Since 1998 Teknion has spearheaded the Blue Sky project, inviting prestigious and young designers to brainstorm on visionary ideas to inspire new thinking and develop products. Securing a multi-million-dollar contract from Boeing in Seattle became Teknion's entrée into the American market. Ability, the company's new mobile furniture system, was launched in 1997 and won three awards from the 1998 NeoCon World's Trade Fair in Chicago, including the award for most innovative. In 1998 *ID* magazine selected Teknion as one of the top forty most design-sensitive companies. The following year, revenues were over $600 million.[64]

TOASTESS

1945–2000, Pointe-Claire, Quebec

For more than fifty years, privately owned Toastess produced small appliances for the Canadian market. Originally named H & S Products for its founders

design departments in Canada (now closed). On staff were psychologists, marketing experts and engineers as well as designers.

Tyson collaborated on many breakthrough products, including the Contempra telephone (1967) and the wedge-shaped, popular, coloured SL-1 business telephone set (1975), which offered advanced features like speed dialing. In 1977 Tyson's team worked on the Imagination series, a group of novelty telephones for the residential market. The Harmony phone (1983) was the company's first fully electronic push-button phone made by computer-aided manufacturing. Considerably lighter in weight than earlier phones, it was also less expensive to produce because it required fewer parts.

Tyson oversaw the development of the first business telephone with an LCD window (1992) and the first digital payphone (1993). In addition to working at Design Interpretative, he served as vice-president of marketing in the U.S. between 1985 and 1991. After his return to Ottawa in 1991, he was vice-president of advanced technology planning, still with his design responsibility. He worked on the production of the wide-screen Vista telephone series. Tyson retired in 2000 and his department, Design Interpretive, closed the following year.[65]

UMBRA

Founded 1979, Scarborough, Ontario

Umbra, one of North America's premier housewares manufacturers, designs for casual living. Such lowly products as blinds, garbage cans and soap dishes become chic best-sellers. Recognized for its sophisticated use of colourful plastics, Umbra also manufactures in aluminum, ceramic, glass and wood.

In the seventies, the graphic designer Paul Rowan designed catalogues for his friend Les Mandelbaum, owner of THC (Trans Canada Hardware), Toronto, a manufacturer of metal cases for rock band equipment. Rowan convinced Mandelbaum that there was an untapped market for decorative paper window shades. In 1979 they founded Umbra Shades, which sold to independent stores like Urban Mode in Toronto. The firm quickly expanded sales to the U.S.

In the 1982 recession, Umbra moved into plastics by persuading underutilized automotive component manufacturing plants to produce their household products like swing-top garbage cans (Umbra's best-seller). In 1984 Umbra moved part of its operation to Buffalo, New York. Private labelling became a major revenue source for the company, and supplying coasters for Starbucks was one of its earliest contracts.

Umbra re-energized the drapery market in 1992 with cash-and-carry kits featuring linen panels and decorative hardware. In 1999 it began producing furniture, including side tables and chairs.

Paul Rowan trained as a graphic designer at Toronto's George Brown College, graduating in 1973. As vice-president of design, he runs an in-house department with the staff designers David Quan, Henry Huang and Liz Crawford. Independent designers who have shaped Umbra's

Harry and Louis Solomon, the firm initially manufactured hand mixers. In 1949 they entered the portable appliance field with an electric toaster and by 1962 were producing electric kettles, electric frying pans and waffle irons.

In the nineties, Toastess changed management and introduced computerized injection-moulded plastic machinery to manufacture translucent electric kettles. As for the toaster, the product that gave Toastess its name, models were imported from China. Despite the media attention for its candy-coloured kettles and a contract from the American housewares distributor Williams-Sonoma, Toastess closed its doors in 2000.

TROTT, WILLIAM

Born 1911, Vancouver; died 1987

An electrical engineer based in Winnipeg, William Trott excelled in designing contract lighting, setting industry standards for modular ceiling and lighting systems. He was a member of the Association of Canadian Industrial Designers and served as chairman of the NIDC between 1950 and 1952. For his lifelong work, he won a National Design Council Award for Excellence in 1967.

Trott graduated from the University of Manitoba in 1933 and worked at Canadian General Electric in Toronto and Winnipeg from 1936 to 1940. In the early fifties, he was a special lecturer on lighting for the University of Manitoba School of Architecture.

In 1940 Trott established Lighting Materials, Winnipeg, to manufacture steel and aluminum grid lighting systems for ceilings. For a brief period, he produced modern aluminum lighting for the Canadian household. Some of his designs were included in the NIDC's *Design Index*. In 1957 he founded Dawn Plastics, Winnipeg, to custom-manufacture components for commercial lighting. His most notable lighting commissions were the Monarch Life building in Winnipeg by Smith, Carter, Parkin, architects and the B.C. Electric building (1957) in Vancouver by Thompson, Berwick, Pratt architects. Both Northern Electric Company and CGE distributed his contract products. Trott retired and sold his companies in 1970.

TYSON, JOHN

Born 1942, Ottawa

As design executive at the Northern Electric Company (now Nortel Networks), John Tyson helped to turn design into a collaborative process, incorporating cutting-edge technology and the latest research in behavioural and consumption patterns.

Tyson graduated in industrial design from the Ontario College of Art in 1966 and joined Northern Electric's engineering department as the company's first in-house industrial designer. As director (and after 1985 vice-president), he shaped the small studio into Design Interpretative (also known as the Corporate Design Group), one of the largest corporate

product line include Helen Kerr, Miles Keller, Karim Rashid, Scot Laughton and Paul Epp. Mandelbaum manages the client base, which includes Pottery Barn, Bed, Bath & Beyond and Crate & Barrel, all in the U.S. The $100 million company exports 85 per cent of its products to the U.S.

VEN-REZ PRODUCTS COMPANY

Founded 1947, Shelburne, Nova Scotia

Like Walter Nugent, the cousins Archibald King and Balfour Swim built a business on the foundation of a single chair design—in their case, a school chair. The company continues to operate more than fifty years later.

After the Second World War, veterans King and Swim wanted to enter the plastics industry. Discouraged by experts, they focused on moulded plywood instead, which they were familiar with because it had appeared in Mosquito bombers. In 1947, with the help of the National Research Council of Canada in Ottawa, the pair designed veneered chair legs that could resist five tons of pressure before breaking—about the same strength as steel. The following year, they took over a former navy plant in Shelburne, Nova Scotia, and founded Ven-Rez Products (short for veneers and resins).

In addition to chairs, the company has also manufactured matching tables, desks and other furniture for the institutional and commercial markets, all shipped flat. Ven-Rez converted to steel and plastic furniture in the late sixties. A modern plant continues to serve institutional customers worldwide.[66]

VILSONS, VELTA

Born 1919, Talsi, Latvia

Velta Vilsons produced custom weaving for the interior design trade. Both craft and design institutions have celebrated her work.

Vilsons arrived in Montreal in 1952, after attending agricultural college in Sweden. She was a piece weaver for Karen Bulow for two and a half years, where she learned to work quickly and efficiently. Moving to Toronto, she founded the Handweaving Studio in 1956 with the former Bulow employee Ilza Polma. By 1960 Vilsons was working independently and renamed the studio in her own name.

Vilsons attended a summer course with the renowned textiles designer Jack Lenor Larsen in Maine in 1962. She competed with Karen Bulow on a smaller scale, specializing in fabric for upholstery and drapery for clients such as Alison Bain, J & J Brook and Leif Jacobsen. As Vilsons's reputation grew, she received commissions for screens and murals. Major projects were the Metro chairman's office in Toronto's New City Hall and screens for the offices of the law firm McCarthy and McCarthy. Toronto Hydro was her last major work, in 1974 (since refurbished). Vilsons exhibited at the first national Crafts Exhibition at the National Gallery in 1957 and won a 1967 Design Award from the National Design Council.[67]

NOTES

1 Modernism

1 Detlef Mertins, "Mountain of Lights," *Toronto Modern Architecture: 1945–1965* (Toronto: Coach House Press, 1987).
2 The most influential example was the 1948 exhibition of the Museum of Modern Art, New York, *International Competition of Low-Cost Furniture Design.*
3 Rhodri Windsor Liscombe, *The New Spirit: Modern Architecture in Vancouver, 1938–1963* (Montreal: Canadian Centre for Architecture; Vancouver: Douglas and McIntyre, 1997).
4 Unpublished Egmont Arens speech, given at Harvard University, 14 February 1950.
5 Quoted by Margaret Ness in "With Our Home Seers," *Saturday Night*, 4 April 1950.
6 Art Gallery of Ontario, *Design in the Household* exhibition, curatorial file.
7 James Ferguson and John Low-Beer, "Survey of Design Requirements and Conditions in the Canadian Furniture Industry," 1950, Appendix, 2. National Gallery of Canada, Box 7.4D *Design in Industry*, File 2, Outside Activities/Organizations. No references are available for Low-Beer or Ferguson, but a J. D. Ferguson of Spencer Supports (Canada),Rock Island, Quebec, was on the NIDC board. By all accounts, the two men sent out surveys and visited plants in a somewhat informal manner.
8 George Soulis, *Canadian Art*, Spring 1955: 125.
9 British Columbia Lumber Manufacturers' Association; the Plywood Manufacturers' Association of B.C.; and the Consolidated Red Cedar Shingle Association of B.C., all headquartered in Vancouver.
10 Allan Collier, research report on the Trend House program, *SSAC Bulletin*, n.d.: 51–54.
11 Text from the conference "How Can We Sell More Modern Furniture?"
12 Liscombe, *The New Spirit.*
13 Denise Piché, *The Governor General's Awards for Architecture 1992* (Toronto: The Royal Architectural Institute of Canada).
14 Linda Lewis, "The Great Furniture Fiasco," *ARIDO* magazine, July/August 1992, *Canadian Architect*, April 1959, Sara Bower, p.47; *Canadian Architect*, June 1965, *Canadian Interiors*, May 1966.
15 Robert Fulford, *This Was Expo* (Toronto: McClelland and Stewart, 1968); Peter Day, "The Future That Can Be Ours," *Art in Everyday Life: Observations in Contemporary Canadian Design* (Toronto: Summerhill Press/The Power Plant, 1988), 137.
16 Lydia Ferrabee, "Jacques Guillon Designers," *Canadian Interiors*, frontispiece, April 1972; Virginia Wright, *Seduced and Abandoned: Modern Furniture Designers in Canada—The First Fifty Years* (Toronto: The Art Gallery at Harbourfront), 5.

2 New Materials and Processes

1 D. Morey Taylor, "Tomorrow's Plastic Today," *Canadian Homes and Gardens*, September 1945: 13, 74.
2 Englesmith taught this industrial design course as a three-week segment, but he was also interested in bringing the architecture school closer to industry: "...it is being recognised that future departments of industrial design might very well be contained in faculties of architecture near industrial centres." University of Toronto, School of Architecture, Third Year: 1948–9, Introduction to Design Problem #2. "An Industrial Design Problem," National Archives of Canada RG 20 A4 1433 file George Englesmith.
3 Ferguson and Low-Beer observed this trend in their 1950 "Survey of Design Requirements and Conditions in the Canadian Furniture Industry."
4 For descriptions of *Design in Industry*, see Virginia Wright, *Modern Furniture in Canada: 1920–1970* (Toronto: University of Toronto Press, 1997, 94–97); Bruce Collins, "Design for Millions," M.A. thesis, Carleton University, 1987.
5 The retail display at Simpson's is described in *Canadian Plastics*, November 1948.
6 For a history of plastics, see *The Plastics Age*, Penny Sparke, ed. (London: Victoria and Albert Museum, 1990); Jeffrey L. Meikle, "Plastic, Material of a Thousand Uses," *Imagining Tomorrow*, ed. Joseph J. Corn (Cambridge, Mass.: The MIT Press, 1986); Andrea di Noto, *Art Plastic, Designed for Living* (New York: Abbeville Press, 1984).
7 *Canadian Plastics*, February 1948; *Canadian Plastics*, July–August 1954: 54; "Plastic Dream Suite," *Canadian Homes and Gardens*, November 1945: 82.
8 Donald Buchanan also comments on the problems of new plastic fabricators in *Design for Use in Canadian Products* (Ottawa: National Gallery of Canada, 1947); *Canadian Plastics*, February 1946; *Canadian Plastics*, October 1959.
9 *Canadian Plastics*, March 1953.
10 "How much better would be the consumer attitude if our merchandise were both DESIGNED and made in Canada" cites one editorial from *Canadian Plastics*, April 1946. "The designer is an investment," notes *Canadian Plastics*, May 1952. "How the Industrial Designer Can Help the Molder—and His Clients," *Canadian Plastics*, November 1966. Sid Bersudsky advertised his services in *Canadian Plastics*, December 1947 and December 1951.
11 Letter from Sid Bersudsky to G. D. Mallory, Industrial Development Division, Department of Trade and Commerce, 4 February 1955, NAC RG 20, Vol. 198, File 13-2, Vol. 5.
12 Signed photocopies of the design by Julian Rowan, dated 29 November 1960, are in the Design Exchange archives; interview with Julian Rowan, June 1996.
13 Author telephone interview with André Morin, Montreal, June 1999.
14 The patent was granted in 1948. Sid Bersudsky Archives, Design Exchange.
15 Egmont Arens cited in *Canadian Plastics*, March 1945.
16 Buchanan describes the experiment in *Design for Use*, 9. The 1945–46 (29th) Annual Report of the National Research Council mentions samples of furniture made from moulded plywood and impregnated fabric on page 44. L.A. Beaulieu produced a comprehensive report on synthetic resin adhesives in 1948, pointed out by Lynn Delgaty, National Research Council Archives (August 1999). James Donahue's annotated sketches describing the process, 15 August 1992, freelance curator Allan Collier Collection, Victoria. Wright discusses the Donahue/ Simpson chair in *Modern Furniture in Canada*, 96–101.
17 The following articles discuss plastic component furniture, "Quebec Molder Develops Hi Density Polyethylene Chair," *Canadian Plastics*, December 1960: 61; "Cost of Large Complex Products Drops Now That They Can Be Mass Produced," *Canadian Plastics*, July 1962: 36–37; Reinforced polyester vinyl bucket chairs by Canadian Seating Company appear in *Canadian Plastics*, September 1962: 76–77, 103.
18 Meikle, "Plastic, Material of a Thousand Uses," 238.
19 "The Impact of Plastic's Technology on Furniture," Furniture Seminar, October 1969, the Manitoba Design Institute, the Manitoba Department of Industry and Commerce, NAC RG20 vol. 2142, File Furniture Seminar Manitoba, Design Institution, 1969 ARC 20/2142/9.
20 *Canadian Interiors*, October 1970; author interview with Victor Prus, Senneville, July 1999.
21 Virginia Wright, "A Modern Furniture Classic: The Furniture of Thomas Lamb," *Ontario Craft*, September 1986: 14–18.
22 "They're Moulding Furniture in Quebec," *Canadian Plastics*, May 1972: 23–24; "Business Success, Jean Fournier, Treco," *Canadian Interiors*, May 1975: 13.
23 Author telephone interview with Paul Boulva, Montreal, January 2000.
24 Robert Forrest of L'image Design observed in "Industry Comments" in *Canadian Interiors*, "Due to the high cost of tooling, moulded furniture must be made in exceptionally large runs to be economical. For this reason the Canadian manufacturer will be at a disadvantage unless he is able to make inroads into the huge market to the South," September 1970: 20; Jo-Anne Thompson, "Plastics Furniture in Canada: Complacency Is a Difficult Thing to Buck," *Canadian Plastics*, April 1973: 47.
25 Andrea Codrington, "Poetry in Plastic," *ID* magazine, November 1997.

26 See Sarah Nichols, ed., *Aluminum by Design* (Pittsburgh: Carnegie Museum of Art, Carnegie Institute, 2000).
27 *Meet the Company: Welcome to Kingston Works!* (Montreal: The Herald Press Limited, 1946).
28 "Because of Its Strength, Appearance and Lightness Aluminum Finds High Spot in Furniture Industry," *Aluminum Ingot*, 16 November 1945; *Aluminum Ingot*, 13 August 1948; "Aluminum Summer Furniture," *Alcan Ingot*, 23 May 1952: 4–7; author telephone interview with William Holtzman, Montreal, August 1998.
29 Author telephone interview with B. F. Harber, Fort Erie, Ontario, August 1998; "Popular Line of Aluminium Furniture Produced by Growing Fort Erie Company," *Aluminium Ingot*, 1947.
30 David Fulton, "Ernest Orr, Straight Talker," *Industrial Canada*, September 1962: 35–36.
31 Jack Luck, "Kitchen Utensils, Aluminum, Description and Photography," Aluminum Laboratories, Montreal, 1944. Design Exchange archives.
32 Jack Luck in a lecture on industrial design to the University of Toronto, Department of Extension, 30 November 1954: 12, unpublished, Design Exchange archives.
33 Dennis P. Doordan, "Promoting Aluminum: Designers and the American Aluminum Industry," *Design History: An Anthology* (Cambridge, Mass., and London: MIT Press), 1995: 158–64.
34 For more information on Jean Raymond Manufacturing, see Monte Kwinter, "Design for Production with Aluminum," *Product Design and Materials*, April 1956.
35 "It's Not Done with Mirrors...It's Aluminum," *Canadian Homes and Gardens*, April 1949: 21–32; *Canadian Homes and Gardens*, June 1949: 42–4.
36 Advertisement promoting aluminum curtain walls, *Canadian Architect*, June 1956; Mertens, *Toronto Modern Architecture*, 18.
37 *Alcan News*, September 1961; Lydia Ferrabee, "Alcan Offices Unique Design in Aluminum," *Canadian Interiors*, November 1964: 20–26.
38 Author telephone interview with Gord Hoselton, Colborne, Ontario, January 1999.
39 Paola Antonelli, *Mutant Materials in Contemporary Design*, (New York: The Museum of Modern Art, 1995), 116.
40 Author telephone interview with Mark Müller, Toronto, October 1999.
41 Wright, *Modern Furniture in Canada: 1920–1970*, 4.
42 "Furniture Makers Work Together to Organize Aircraft Components Job," *Canadian Aviation*, 15 October 1942: 84; John Collins, "Turning Bombers into Lounge Chairs," *Material History Bulletin*, Spring 1988.
43 Ferguson and Low-Beer, "Survey of Design Requirements...," Appendix, 2.
44 Donald Buchanan in a letter to George Englesmith, 21 February 1949, stated that "this is the kind of chair we had demonstrated in the kitchen of the Design Centre." NAC RG 20 A4 Vol. 1433 file George Englesmith architect.
45 "Curved Plywood with Hi-Frequency Glue-Line Curing," *Canadian Woodworker*, March 1995: 34–35, 62, 66; Murray Waghorne, "The Future for Moulded Plywood," *Canadian Woodworker*, March 1956: 31, 62.
46 *Canadian Woodworker*, March 1957; *Canadian Woodworker*, July 1957.
47 Author interview with Ted Samuel, Toronto, the founder and president of Curvply, November 1998; Joe Reddy, "Diversity of Products Key to Increased Production and Expansion at Curvply," *CWPI Furniture Production*, February 1966: 29–31.
48 Antonelli, *Mutant Materials in Contemporary Design*, 96.

3 Craft, Design and Industry

1 For a history of crafts in Canada, see Sandra Flood, "Canadian Craft and Museum Practice 1900–1950," Ph.D. thesis, University of Manchester, 1998; Gail Crawford, *A Fine Line: Studio Crafts in Ontario from 1930 to the Present* (Toronto: Dundurn Press, 1998).
2 Gloria Lesser, *École du Meuble 1930–50* (Montreal: Le Château Dufresne, Musée des Beaux-Arts, 1989).
3 Ian McKay, "Mary Black and the Invention of Handicrafts," *The Quest of the Folk: Antimodernism and Cultural Selection in Twentieth-Century Nova Scotia* (Montreal/Buffalo: McGill-Queen's University Press, 1994).
4 *Design in Industry* invitation/flyer, ROM Library RG 10 box file 5.
5 Donald Buchanan, "Design in Industry—A Misnomer," *Canadian Art*, summer 1945.
6 *Design in Industry* Internal Memo. ROM Library RG 10 box file 5.
7 Author interview with Michael Fortune, Toronto, July 1999; Donald McKinley, "The School of Crafts and Design," in James Strecker, *Sheridan: The Cutting Edge in Crafts* (Erin, Ont.: Boston Mills Press, 1999), 66–67.
8 Author telephone interview with Patty Johnson, Toronto, January 2000.
9 Adele Weder, "Making It," *Azure*, September/October 1997.
10 Author telephone interview with Geoffrey Lilge, Edmonton, January 2000.

4 Canadian Design in the Pop Era

1 For a discussion on Pop design, see Nigel Whiteley, *Pop Design—Modernism to Mod* (London: Design Council, 1987); Lesley Jackson, *The Sixties* (London: Phaidon, 1999).
2 The Unhouse was published in *Art in America*, April 1964, *Domus* 1966, and was included as well in Charles Jencks, *Architecture 2000* (New York: Praeger Publishers, 1971).
3 Much of the information presented here is from the exhibition *Pop in Orbit: Design from the Space Age*, Design Exchange, 7 September 1995 to 26 February 1996.
4 "Turned on Town-House," *Chatelaine*, September 1969.
5 Big Steel, Fairview Mall, *Canadian Interiors*, January 1971.
6 Author interview with Jerome Markson, Toronto, August 1999; *Canadian Interiors*, February 1969: 35.
7 Author conversation with Alison Hymas, Toronto, July 1999.
8 Paula McCullough, "Architecture Meets Psychedelia," *Azure*, August/September 1991: 12–13; David Culsiau/Marco Polo, "Uno Prii," *The Newsletter of the Toronto Society of Architects*, Fall 1991: 21–41.
9 *Canadian Interiors*, August 1971.
10 Author telephone interviews with Gary Smith and Steve Smith, October 1999.
11 *Canadian Interiors*, January 1968; *Canadian Interiors*, September 1970; "The 1970 Look," *Canadian Home*, January 1970: 2–5.
12 Sidney Gibson, "Plastics and Design," *CWPI Furniture Wood Products*, November/December 1969: 22, 26–27.
13 Reyner Banham, *Design by Choice* (London: Academy Editions, 1981).
14 "Ottawa Closes Design Centres," *Canadian Interiors*, February 1970: 37–42; author conversation with Jan Kuypers, 1991; Sigrun Bülow-Hübe Archives, Canadian Architecture Collection, McGill University, Montreal.
15 Letter to Aubrey Pugsley, Business Manager, Dominion Chair Company, December 1975, NAC RG 20 Vol. 2132, file Nova Scotia Products, Development Cases 1975–79.
16 Author interview with Douglas Ball, Montreal, July 1998.

5 From Postmodernism to Pluralism

1 Gazelle Chair 10th anniversary exhibit, Design Exchange, 15 April to 28 September 1997.
2 See Harold Kalman, *A History of Canadian Architecture* Vol. 2 (Toronto: Oxford University Press, 1994).
3 Virtu 3 catalogue, 1987.
4 "Federal Government to Encourage Private Sector in Design Field," Regional Industrial Expansion news release, Government of Canada, May 1985, Design Exchange.
5 Piché, *The Governor General's Awards for Architecture 1992*, 59.
6 Author interview with Tom Deacon, Toronto, March 1999.
7 Author telephone interview with David Burry, Montreal, July 1999.
8 Author telephone interview with Karim Rashid, New York City, May 1999. Ron Arad discusses the importance of individuality in design in *International Design Studio 9*, Jeremy Myerson, ed., 1994.
9 Author telephone interview with John Hellwig, Toronto, July 2000.
10 Design Exchange staff interview with Umbra, Scarborough, March 2000.
11 Author interview with Mike Keilhauer, Scarborough, May 2000.
12 Author conversation with Klaus Nienkämper, Toronto, October 1999.
13 Jean Nouvel, "Design for the Present," *International Design Yearbook 10*, Paul Jodard, ed., 1995.

6 Furniture

1 Load-Test Report, Guillon Chair, 6 March 1953, Milton Hersey Company, courtesy of Jacques Guillon.
2 Author interview with Jacques Guillon, Montreal, 7 July 1998.
3 A. C. Parks, March 1959, "Furniture Manufacturing in the Atlantic Provinces," Atlantic Provinces Council, Halifax.
4 Letter from Jan Kuypers to Joy Parr, professor, Department of History, Innis College, University of Toronto, 15 January 1997, Kuypers file, Design Exchange.
5 John Collins, "Turning Bombers into Lounge Chairs," *Material History Bulletin*, spring 1988.
6 Raymond Stanton, *Visionary Thinking: The Story of Canada's Electrohome* (Kitchener: Canadian Corporate Histories, 1997), 55–57.
7 Series of Eaton's interdepartmental memos, December 1946 to April 1948, Eaton's Archives, binder Swedish Design Furniture (Magnus Werner).
8 Robert Fones, *A Spanner in the Works: The Furniture of Russell Spanner 1950–1953* (Toronto: The Power Plant, 1990); personal archives of Robert Fones, Toronto.
9 Wright, *Modern Furniture in Canada: 1920–1970*.
10 Author interview with Harvey Meighan, former sales manager, Leif Jacobsen, Toronto, 27 July 1999.
11 Author interview with Janis Kravis, former owner, Karelia International, Toronto, 10 November 1999.
12 Author interview with Robin Lauer, president of Brunswick Contract Furniture, Toronto, 4 February 1999.
13 Fones, *A Spanner in the Works*; Fones's personal archives.
14 Fones, *A Spanner in the Works*.
15 Author interview with James Snyder and Jamie Snyder, grandson and great-grandson of the founder of Snyder's, Waterloo, Ontario, January 1999. Jamie Snyder is president of James C. Snyder, a continuation of the firm.
16 Peter Cotton Collection, British Columbia Archives, Victoria.
17 Wright, *Modern Furniture in Canada*, 98
18 Author telephone interviews with Wayne King (Archibald King's son) with notes and recollections from Joan King (Archibald King's widow) and Wallace Buchanan (Ven-Rez Products), Shelburne, Nova Scotia, July to November 1999.
19 Author telephone interview with former Donahue student Grant Marshall, professor emeritus, Interior Design Department, University of Manitoba, Winnipeg, August 1997.
20 Author interview with Lawrie McIntosh, Toronto, January 1994; "Second Product Design Competition Report" submitted by the judges, Box 74D Outside Activities/Organizations, *Design in Industry* file no. 4, National Gallery of Canada Library.
21 Author interview with Robert Kaiser, Toronto, 13 December 1999.
22 Imperial furniture brochure, n.d., ACIDO archives, Jan Kuypers file.
23 Author correspondence with Court Noxon, Bloomfield, Ontario, 23 February 2000.
24 Sigrun Bülow-Hübe archive, Canadian Architecture Collection, McGill University, Montreal.
25 Author interview with John Stene, Toronto, 30 October 1999.
26 Author interview with Robin Lauer, Toronto, 4 February 1999.
27 Norman Hay, "Robin Bush," *Canadian Art*, May 1959: 120.
28 Author interview with Walter Nugent, Oakville, Ontario, 11 June 1998.
29 Lydia Ferrabee, "Toronto Airport Interior Design," *Canadian Architect*, February 1964: 63–67; M. Michaelson, "Focus on Canadian Design," Office Administration, August 1966: 8–10.
30 Stefan Siwinski file, Design Exchange, gift of Stefan Siwinski.
31 Minutes of Executive Meeting, 26 October 1962, COSF, The Percy R. Hilborn Papers, Cambridge Public Archives, Cambridge, Ontario; Ferrabee, "Toronto Airport Interior Design."
32 Author interview with Jack Dixon, Port Perry, Ontario, 13 May 1999.
33 Author interview with Svend Nielsen, former vice-president, Leif Jacobsen, Toronto, 20 December 1999.
34 Author interview with Christen Sorensen, Toronto, 7 October 1999.
35 The 1+1 series appears in *Canadian Interiors*, September 1970.
36 Author interview with Jacques Guillon, Montreal, 7 July 1998.
37 Author correspondence with Court Noxon, 23 February 2000.
38 Author interview with Donald Lapp, Kitchener, January 1999.
39 Author telephone interview with James Murray, Ayr, Ontario, 26 September 2000.
40 Author interview with Douglas Ball, Montreal, July 1998.
41 Author interview with Jerry Adamson, Toronto, 12 May 1998; Barbara Plumb, "Inside Habitat," *The New York Times Magazine*, 30 July 1967; Ann Barr, "A Lesson from Habitat," *House and Garden*, July/August 1967: 22–23; "Exciting Uses of Plastic," *Canadian Interiors*, November 1967: 47–48.
42 Author telephone interview with Michael Stewart, May 1999.
43 Memo to Dick Hilborn from Robin Bush, ROS Manufacturing and Marketing program, 22 August 1962, Percy R. Hilborn Papers.
44 *Canadian Interiors*, July 1965: 51; Robin Bush Archive, Design Exchange.
45 Author interview and correspondence with Anna Chester (Hugh Spencer's former wife), Battersea, Ontario, December 1999 to October 2000.
46 Letter to Walter Nugent from Donald E. Hewson, Dennison Associates, 20 November 1972; letter to Walter Nuggent (*sic*) from George Vondrejs, 18 September 1975, Design Exchange Archives.
47 Author interview with Bob Forrest, Toronto, 26 May and 1 June 2000.
48 Author interview with Philip Salmon, Toronto, 1999.
49 Author interview with Bill Lishman, Blackstock, Ontario, 21 December 1999.
50 Author interview with Linda Lewis of Ryerson Polytechnic University, Toronto, September 2000, who is cataloguing Thomas Lamb's archives.
51 Author conversation with Thomas Lamb, September 1991.
52 The MOMA accepted Neil Bendtsen's Ribbon chair in its study collection several years before Lamb's design but considers it Danish rather than Canadian as Bendtsen was living in Denmark when he developed the design and is a citizen of Denmark.
53 Vello Hubel file, Design Exchange, gift of Vello Hubel.
54 Gazelle chair, 10th Anniversary Exhibit, Design Exchange.
55 Author telephone interview with Mark Müller, Toronto, March 2000.
56 Author interview with Tom Deacon, Toronto, March 1999; author interview with Mike Keilhauer, Scarborough, May 2000.

7 Lighting

1 W. A. Trott, "1953 Design Award NIDC Canada, Introducing the Winners," *Industrial Canada*, April 1953: 53, 56.
2 For a discussion of twentieth-century lighting design, see Jeremy Myerson and Sylvia Katz, *Conran Design Guides Lamps and Lighting* (London: Conran Octopus, 1990).
3 Buchanan, *Design for Use in Canadian Products*, 21; "Top Lighting—a New Trend in Building," *Canadian Plastics*, December 1959: 28–29; W. E. Harper, "Lighting Fixtures—and Acrylic Future," *Canadian Plastics*, January 1950: 12, 41.
4 Modulite advertised in *Canadian Architect*, May 1956: 61; Canadian Architect, July 1956: 55; author telephone interview with Norman Slater, August 1998.
5 For information on J. A. Wilson Lighting, see Monte Kwinter, "Design for World Markets," *Product Design and Materials*, June 1956: 8; D. C. McCormack, "Industrial Design for the Small Manufacturer," *Industrial Canada*, September 1957: 59, 61.
6 Frank Reed discusses his lighting designs in "Industrial Design Conference," *Industrial Canada*, July 1952: 269–93.
7 Kurt Versen "Originals" advertised by Simpson's in *Canadian Homes and Gardens*, April 1951: 5.
8 John Merriman, "Crown Electric Was First of Its Kind," *The Brantford News*, 13 February 1980; Douglas Reville, *History of the County of Brant* Vol. 2, (Brantford, Ont.: The Hurley Printing Company, 1920).
9 Bruce Douglas, Medalta Lamps, 1996. Self-published.
10 L'Atelier L'Argile Vivante/Céramique de Beauce catalogue, 1963, private collection of Daniel Cogné.
11 "Perception...Contemporary Lamp Group" by Deilcraft, n.d., catalogue courtesy of Electrohome.
12 Author conversation with Tony Caruso, former president of Sverige, Montreal, October 2000, and Michel Morelli, Montreal, October 1999.
13 Kate McIntyre, *Desk Lamp* (London: Aurum Press, 1998), 29.
14 Author interview with Marten Bostlund, Oak Ridges, Ontario, April 1998; "Lotte Stoneware Lamps Fiberglass Shades" catalogue (Bostlund Industries, n.d.); Cash Mahaffy, "Live with Lamps," *Canadian Homes*, January 1966: 10, 13; Ray Magladry, "From Bostlund's Barns to the World Markets," *Toronto Daily Star*, 3 April 1963; Erid Haworth, "Filament Winding Creates Quality Housewares Line," *Progressive Plastics*, January 1963: 25.
15 Author conversation with Francesco Piccaluga, June 2000; The Opus Architectural Lighting Collection, Sonnenberg Industries catalogue, n.d.

8 Textiles

1 Lesser, *École du Meuble*, 75–78; Joyce Bainard, "We Explore Canadian Painting," *Canadian Homes and Gardens*, April 1950: 39–40; Memo to Mr. B. E. Mercer H. F. Merchandise Office, Re Canadart Fabrics, from F. J. Carter, 4 June 1950. Carter defended Eaton's lack of involvement in the project: "The Canadart group has a very limited use…they would require a promotion of such size to show even minor results.…The writer feels extremely dubious as to whether these patterns as a group would have been selected by us had they been offered in the regular way. The type of design is well covered by our present stock of both screen and roller prints." Ontario Archives, F229 Series 69 Box 2 File 267, Draperies.
2 A. Guerin, Canadian Textiles Industry, 16 July 1964, NAC RG 20 Volume 2057, File 6-1335-01.
3 Donald Rosser, 1948 NIDC report, segments reprinted in "Who Designs Canadian Textiles", *Canadian Art*, December/January 1949–50: 50–53.
4 NGC 74-D Outside Activities/Organizations, *Design in Industry*, File 7.
5 Nova Scotia Design Institute, PDDP Management Committee, P. A. Brett RE Designcraft, April 6, 1976. RG 20 2132, File Nova Scotia Products, NAC.
6 Crawford, *A Fine Line*, 35.
7 Ibid., 63–65.
8 Allan Collier, "Modernism at Home, 9. Modern Design in B.C. 1945–6," Charles H. Scott Gallery, Emily Carr College of Art and Design, Vancouver, 1995.
9 Although Borduas used the drapery in his own home, he later claimed, "I have never made designs for fabrics, nor for anything at all with any practical or utilitarian application and I hope I never shall." François-Marc Gagnon, *Paul-Émile Borduas* (Montreal: Musée des Beaux-Arts, 1988), 178.
10 Mayfair, February 1951; J. H. Thompson, *Canadian Textiles* (Leigh-on-Sea: F. Lewis Publishers, 1963); Thor Hansen, *Canadian Crafts in Industry* (Toronto: British American Oil Company, n.d.); *B-A Canadiana Arts & Crafts in Industry* (Toronto: British American Oil Company, n.d.); Huronia Museum vertical files; "Celebrating Thor Hansen," *Craftnews*, November 1989.
11 Pearl McCarthy, "Hand-Printed Textiles Made in Barn at Bronte," *The Globe and Mail*, Toronto, 8 November 1947; Jo Carson, "Fabrics Become Family Industry," *The Globe and Mail*, Toronto, 29 July 1961.
12 Sarah Baker, Design Development, Knoll Textiles, letter to Design Exchange, July 1999.

9 Consumer Electronics

1 Author interview and correspondence with Bill McGregor, senior vice-president, Electrohome, Kitchener, January 1999 to October 2000.
2 Author correspondence with Bryan Dewalt, curator, communications, National Museum of Science and Technology, Ottawa, 23 May 2000.
3 Ian Anthony, historian, Rogers Communications, Toronto. Anthony compiled the internal company document, "Complete Retrospectus of Rogers Telecommunications Limited: Spanning the Years 1911–1939 and 1959–1999."
4 Annual service "work sheets" circulated by the Radio College of Canada, Toronto.
5 Author telephone interview with Lloyd Swackhammer, former president and current member of the London Vintage Radio Club, London, Ontario, 22 October 1999. Swackhammer is preparing a list of radio manufacturers in Canada.
6 Author interview with André Morin, former industrial designer, RCA Victor Company, Montreal, 28 October 1999.
7 Eaton's archives, Archives of Ontario, Toronto.
8 Author telephone interview with André Morin, 28 October 1999.
9 *Product Design and Materials*, April 1956, Summary of Royal Commission on Canada's Economic Prospects, Economic Effects on the Electronics Industry.
10 Author interview with Mike Batch, owner, Vintage Radio and Gramophone, Toronto, 14 October 1999.
11 Interview with a former chief designer, consumer electronics manufacturer, Toronto, 5 November 1999.
12 *Canadian Homes and Gardens*, January 1951.
13 Author telephone interview with André Morin, Montreal, 28 October 1999.
14 Peter Day and Linda Lewis, *Art in Everyday Life: Observations on Contemporary Canadian Designs* (Toronto: Summerhill Press/The Power Plant, 1998).
15 Jeffrey Liss and Sandra Shaul, eds., *Watching TV: Historic Televisions and Memorabilia from the MZTV Museum* (Toronto: Royal Ontario Museum and MZTV Museum, 1995).
16 *Product Design and Materials*, April 1956, Summary of a Royal Commission on Canada's Economic Prospects, Economic Effects on the Electronics Industry.
17 John Sideli, *Classic Plastic Radios of the 1930s and 1940s, A Collector's Guide to Catalin Models*. (New York: E. P. Dutton, 1990); author interview with an Addison radio collector, Kim Stephenson, October 1998.
18 Author interview with George Found, Grandpa's Radios, Kitchener, 6 November 1999.
19 Author telephone interview with André Morin, Montreal, 28 October 1999.
20 Author interview and correspondence with Anna Chester (former wife of Hugh Spencer), Battersea, Ontario, December 1999 to October 2000.
21 Author telephone interview with Gordon Duern, August 1993.
22 Author correspondence with Keith McQuarrie, 2 May 1995, and author interview with McQuarrie, Waterloo, 8 November 1999.
23 Author interview and correspondence with Bill McGregor.
24 Ibid.

10 Ceramics

1 Donald Webster, *Early Canadian Pottery* (Toronto: McClelland and Stewart, 1971); Elizabeth Collard, *Nineteenth-Century Porcelain in Canada* (Montreal: McGill University Press, 1967).
2 Department of the Interior, National Development Bureau, "The Ceramic Industry of Canada: with special reference to the manufacture of white ware." (National Development Bureau, Department of the Interior, Ottawa, 1931.)
3 In *Design for Use*, Buchanan praised Beauceware's utility items, such as cereal bowls, cups and beer mugs; Daniel Cogné, "Du Beauce sur nos tables," *Cap-aux-Diamants* 44, hiver 1996: 40–43.
4 Ruth Home, "Pottery in Canada," *Canadian Geographic Journal*, February 1944: 71.
5 In 1926 Bertram Watkin was a modeller at Onondaga Pottery (later known as Syracuse China Company) in Syracuse, New York. See Paul Atterbury, *Miller's Twentieth-Century Ceramics* (London: Octopus Publishing Group, 1999), 147. In the thirties, the peripatetic Watkin worked at Ecanada Art Pottery, Hamilton.
6 Anne Hayward, *The Alberta Pottery Industry, 1912–1990* (Hull: Canadian Museum of Civilization, forthcoming publication). *The Clay Products News and Ceramic Record*, August 1947, noted that "export markets have been open and waiting for Canadian pottery ware but to date the Medicine Hat Potteries have been unable to produce any surplus goods to enable them to enter this field."
7 Fraser Robertson, "Ontario Potter Sends Teapots to Newcastle," *The Clay Products News and Ceramic Record*, August 1964: 14; "Laurentian Art Pottery," *The Clay Products News and Ceramic Record*, June 1963: 20.
8 For an overview on Quebec pottery, see Paul Bourassa, *Trajectoires: La Céramique au Québec des Années 1930 à Nos Jours* (Quebec City: Musée du Québec, 1999); L'Atelier L'Argile Vivante/Céramique de Beauce catalogue, 1963, private collection of Daniel Cogné.
9 For an overview of Canadian studio pottery, see Home, "Pottery in Canada," 65–77; Evelyn Charles, "A History of the Guild," *Tactile*, August 1976: 3–6; Judith Ross Thompson, *Down to Earth Canadian Potters at Work* (Toronto: Nelson Canada, 1981).
10 Crawford, *A Fine Line*, 42.
11 Stephen Inglis, *The Turning Point: The Deichmann Pottery, 1935–63* (Hull: Canadian Museum of Civilization, 1991).
12 "Lambert Potteries," *The Clay Products News and Ceramic Record*, September 1951: 7–9.
13 Nancy Townsend, *The History of Ceramics in Alberta* (Edmonton: Edmonton Art Gallery/Alberta College of Art, 1975).
14 Sue Jefferies, "A Legacy of Commitment," *Ontario Craft*, Summer 1987: 19–22.
15 "A Visit with Two Talented Potters," *Canadian Homes*, September 1961; *Form and Fantasy: Ceramics by Theo and Susan Harlander* (Whitby, Ont.: The Station Gallery, 1995).

11 Glass and Miscellany

1 Lynda Parker, "Depression Glass," *Canadian Collector*, January/February 1987: 34–37; with notes from glass collector Walter Lemiski.
2 For an overview on Canadian glass, see Thomas King, *Glass in Canada* (Erin, Ont.: Boston Mills Press, 1987); Ian Warner, "Canadian Swirl," *Canadian Depression Glass Association Review*, August/September 1998: 9–12.
3 King, *Glass in Canada*, 192; Sandra Handler, "Tableware Manufactured by Corning Glassworks of Canada," *The Daze Past*, August 1975.
4 Sandra Handler, "Cornflower a Canadian Original," *The Daze Past*, June 1975; Walter Lemiski, "Hughes Corn Flower: An Elegant Canadian Tradition," *Canadian Depression Glass Association Review*, June/July 2000.
5 Mary Coward, *Altaglass, A Guide for Collectors* (Edmonton: self-published, 1999); King, *Glass in Canada*, 229.
6 King, *Glass in Canada*; author interview with Toan Klein, Toronto, August 2000.
7 "Murano Glass Ornamental Glass," *Canadian Jeweller*, April 1962: 18–19; "Ornamental Glassware at Cornwall, Ontario," *The Clay Products News and Ceramic Record*, September 1963: 12–14; King, Glass in Canada, 230.
8 Rosalyn Morrison, *Canadian Glassworks*, 1970–1990 (Toronto: Ontario Crafts Council, 1990).
9 "The Tableware Picture," *Canadian Plastics*, September 1950: 48; "Melamine Dinnerware in Canada," *Canadian Plastics*, April 1955: 56–57, 70–72; *Industrial Canada*, July 1953.
10 Pointed out to authors by the Toronto Melmac collector Lloyd Gray, who is preparing *The Collector's Guide to Canadian Melmac*; see also Donald Emmerson, *Canadian Inventors and Innovators: Pioneering in Plastics 1885–1950* (Toronto: Canadian Plastics Pioneers, 1978).
11 Author conversation with John Tyson, Ottawa, September 2000; "Telephones Use Your Imagination," *Telesis*, June 1977: 95; David Robertson, Terry Thomas and Peter Trussler, "User Value Drives BNR Telephone Design," *Telesis*, July 1991: 111–14; Nortel Networks transcripts interview between Gerald Levitch and Ian Craig, April 2000; Nortel Networks transcripts interview between Gerald Levitch and John Tyson.
12 Author conversation with Linda Lewis, Ryerson Polytechnic University, April and May 2000; invoice for three days' work on covered casseroles from Ruth Gowdy McKinley to Thomas Lamb, 12–14 April 1979, Thomas Lamb archives.
13 Author conversation with Marcel Girard, Montreal, June 2000.
14 Author telephone interview with Michael Santella, Montreal, May 1999.

12 Small Appliances

1 Canadian General Electric brochure, n.d., Simcoe County Archives; described by Fred Moffatt, May 1999.
2 Franz Klingender, "To Lighten the Burden of Womenkind [*sic*], the Mechanization of Domestic Equipment (1890–1960)," unpublished paper, National Museum of Science and Technology, Ottawa, 1994.
3 For an overview on small appliances, see Penny Sparke, *Electrical Appliances* (London: Unwin Hyman, 1987).
4 Joy Parr, *Domestic Goods: The Material, the Moral and the Economic in the Postwar Years* (Toronto: University of Toronto Press, 1999), 76–77.
5 Author interview with Fred Moffatt, Toronto, March 1990; Archibald Johnston, "Canadian General Electric's First Hundred Years" (Canadian General Electric Company brochure, n.d.).
6 Inventor's bonus, internal memo, June 1945, from L. C. Prittie to K. J. Cokery, Simcoe County Archives, Barrie.
7 W. F. McMillan, "The Electrical Manufacturing Industry in Canada," *Industrial Canada*, February 1948: 57; author interview with Fred Moffatt, March 1990.
8 General Tour Notes, General Steel Wares, 19 November 1965, prepared by P. Fredenburgh, NAC RG20 Vol. 2070, p. 8001-270/G47.
9 Author interview with Toastess staff, July 1999.
10 Internal memo, "Live Better Electrically Program," April 1958, F5701, Ontario Hydro Corporate Archives, File 5701.
11 Office of Design, Department of Industry, Trade and Commerce and the National Design Council, Proctor-Lewyt, McGraw-Edison of Canada in "A Series of Canadian Product Case Studies Showing the Benefits of Effective Design Management" (unpublished report, April 1973: 66–71), Design Exchange archives.
12 An internal CGE memo explains the new competition and the need for "shorter lead times and national distribution." Memo, some comments on Housewares Business Environment as related to manufacturing function, 1968, Simcoe County Archives, Barrie. Discussing Canada's deteriorating competitive position in the small appliance industry are the government reports "Electric Products Sector Task Force," 1975, NAC, RG 20 Vol. 2147 File Arc-20/2147/9; "Summary of the Industry Sector Studies for the Innovation Committee Design Canada," December 1977, NAC, RG 20 Vol. 2147 File ARC-20/2147/9.
13 "Kettles Every Day Icons," *Design*, April 1993: 41; John Heskett, *Industrial Design* (New York: Oxford University Press, 1980), 70–71.
14 Peter Dormer, *Design Since 1945* (New York: Thames and Hudson, 1993), 47–48.
15 Author interview with Glenn Moffatt, Toronto, September 1996.
16 Author telephone interview with Jerry Adamson, Toronto, May 1998.

13 Metal Arts

1 Author interview with Callie Stacey (Harold Stacey's daughter), 20 January 2000.
2 Francis Harris, "The Old Silver of Nova Scotia," *Mayfair*, September 1949: 66.
3 René Villeneuve, assistant curator, early Canadian art, National Gallery of Canada, Ottawa, *Vernissage*, Autumn 1999.
4 Robin Patterson, modern history division, British Columbia Provincial Museum, Victoria, notes to a Carmichael exhibition at the museum, February 1980.
5 Author interview with Callie Stacey, 20 January 2000.
6 Joan Fussell (Andrew Fussell's daughter) in "Andrew Fussell: Canadian Silversmith," *The Silver Society of Canada Newsletter*, Autumn 1999.
7 Notes to a donation of Carl Poul Petersen silverware by Conrad Graham, curator of decorative arts, McCord Museum of Canadian History, Montreal, published in *Fontanus IX*, 1996.
8 Author interview with Anne Barros, Toronto, May 1997.
9 Author interview with Helen Kerr, Toronto, July 1999.

Biographies and Corporate Histories

1 Author interview with Mike Batch, Vintage Radio and Gramaphone, Toronto, 14 October 1999; interview with an Addison radio collector, Kim Stephenson, Oakville, Ontario, October 1998.
2 Author telephone interview with Michael Stewart, Toronto, May 1999.
3 Author interview with Douglas Ball, Montreal, July 1998.
4 Author interview with Anne Barros, Toronto, May 1997.
5 B.C. Ceramics undated catalogue from the collection of Allan Collier; see also Allan Collier, "Modernism at Home: Modern Design in B.C. 1945–60," exhibition pamphlet (Vancouver: Charles H. Scott Gallery, Emily Carr College of Art and Design, 1995). Dates for the company's time in business were kindly provided by Alan Elder, who interviewed Herta Gerz, Vancouver, August 2000.
6 Anne Barros, *Ornament and Object: Canadian Jewellery and Metal Art, 1946–1996* (Erin, Ont.: Boston Mills Press, 1997).
7 Crawford, *A Fine Line*.
8 Bob Collins, "Award-Winning Fabric Designers," *Canadian Homes and Gardens*, May 1954: 82–84; "J & J Brook," *Canadian Interiors*, November 1965: 43–47; author interview with Joanne Brook, Toronto, May 1999.
9 Gloria Lesser, "Karen Bulow: Masterweaver," *Canadian Society of Decorative Arts Bulletin*, winter 1990: 8–9; "Seven Canadian Designers for Living," *Canadian Homes and Gardens*, September 1958: 20; Walter Rendell Storey, "Fabrics for Diverse Purposes Come from Karen Bulow's Looms," *Handweaver and Craftsman*, Winter 1953–54: 14–17, 53; notes from Pat Harris file, October 1993, Department of Textiles, Royal Ontario Museum; "Karen Bulow," *Canadian Interiors*, August 1975: 14–16; author interview with Edward Steeves, Toronto, June 1999.
10 Sigrun Bülow-Hübe archive, Canadian Architecture Collection, McGill University, Montreal.
11 The Percy R. Hilborn Papers, Canadian School and Office Furniture, Cambridge Public Archives, Cambridge, Ontario.

12 Author telephone interview with Tom Hanson, director of Canadian operations for Herman Miller, May 1998.
13 With notes from Renald Fortier, curator, National Aviation Museum, Ottawa, 18 February 1999; author interview with Kasia Seydegart, granddaughter of Hilary Stykolt, October 2000.
14 Author interview and correspondence with Anna Chester, Battersea, Ontario, December 1999 to October 2000; Clairtone Sound Corporation archives, Design Exchange, Toronto.
15 Jonathan Crinion file, Design Exchange, gift of Jonathan Crinion.
16 Author interview with Michel Dallaire, 9 July 1998.
17 The Museum of Civilization archives, Hull, Quebec, holds correspondence between Kjeld Deichmann and Bernard Leach.
18 *Canadian Woodworker*, December 1947; *Canadian Plastics*, February 1948.
19 Sandra Handler, "The Dominion Glass Company of Canada," *The Daze Past*, December 1975: 4; Dominion Glass Company Ltd., *Industrial Canada*, May 1967: 231–32; Linda Jacobs, "Libbey-St. Clair," *Canadian Clay & Ceramic*, July/August 1979: 10–12; King, *Glass in Canada*, 196.
20 From letters and drawings submitted to Allan Collier, Victoria, from James Donahue, 15 August 1992, Chester, Nova Scotia. (Collier writes extensively on Western Canadian modern furniture and has curated numerous exhibitions); Wright, *Modern Furniture in Canada*, 98.
21 Peter Kaellgren, "Ecanada Art Pottery," *Canadian Collector*, March/April 1979: 56–63; "Ecanada Art Pottery Produces Ovenware Line," *The Clay Products News and Ceramic Record*, January 1952; "George Emery Canadian Society President Tells Life Story," *The Clay Products News and Ceramic Record*, February 1950: 13–17.
22 Author interview and correspondence with Bill McGregor, Electrohome, January 1999 to October 2000; *CWPI/Furniture Production*, October 1967: 31; *Visionary Thinking: The Story of Electrohome* (Kitchener; Canadian Corporate Histories).
23 Linda Lewis, "Design Is My Life, Al Faux 1931–1978," *ARIDO* (Association of Registered Interior Designers of Ontario) magazine, 1990.
24 Henry Finkel Archives, Musée du Québec; Henry Finkel, "ACID, Association of Canadian Industrial Designers," *Canadian Art*, February 1959: 36–37, 39; David Fulton, "Henry Finkel: Total Solutions," *Industrial Canada*, January 1963: 33–34.
25 Joan Fussell, "Andrew Fussell: Canadian Silversmith," *The Silver Society of Canada Newsletter*, Autumn 1999.
26 Author interview with Jeff Goodman, Toronto, October 1999.
27 Author interview with Jacques Guillon, Montreal, 7 July 1998.
28 Letter from Siegmund-Werner to Julien Hébert, 4 April 1951, Julien Hébert archives, Musée du Québec; David Fulton, "Julien Hébert, 100 Ways to Fold a Chair," *Industrial Canada*, August 1962: 25; additional notes from Paul Bourassa, curator, decorative arts, Musée du Québec, Quebec City.
29 Author interview with Ivan Lacroix, president, Baronet Corporation, Ste-Marie, Quebec, 3 January 2000.
30 Author interview with Robert Kaiser, Toronto, 13 December 1999.
31 Author interview with Jerry Adamson, Toronto, 12 May 1998.
32 Author interview with Mike Keilhauer, Scarborough, May 2000.
33 Author interview with Helen Kerr, Toronto, 21 July 1999.
34 Autobiographical notes of Jan Kuypers, 29 May 1995.
35 Wright, *Ontario Craft*, September 1986; Linda Lewis, Ryerson Polytechnic University, Toronto, September 2000; interviews with Thomas Lamb.
36 *Never Leave Well Enough Alone*, Gallery 76, Ontario College of Art, 22 March–2 April 1988.
37 Author interview with Svend Nielsen, Toronto, 19 December 1999.
38 Author interviews with Bob Forrest, Toronto, 26 May and 1 June 2000.
39 Author interview with Bill Lishman, Blackstock, Ontario, 21 December 1999.
40 Anne Hayward, "Medalta's Art Department: A Strategy for Product Diversification," *Material History Review* 39, Spring 1994: 20; Ronald M. Getty, *Know Your Medalta* (Medicine Hat: ICM Press, 1995).
41 Ronald M. Getty, *The Kilns of South Eastern Alberta* (Medicine Hat: ICM Press, 1994); Ronald Getty, "Collecting Hycroft China: A Window to History" (self-published, n.d.).
42 Author interview and correspondence with Court Noxon, Bloomfield, Ontario, 23 February 2000; Metalsmiths Company corporate archives, Toronto.
43 Author interview with André Morin, Montreal, 28 October 1999.
44 Author interview with James Murray, Ayr, Ontario, 27 September 2000.
45 Barbara Mayer, *Nienkämper at 30, Meeting the Challenges of Change* (Toronto: Nienkämper, 1998); author interview with Klaus Nienkämper, Toronto, October 1999.
46 Letter from William Doig to Henry Finkel at the Association of Canadian Industrial Designers inquiring about membership in ACID, 11 August 1953; author correspondence with Gerald Levitch, 15 May 2000.
47 Author interview with Walter Nugent, Oakville, Ontario, 11 June 1998. Nugent filed a petition in bankruptcy against the company in 1973, but it continued to operate until 1976. Walter Nugent Designs, financial statements, 30 April 1973, and 30 April 1976, Dunwoody and Company.
48 Author interview with Gustavo Martinez, Toronto, July 1998.
49 Gloria Lesser, "Carl Poul Petersen: Master Danish-Canadian Silversmith," *Material History Review* 43 (Spring 1996): 47–53; undated C. P. Petersen & Sons brochure supplied by Sotheby's (Canada), Toronto.
50 Author telephone interview with Francesco Piccaluga, Toronto, June 2000.
51 Author correspondence with Rob Carriere, vice-president, Precidio, Brampton, Ontario, September 2000.
52 Author telephone interview with Geoffrey Lilge, Edmonton, January 2000.
53 Author interview with Ian Anthony, historian, Rogers Communications, Toronto, 28 October 1999; Ian Anthony, "Complete Retrospectus of Rogers Telecommunications Limited: Spanning the Years 1911–1939 and 1959–1999"; notes from Bryan Dewalt, National Museum of Science and Technology, Ottawa, 23 May 2000.
54 Gloria Lesser, Beauregard and Lamarre: *A Textile Collaboration (1975–1999)*, The Museum for Textiles, Toronto, 28 October 1995 to 3 March 1996.
55 Stefan Siwinski file, Design Exchange, gift of Stefan Siwinski.
56 Author interview with James Snyder and Jamie Snyder, Kitchener, January, 1999.
57 Author interview with Christen Sorensen, Toronto, 7 October 1999.
58 "Sovereign Potters Expanding," *The Clay Products News and Ceramic Record*, January 1941: 5–8; "Sovereign Potters Welcome English Potters," *The Clay Products News and Ceramic Record*, October 1947; "Hamilton Potters Turn Out 50,000 Pieces a Day," *Saturday Night*, 23 August 1947; author conversation with Lois Etherington Betteridge and Bruce Etherington (the children of William Etherington), February 1999; Ann Kerr, *The Collector's Encyclopedia of Russel Wright Designs* (Paducah, Ken.: Collector Books, 1990); Russel Wright archives, Syracuse University Library.
59 Fones, *A Spanner in the Works*; Fones's personal archives.
60 Hugh Spencer's notes, Opus International catalogue, 1968; author interview and correspondence with Anna Chester, December 1999 to October 2000.
61 Author interview with and research by Callie Stacey, Toronto, 20 January 2000; Robert Stacey (Harold Stacey's son), "Stacey Sterling," *Metalsmith*, Spring 1985: 11.
62 Author interview with John Stene, Toronto, 30 October 1999.
63 Author telephone interview with Martha Sturdy, Vancouver, July 1999.
64 Author interview with John Hellwig, Toronto, July 2000.
65 Author telephone interview with John Tyson, Ottawa, September 2000.
66 Author interview with Wayne King (Archibald King's son); notes and recollections from Joan King (Archibald's widow) and Wallace Buchanan, Ven-Rez, Shelburne, Nova Scotia, July to November 1999.
67 Author interview with Velta Vilsons, Toronto, March 2000; Canadian Guild of Crafts, "Women in Craft—A Retrospective" (Toronto: Craft Gallery/Canadian Guild of Crafts, Ontario, 1975).

SOURCES

Periodicals

Alcan Ingot

Aluminum Ingot

Azure

Canadian Architect

Canadian Art

Canadian Geographic Journal

Canadian Homes Magazine

Canadian Homes and Gardens

Canadian Interiors

Canadian Jeweller

Canadian Plastics

Canadian Society of Decorative Arts Bulletin

Canadian Textiles Journal

Canadian Wood Product Industries (CWPI) and Furniture Wood Products

Canadian Woodworker

Chatelaine

Clay Products News and Ceramic Record (after 1965 known as *Canadian Clay and Ceramics*)

Craft Ontario

Craftsman (after 1981 known as *Ontario Craft*)

Décormag

Echoes

Furniture and Furnishings

Industrial Canada

Industrial Design

Insite

Journal of the Royal Architectural Institute of Canada

MAGazine: A Publication of the Metal Arts Guild

Material History Bulletin

Mayfair

Modern Plastics

Radio and Appliance Sales (later named *Radio-Television and Appliance Sales*)

Radio College of Canada: Circuit Manuals

Radio Trade Builder (variously known as *Trade Builder, Radio and Electrical Trade Builder, Radio Video Electrical Trade Builder, Radio Appliance Trade Builder*)

The Silver Society of Canada Journal

Tactile

Western Homes and Living

Research Sources

TORONTO

Archives of Ontario
- Eaton's archives
- Ontario Crafts Council archives

Art Gallery of Ontario, E.P. Taylor Reference Library and Archives

Canadian Ceramic Society

Cambridge Public Archives, Cambridge, Ontario
- The Percy R. Hilborn Papers, archives
- Canadian School and Office Furniture

Design Exchange Resource Centre
- Vertical Files
- Association of Chartered Industrial Designers of Ontario (ACIDO), archives
- Bersudsky, Sid, archives
- Bush, Robin, archives
- Clairtone Sound Corporation, archives
- KAN Industrial Design, archives
- Luck, Jack, archives

Hamilton Public Library, Special Collections Department

Huronia Museum, Midland, Ontario

Hansen, Thor, archives

Metal Arts Guild of Ontario

Toronto Reference Library, Canadian Artist Files

Royal Ontario Museum Library and Archives
- Bulow, Karen (Pat Harris File, Textile Department)

County of Simcoe Archives, Minesing, Ontario
- Canadian General Electric Company archives

OTTAWA/HULL

Canadian Museum of Civilization Library and Archives

National Archives of Canada

National Gallery of Canada Library and Archives

National Museum of Science and Technology Library and Archives

MONTREAL

Alcan Archives

Bell Canada Archives

Canadian Architecture Collection, McGill University
- Bülow-Hübe, Sigrun, archives

QUEBEC CITY
Musée du Québec
Hébert, Julien, archives

VANCOUVER
British Columbia Archives
Cotton, Peter, archives

City of Vancouver Archives and Records

University of British Columbia Special Collections

Vancouver Art Gallery Library and Archives

Vancouver Public Library Fine Arts Library

Corporate and Private Archives

BALL, DOUGLAS/SUNAR INDUSTRIES (designer archives), Montreal

BLEASDALE, ADRIAN (private archives), Vancouver

BOSTLUND INDUSTRIES (company archives), Oak Ridges, Ontario

BROOK, JOANNE (designer archives), Toronto

COGNÉ, DANIEL (private archives), Hull, Quebec

COLLIER, ALLAN (private archives), Victoria

ELECTROHOME (corporate archives), Kitchener

FORREST, BOB/L'IMAGE DESIGN (designer/company archives), Toronto

FOUND, GEORGE/GRANDPA'S RADIOS (private archives), Kitchener, Ontario

GUILLON, JACQUES (designer archives), Montreal

KING, JOAN/VEN-REZ (designer/company archives), Shelburne, N.S.

KRAVIS, JANIS/KARELIA (designer archives), Toronto

METROPOLITAN HOME (private archives), Vancouver

NIELSEN, SVEND/LEIF JACOBSEN (designer archives), Toronto

NOXON, COURT/METALSMITHS (designer/company archives), Bloomfield and Markham, Ontario

NUGENT, WALTER (designer archives), Oakville, Ontario

ROGERS COMMUNICATIONS (corporate archives), Toronto

SNYDER'S (corporate archives), Waterloo

SORENSEN, CHRISTEN (designer archives), Hudson, Quebec

SPANNER, RUSSELL (Robert Fones research/private archives), Toronto

STACEY, HAROLD (Callie Stacey private archives), Toronto

STENE, JOHN/BRUNSWICK CONTRACT FURNITURE (designer/company archives), Toronto

VIRTU VINTAGE OFFICE FURNITURE (vintage catalogues), Toronto

Theses/Unpublished Papers

BOYKIW, ALAN, 1993. "Centering the Margins: The Realities and Possibilities of Industrial Design in Canada," M.A. thesis, Carleton University, Ottawa.

CHRISTIE, ROBERT, 1964. "Development of the Furniture Industry in the Southwestern Ontario Furniture Manufacturing Region," M.A.thesis, University of Western Ontario, London.

COLLINS, JOHN BRUCE, 1986. "Design for Use, Design for Millions: Proposals and Options of the National Industrial Design Council, 1948–1960," M.A.thesis, Carleton University, Ottawa.

ELDER, ALAN, 2000. "On the Home Front: Representing Canada at the Triennale di Milano, 1957," M.A.thesis, University of British Columbia, Vancouver.

FLOOD, SANDRA, 1998. "Canadian Craft and Museum Practice 1900–1950," Ph.D. thesis, University of Manchester, England.

HAYWARD, ANNE, 1997. "The Alberta Pottery Industry, 1912–1990." Hull: Canadian Museum of Civilization, forthcoming.

HODGES, MARGARET, 1996. "Sigrun Bülow-Hübe: Scandinavian Modernism in Canada," M.A. thesis, Concordia University, Montreal.

KLINGENDER, FRANZ, 1994. "To Lighten the Burden of Womenkind: The Mechanization of Domestic Equipment (1890–1960)," National Museum of Science and Technology, Ottawa.

WRIGHT, CYNTHIA JANE, 1992. "The Most Prominent Rendezvous of the Feminine Toronto: Eaton's College Street and the Organization of Shopping in Toronto, 1920–1950," Ph.D. thesis, University of Toronto.

Reports and Directories

Acton, James, 1923. *Canadian Book of Furniture.* Toronto: Acton Publishing Company.

Bond, David E., and Ronald J. Wonnacutt, 1968. *Trade Liberalization and the Canadian Furniture Industry.* Toronto: University of Toronto Press.

Canada Design '67 Catalogue *Products for Buildings*, 1967. Ottawa: The National Design Branch, Department of Industry, Ottawa, Canada.

Canadian Business Equipment Manufacturers Association, 1964. "Discussions re steel vs. wood office furniture." (Curtis, Harter, Royal Metal, Sunshine, Office Specialty, et al.).

Ferguson, James, and John Low-Beer, 1950. "Survey of Design Requirements and Conditions in the Canadian Furniture Industry," National Gallery of Canada. Box 7.4 D Design in Industry File 2, Outside Activities/Organizations Appendix.

Furniture Manufacturing in the Atlantic Provinces, March 1959. Halifax: Atlantic Provinces Economic Council.

Government grants to furniture makers: GAAP (General Adjustment Assistance Program), PEP (Program to Enhance Productivity) and CASE (Counselling Assistance to Small Enterprises).

Household Furniture, Part I, Overview and Prospects. Industry Canada, 1996. From series titled *Sector Competitiveness Frameworks*, 1996.

"How Can We Sell More Modern Furniture?", October 1954. National Industrial Design Council conference. (National Gallery of Canada Library).

International Home Furnishings Fair, New York, Toronto Trade mission (Du Barry, Galaxi, Canada Cabinets, Kaufman, Reff, etc.).

The Manitoba Design Institute Department of Industry and Commerce, October 1969. "The Impact of Plastic's Technology on Furniture, Furniture Seminar." National Gallery of Canada, Box RG20 Vol. 2142.

National Furniture Conference, Toronto, 30 October 1954. Jean McKinley, editor, *Canadian Homes and Gardens*, keynote address.

Report of a Study to Determine the Market in the United States for Hardwood Furniture Components. Ottawa: Information Canada.

Spalding, Jeffrey J., 1979. *Silversmithing in Canadian History.* Calgary: Glenbow-Alberta Institute.

The Wood Furniture Components Industry, A British Columbia Industry Study, 1966.

Canadian Monographs and Exhibition Catalogues

ANTHONY, IAN, 1998. "Complete Retrospectus of Rogers Telecommunications Limited: Spanning the Years 1911–1939 and 1959–1999." Internal company document.

BARROS, ANNE, 1997. *Ornament and Object: Canadian Jewellery and Metal Art, 1946–1996.* Toronto: Boston Mills Press.

BOTHWELL, ROBERT, IAN DRUMMOND AND JOHN ENGLISH, 1981. *Power, Politics and Provincialism.* Toronto: University of Toronto Press.

BOURASSA, PAUL, 1999. *Trajectoires: La Céramique au Québec des années 1930 à nos jours.* Quebec City: Musée du Québec.

BUCHANAN, DONALD, 1947. *Design for Use in Canadian Products.* Ottawa: National Gallery of Canada.

CANADIAN MUSEUM OF CIVILIZATION, 1999. *Common Ground: Contemporary Craft, Architecture and the Decorative Arts.* Hull: Canadian Museum of Civilization.

Celebration 1946–1996, 50th Anniversary of the Metal Arts Guild of Ontario. Toronto: The Metal Arts Guild of Ontario.

COLLARD, ELIZABETH, 1967. *Nineteenth-Century Porcelain in Canada.* Montreal: McGill University Press.

COLLIER, ALLAN, 1995. "Modernism at Home: Modern Design in B.C. 1945–60." Exhibition pamphlet. Vancouver: Charles H. Scott Gallery, Emily Carr Institute of Art and Design.

———, 1988. *West Coast Modern: Furniture 1945–1960.* Exhibition pamphlet. Vancouver Art Gallery, October–November.

CRAWFORD, GAIL, 1998. *A Fine Line: Studio Crafts in Ontario from 1930 to the Present.* Toronto: Dundurn Press/Metal Arts Guild.

DAY, PETER, AND LINDA LEWIS, 1988. *Art in Everyday Life: Observations on Contemporary Canadian Design.* Toronto: Summerhill Press/The Power Plant.

Design for Living, 1949. Exhibition catalogue. Vancouver: Community Arts Council and the Vancouver Art Gallery.

EMMERSON, DONALD, 1978. *Canadian Inventors and Innovators: Pioneering in Plastics 1885–1950.* Canadian Plastics Pioneers, Toronto.

FLANDERS, JOHN, 1981. *The Craftsman's Way.* Toronto: University of Toronto Press/Massey Foundation.

FONES, ROBERT, 1990. *A Spanner in the Works: The Furniture of Russell Spanner 1950–1953.* Toronto: The Power Plant.

FREEDMAN, ADELE, 1990. *Sight Lines: Looking at Architecture and Design in Canada.* Toronto: Oxford University Press.

FULFORD, ROBERT, 1968. *This Was Expo.* Toronto: McClelland and Stewart.

GALERIE ELENA LEE, 1996. *20 Years.* Montreal: Galerie Elena Lee Verre d'Art.

GETTY, RONALD, 1994. *The Kilns of South Eastern Alberta.* Medicine Hat: ICM Press.

———, 1995. *Know Your Medalta.* Medicine Hat: ICM Press.

HOLTZ, FRED, 1999. *Eleanor Beveridge, Foster Beveridge, Studio Ceramics 1957–85.* Halifax: Nova Scotia College of Art and Design.

HOPKINS, GARTH, 1978. *Clairtone: The Rise and Fall of a Business Empire.* Toronto: McClelland and Stewart.

INGLIS, STEPHEN, 1991. *The Turning Point: The Deichmann Pottery, 1935–63.* Hull: Canadian Museum of Civilization.

KALMAN, HAROLD, 1994. *A History of Canadian Architecture*, Vol. 2. Don Mills, Ont.: Oxford University Press.
KING, THOMAS, 1987. *Glass in Canada.* Erin, Ont.: Boston Mills Press.

LESSER, GLORIA, 1996. *Beauregard and Lamarre: A Textile Collaboration (1975–1995).* The Museum for Textiles, Toronto, October 28, 1995–March 3, 1996.

———, 1989. *École du Meuble 1930–50: Interior Design and Decorative Art in Montreal.* Montreal: Le Château Dufresne Inc./ Musée des Arts Décoratifs.

LISCOMBE, RHODRI WINDSOR, 1997. *The New Spirit: Modern Architecture in Vancouver 1938–1963.* Montreal: Canadian Centre for Architecture; Vancouver: Douglas & McIntyre.

LISS, JEFFREY, AND SANDRA SHAUL, eds., 1995. *Watching TV: Historic Televisions and Memorabilia from the MZTV Museum.* Toronto: Royal Ontario Museum and MZTV Museum.

LYNWOOD ARTS CENTRE, 1987. *Chairs: Designed and Made in Ontario.* Simcoe, Ont.: Lynwood Arts Centre.

MAYER, BARBARA, 1998. *Nienkämper at 30, Meeting the Challenge of Change.* Toronto: Nienkämper.

McKAY, IAN, 1994. *The Quest of the Folk: Antimodernism and Cultural Selection in Twentieth-Century Nova Scotia.* Montreal/Buffalo: McGill-Queen's University Press.

McRAY, D.G.W., 1944. *The Arts and Crafts of Canada.* Toronto: The Macmillan Company of Canada.

MERTENS, D., M. BARANESS, R. CAWKER, B. SHIM AND G. KAPELOS, 1987. *Toronto Modern Architecture 1946–1965.* Toronto: Bureau of Architecture and Urbanism and Coach House Press.

METAL ARTS GUILD OF ONTARIO, 1996. *Celebration: 1946–1996. 50th Anniversary of the Metal Arts Guild.* Exhibition catalogue. Toronto: Metal Arts Guild of Ontario.

MORRISON, ROSALYN, 1990. *Canadian Glassworks*, 1970–1990. Toronto: Ontario Crafts Council.

MURRAY ZANTOVSKA, IRENA, ed., 1997. *Sigrun Bülow-Hübe: A Guide to the Archive.* Montreal: Canadian Architecture Collection, McGill University.

NEWMAN, PETER C., 1995. *Nortel, Northern Telecom, Past, Present, Future.* Toronto: Northern Telecom and Power Reporting.

PARR, JOY, 1999. *Domestic Goods: The Material, the Moral, and the Economic in Postwar Years.* Toronto: University of Toronto Press.

ROSS, JUDITH THOMPSON, 1981. *Down to Earth Canadian Potters at Work.* Don Mills, Ont.: Nelson Canada Limited.

THE ROYAL ARCHITECTURAL INSTITUTE OF CANADA, 1992. *The Governor General's Awards for Architecture 1992.* Ottawa: RAIC.

SHENSTONE, DOUGLAS, 1990. *For the Love of Pewter: Pewterers of Canada.* Ottawa: self-published.

STACEY, ROBERT, ed., 1992. *Achieving the Modern: Canadian Abstract Painting and Design in the 1950s.* Winnipeg: Winnipeg Art Gallery.

STANTON, RAYMOND, 1997. *Visionary Thinking: The Story of Canada's Electrohome.* Kitchener: Canadian Corporate Histories.
STEVENS, GERALD, 1961. *Early Canadian Glass.* Toronto: Ryerson Press.

STRECKER, JAMES, 1999. Sheridan: *The Cutting Edge in Crafts.* Erin, Ont.: Boston Mills Press.

WEAVER, RONALD, 1997. *Modern Furniture Made in Canada: 1945–1960.* Vancouver: self-published.

WEBSTER, DONALD, 1971. *Early Canadian Pottery.* Toronto: McClelland and Stewart.

WHHITBY ARTS INCORPORATED, The Station Gallery, 1995. *Form and Fantasy: Ceramics by Theo and Susan Harlander.*

WRIGHT, VIRGINIA, 1997. *Modern Furniture in Canada: 1920 to 1970.* Toronto: University of Toronto Press.

———, 1985. *Seduced and Abandoned: Modern Furniture Designers in Canada—The First Fifty Years.* Toronto: The Art Gallery at Harbourfront.

General

ANTONELLI, PAOLA, 1995. *Mutant Materials in Contemporary Design.* New York: The Museum of Modern Art.

ATTERBURY, PAUL, 1999. *Miller's Twentieth-Century Ceramics.* London: Octopus Publishing Group.

CASTELLI, CLINO, 1990. *Transitive Design.* Milan: Electa.

COLLINS, PHILIP, 1987. *Radios: The Golden Age.* San Francisco: Chronicle Books.

———, 1991. *Radios Redux: Listening in Style.* San Francisco: Chronicle Books.

THE DETROIT INSTITUTE OF ARTS/THE METROPOLITAN MUSEUM OF ART, 1983. *Design in America.* New York: Harry N. Abrams in association with the Detroit Institute of Arts and the Metropolitan Museum.

DI NOTO, ANDREA, 1984. *Art Plastic, Designed for Living.* New York: Abbeville Press.

DORMER, PETER, 1993. *Design Since 1945.* London: Thames and Hudson.

EIDELBERG, MARTIN, ed., 1991. *Design 1935–1965—What Modern Was.* New York: Harry N. Abrams.

GARNER, PHILIPPE, 1980. *Twentieth-Century Furniture.* Oxford: Phaidon Press.

GREENBERG, CARA, 1984. *Mid-Century Modern: Furniture of the Fifties.* New York: Harmony Books.

HABEGGER, JERRYLL, AND JOSEPH H. OSMAN, 1997. *Sourcebook of Modern Furniture.*, 2d ed. New York and London: W. W. Norton & Company.

HAWES, ROBERT, 1996. *Bakelite Radios.* Edison, N.J.: Chartwell Books.

HESKETT, JOHN, 1980. *Industrial Design.* New York and Toronto: Oxford University Press.

HIESINGER, KATHRYN, 1983. *Design Since 1945.* Philadelphia: Philadelphia Museum of Art.

HIESINGER, KATHRYN, AND GEORGE MARCUS, 1993. *Landmarks of Twentieth-Century Design: An Illustrated Handbook.* New York: Abbeville Press.

IKOKU, NGOZI, 1999. *Victoria and Albert Museum's Textile Collection, British Textile Design From 1940 to the Present.* London: V&A Publications.

JACKSON, LESLEY, 1994. *Contemporary: Architecture and Interiors of the 1950s.* London: Phaidon Press.

———, 1998. *The New Look: Design in the Fifties. Manchester:* Manchester City Art Galleries; London: Thames and Hudson.

———, 1998. *The Sixties: Decade of Design Revolution.* London: Phaidon Press.

JULIER, GUY, 1993. *20th Century Design Encyclopaedia.* London: Thames and Hudson.

MARCUS, GEORGE H., 1998. *Design in the Fifties: When Everyone Went Modern.* Munich/New York: Prestel-Verlag.

MEIKLE, JEFFREY, 1986. "Plastic, Material of a Thousand Uses," in *Imagining Tomorrow,* Joseph J. Corn, ed. Cambridge, Mass.: The MIT Press.

NICHOLS, SARAH, 2000. *Aluminum by Design.* Pittsburgh: Carnegie Museum of Art.

RILEY, TERENCE, 1995. *Light Construction.* New York: The Museum of Modern Art.

SIDELI, JOHN, 1980. *Classic Plastic Radios of the 1930s and 1940s: A Collector's Guide to Catalin Models.* New York: E.P. Dutton.

SPARKE, PENNY, 1987. *Electrical Appliances.* London: Unwin Hyman.

———, ed., 1990. *The Plastics Age.* London: Victoria and Albert Museum.

THACKARA, JOHN, 1988. *Design After Modernism.* New York: Thames and Hudson.

VENABLE, CHARLES, 2000. *China and Glass in America 1880–1990.* Dallas: Dallas Museum of Art.

WATSON, OLIVER, 1990. *Studio Pottery: Twentieth Century British Ceramics in the Victoria and Albert Museum Collection.* London: Phaidon.

WHITELEY, NIGEL, 1987. *Pop Design—Modernism to Mod.* London: London Design Council.

WOODHAM, JONATHAN, 1997. *Twentieth-Century Design.* Oxford: Oxford University Press.

INTERVIEW LIST

Modernism

Arthur Erickson, September 1998

Alison Hymas, July 1999

Linda Lewis, September 2000

Jerome Markson, July 1999

Peter and Cornelia Oberlander, September 1998

Murray Oliver, July 1999

Blythe Rogers, September 1998

Furniture

Jerry Adamson, May 1998

Douglas Ball, July 1998

Adrian Bleasdale, September 1998

David Burry, July 1999

Allan Collier, September 1998

Tom Deacon, March 1999

Jack Dixon, 12 July 1999

Frank Dudas, 28 December 1999

Murray Dunne, September 1998

Robert Fones, January 2000

Bob Forrest, 26 May and 1 June 2000

Jacques Guillon, July 1998

John Hellwig, July 2000

Mary Honderich, 14 December 1999

Robert Kaiser, 13 December 1999

Mike Keilhauer, May 2000

Miles Keller, March 2000

Wayne King, Joan King and Wallace Buchanan, July and November 1999

Janis Kravis, 10 November 1999

Ivan Lacroix, 3 January 2000

Donald Lapp, January 1999

Robin Lauer, 4 February 1999

Linda Lewis, January 1999

Bill Lishman, 21 December 1999

Harvey Meighan, 27 July 1999

Mark Müller, March 2000

James Murray, 27 September 2000

Svend Nielsen, 19 December 1999

Klaus Nienkämper, October 1999

Court Noxon, 23 February 2000

Walter Nugent, June 1998

Ted Samuel, November 1998

James Snyder and Jamie Snyder, January 1999

Christen Sorensen, 7 October 1999

John Stene, 30 October 1999

Michael Stewart, May 1998

Lighting

Simon Ben Ghozi, October 1999

Marten Bostlund, April 1998

Gustavo Martinez, July 1998

Michel Morelli, October 1999

Textiles

Joanne Brook, May 1999

Skye Morrison, January 2000

Gary and Steve Smith, October 1999

Edward Steeves, June 1999

Velta Vilsons, March 2000

Consumer Electronics

Ian Anthony, 28 October 1999

Iain Baird, October 1998

Mike Batch, 14 October 1999

Gary Borton,7January 1998

George Bunda, 10 November 1999

Anna Chester, December 1999

Peter Denman, 22 October 1999

Bryan Dewalt, 23 May 2000

Gordon Duern, August 1993

George Found, 6 November 1999

Arnold Kenton, 26 October 1999

Bill McGregor, January 1999

Keith McQuarrie, August 1990, 8 November 1999

André Morin, 28 October 1999

Kim Stephenson, October 1998

Lloyd Swackhammer, 22 October 1999

Ceramics

Scot Barnim, July 1999

Gaétan Beaudin, May 2000

Alain Bonneau, May 1998

Louise Bousquet, December 2000

Daniel Cogné, April 1999

Glass and Miscellany

Michel Dallaire, July 1998

Koen de Winter, July 1998

Marcel Girard, June 2000

Jeff Goodman, October 1999

Lloyd Gray, July 1999

Toan Klein, August 2000

Walter Lemiski, September 2000

Les Mandelbaum and Paul Rowan, March 2000

Karim Rashid, May 1999

Angelo Rossi, August 2000

Martha Sturdy, July 1999

John Tyson, September 2000

Willa Wong, August 2000

Small Appliances

Fred Moffatt, March 1990, 15 December 1999

Glenn Moffatt, September 1996

Harold Shifman, November 1999

Toastess staff, July 1999

Metal Arts

Anne Barros, May 1997

Murray James, September 1998

Helen Kerr, 21 July 1999

Gabriel Robitaille, 16 December 1999

Callie Stacey, 20 January 2000

NOTE: THIS LIST INCLUDES EXTENSIVE INTERVIEWS THAT RELATE TO FEATURED OBJECTS. HUNDREDS MORE PEOPLE WERE SPOKEN TO CASUALLY.

PHOTOGRAPHY CREDITS

Jacket front image: 7.2 Reed lamp. Design Exchange Collection. Photo by Pete Paterson.
Jacket back image: 6.64 Deacon Tom chair. Design Exchange Collection. Photo courtesy of Keilhauer. Photo by Karen Levy.
End papers: 8.8 Gallop Meadow. Design Exchange Collection (001.7.1). Gift of Joanne Brook. Photo by Pete Paterson.
Preface: 12.1 Moffatt dome kettle. Design Exchange Collection (999.3.10). Photo courtesy of Canada Post Corporation/ Guy Lavigueur, Productions Punch Inc.

1 Modernism

1.1 [Hubel] Playsphere. Vello Hubel file, Design Exchange.
1.2 Canadian Pavilion, Milan Triennale. Design Exchange. Gift of Linda Lewis in memory of Norman Hay. Photo by the National Film Board of Canada.
1.3 Stene dining set. *Contemporary Furnishings* exhibit. Photo courtesy of John Stene.
1.4 Noxon rack. Photo courtesy of Metalsmiths.
1.5 Kuypers chair. KAN Industrial Design archives, Design Exchange. Gift of KAN Industrial Design.
1.6 Three Small Rooms restaurant. Photo courtesy of Janis Kravis.
1.7 Expo 67. Photo courtesy of Roderick Robbie.

2 New Materials and Processes

2.1 Autoclave. Photo *Canadian Art* VII, no.2 (February 1945): 72.
2.2 Sid Bersudsky. Sid Bersudsky archives. Design Exchange. Gift of Joanne Bersudsky.
2.3 Cord Chair. Morgan's display. Photo courtesy Jacques Guillon.
2.4 Rowan thermos. KAN Industrial Design archives, Design Exchange. Gift of KAN Industrial Design.
2.5 Treco/Maur chair. Chair courtesy of period (.) gallery. Photo by Pete Paterson.
2.6 Bersudsky acrylic chair. Sid Bersudsky archives, Design Exchange. Gift of Joanne Bersudsky.
2.7 Alcan headquarters. Photo courtesy of Alcan.
2.8 De Winter light. Photo courtesy of Axis Lighting.
2.9 Laughton stool. Photo courtesy of Pure Design.
2.10 Keller chair. Photo courtesy of Allseating Corporation.
2.11 Wood garden tools. Design Exchange Collection (992.4). Photo courtesy of Canada Post Corporation/ Guy Lavigueur, Productions Punch Inc.
2.12 Dixon plywood chair. Photo courtesy of Jack Dixon.
2.13 Lamb Roo chair. Photo courtesy of Thomas Lamb archives.
2.14 Jones wall desk. Photo courtesy of Andrew Jones.

3 Craft, Design and Industry

3.1 Fussell detail. Photo courtesy of Joan Fussell.
3.2 Pure Design CD rack. Photo courtesy of Pure Design.
3.3 Johnson stool. Photo courtesy of Patty Johnson.

4 Canadian Design in the Pop Era

4.1 Faux G2 stereo. Clairtone Sound Corporation archives, Design Exchange. Gift of Peter Munk.
4.2 Dallegret interior. Photo courtesy of François Dallegret/SODRAC (Montreal 2000). Photo by Bruno Massenet.
4.3 Wiggins barbecue. Design Exchange Collection (996.1). Photo courtesy of Canada Post Corporation/Guy Lavigueur, Productions Punch Inc.
4.4 Noxon chair. Photo courtesy of Metalsmiths Company.
4.5 Kuypers Muffin chair. Design Exchange archives. Gift of KAN Industrial Design.
4.6 Siwinski acrylic chair. Photo by Wim Vanderkooy.

5 From Postmodernism to Pluralism

5.1 Rocket Pepper Mills. Photo courtesy of Umbra.
5.2 Mississauga City Hall. Photo courtesy of Kirkland Partnership. Photo by Robert Burley.
5.3 Hubel dresser. Design Exchange archives. Photo courtesy of Vello Hubel.
5.4 Burry lounge. Photo courtesy of David Burry.
5.5 Yabu Pushelberg table. Photo courtesy of Nienkämper.
5.6 Ability system. Photo courtesy of Teknion.
5.7 Rashid dresser. Photo courtesy of Karim Rashid.
5.8 Max chair. Photo courtesy of Nienkämper.
5.9 Rashid chair. Photo courtesy of Umbra.

6 Furniture

6.1 Aluminum chairs. Photo courtesy of Alcan.
6.2 Guillon cord chair. Photo courtesy of Canada Post Corporation/Guy Lavigueur, Productions Punch Inc.
6.3 Canadian Wooden Aircraft cabinet and chair. Furniture courtesy of Kasia Seydegart. Photo by Sophie Hogan.
6.4 Spanner table. Courtesy of Robert Fones. Photo by Robert Fones.
6.5 Ball RACE. Design Exchange archive. Gift of Linda Lewis.
6.6 Spencer chair. Hugh Spencer file, Design Exchange. Gift of Anna Chester.
6.7 Laughton storage. Photo courtesy of Umbra.
6.8 De Blois bed. Photo courtesy of Baronet.
6.9 Czerwinki/Stykolt dining chair. Design Exchange Collection (998.4). Gift of Marie DunSeith. Photo by Pete Paterson.
6.10 Czerwinski/Stykolt dining chair (inset). Chair courtesy of Kasia Seydegart. Photo by Sophie Hogan.
6.11 Czerwinski/Stykolt lounge. Chair courtesy of Nancy Watt. Photo by Pete Paterson.
6.12 Czerwinski/Stykolt lounge. Chair courtesy of Douglas Richardson. Photo by Pete Paterson.
6.13 Bush/Morrison Airfoam sofa. Robin Bush archive, Design Exchange. Gift of Matthew Bush. Photo by Graham Warrington Photo Studios.
6.14 Ruspan Originals lounge. Chair courtesy of Robert Fones. Photo by Robert Fones.
6.15 Spanner Catalina chair. Chair courtesy of Morba. Photo by Pete Paterson.
6.16 Snyder's sofa and table. Photos courtesy of James C. Snyder Limited.
6.17 Cotton chairs. Robin Bush archive, Design Exchange. Gift of Matthew Bush. Photo by Graham Warrington Photo Studios.
6.18 Donahue/Simpson plastic chair. Photo courtesy of National Archives of Canada. PA205279.
6.19 King/Swim chair. Photo courtesy of Joan King. Photo by Eldon Whynot.
6.20 Donahue Winnipeg chair. Design Exchange Collection (997.7). Gift of Yabu Pushelberg. Photo by Pete Paterson.
6.21 Dodds stacker. Photo courtesy of ACIDO archives, Design Exchange. Photo by David K. Galloway, Toronto.
6.22 McIntosh chair. Photo courtesy of ACIDO archives, Design Exchange.
6.23 Hébert lounge. Photo courtesy of Alcan. Inset photo courtesy of ACIDO archives, Design Exchange.
6.24 Kaiser armchair. Photo courtesy of the Royal Ontario Museum, ROM (998.281).
6.25 Kuypers Nipigon chair. Design Exchange Collection (997.4). Photo by Pete Paterson.
6.26 Kuypers vanity desk. Design Exchange Collection (997.3). Photo by Pete Paterson.
6.27 Noxon lounge. Photo courtesy of Court Noxon.
6.28 Bülow-Hübe cocktail table. Photo courtesy of Sigrun Bülow-Hübe Archive, Canadian Architecture Collection, McGill University.
6.29 Stene chair. Photo courtesy of John Stene. Photo by Rapid, Grip & Batten, Toronto.
6.30 Stene stool. Photo courtesy of John Stene. Photo by the National Film Board of Canada.
6.31 Stene table. Photo courtesy of John Stene. Photo by Rapid, Grip & Batten, Toronto.

6.32 Bush Prismasteel. Robin Bush archive, Design Exchange. Gift of Matthew Bush.
6.33 Nugent chair. Walter Nugent file, Design Exchange. Gift of Walter Nugent.
6.34 Nugent table. Walter Nugent file, Design Exchange. Gift of Walter Nugent.
6.35 Siwinski lounge. Photo courtesy of Stefan Siwinski.
6.36 Siwinski three-legged dining chair. Design Exchange Collection (994.8). Photo courtesy of Canada Post Corporation/Guy Lavigueur, Productions Punch Inc.
6.37 Bush Lollipop. Robin Bush archive, Design Exchange. Gift of Matthew Bush.
6.38 Dixon chair. Photo courtesy of Jack Dixon.
6.39 Nielsen/Jacobsen lounge chair. Photo courtesy of Svend Nielsen. Photo by Leif Jacobsen.
6.40 Sorensen 1+1. Photo courtesy of Christen Sorensen..
6.41 Guillon Alumna. Photo courtesy Musée du Québec (98.182). Gift of Fred Forman. Photo by Patrick Altman.
6.42 Noxon tub chair. Photo courtesy of Metalsmiths.
6.43 Lapp chair. Photo courtesy of Donald Lapp.
6.44 Murray GT-3A armchair. Photo courtesy of James Murray.
6.45 Ball System S desk. Photo courtesy of Douglas Ball. Photo by Len Korean.
6.46 Siwinski plastic chair. Photo by Wim Vanderkooy.
6.47 Adamson Habitat. KAN Industrial Design archives, Design Exchange. Gift of KAN Industrial Design. Photo by Ray Weber.
6.48 Muller/Stewart stacker. Design Exchange Collection (996.6). Gift of Keith Muller. Photo courtesy of Canada Post Corporation/ Guy Lavigueur, Productions Punch Inc.
6.49 Muller/Stewart Image. Photo courtesy of Michael Stewart.
6.50 Bush radial. Robin Bush archive, Design Exchange. Gift of Matthew Bush.
6.51 Bush steel shell. Robin Bush archive, Design Exchange. Gift of Matthew Bush.
6.52 Spencer Slinger lounge. Hugh Spencer file, Design Exchange. Gift of Anna Chester.
6.53 Boulva Lotus chair. Photo courtesy of Paul Boulva. Photo by Photo Studio 70 Inc., Montreal.
6.54 Forrest 2001 chair. Design Exchange Collection (001.2). Gift of Honey and Alan Stark. Photo by Pete Paterson.
6.55 Salmon/Hamilton stools. Design Exchange. Gift of Linda Lewis.
6.56 Lishman rocker. Photo courtesy of Bill Lishman.
6.57 Lamb Steamer chair. Furniture courtesy of Marianne Lamb. Photo courtesy of Canada Post Corporation/Guy Lavigueur, Productions Punch Inc.
6.58 Fortune chair. Photo courtesy of Michael Fortune.
6.59 Epp chair. Design Exchange. Gift of Linda Lewis.
6.60 Hubel Clover Leaf tables. Photo courtesy of Creative Space.
6.61 Crinion Gazelle. Design Exchange Collection (992.14). Gift of manufacturer. Photo courtesy of Canada Post Corporation/ Guy Lavigueur, Productions Punch Inc.
6.62 Ball Clipper. Photo by Douglas Ball.
6.63 Müller Parabola Tangent shelving. Photo courtesy of Nienkämper.
6.64 Deacon Tom chair. Photo courtesy of Keilhauer. Design Exchange collection. Photo by Karen Levy.

7 Lighting

7.1 Trott lamp. Design Exchange Collection (997.5). Gift of 20th Century Gallery. Photo by Pete Paterson.
7.2 Reed lamp. Design Exchange Collection. Photo by Pete Paterson.
7.3 Baldwin mushroom. Electrohome file, Design Exchange. Gift of Electrohome.
7.4 Morelli Rappola. Photo courtesy of Michel Morelli.
7.5 Bazz lamp. Photo courtesy of Bazz.
7.6 Copeland Tango. Photo courtesy of Arteluce.
7.7 Addison Peruse. Photo courtesy of Addison Lanier.
7.8 Bostlund group. Photo courtesy of Marten Bostlund.
7.9 Bostlund table lamp. Design Exchange Collection (000.1). Photo by Pete Paterson.
7.10 Cotton lamp. Photo courtesy of British Columbia Archives
7.11 Campbell lamp. Design Exchange Collection (996.3). Gift of Mary-Ann Metrick. Photo by Roman Pylypczak.
7.12 Meggit/Gott lamp. Photo *Canadian Art* V, no. 2 (1947-48): 86. (1-61502).
7.13 Dallegret KiiK. Photo courtesy of François Dallegret.
7.14 Ball Glo-ups. Photo courtesy of Douglas Ball.
7.15 Martinez cylinder. Lamp courtesy of Gustavo Martinez. Photo by Pete Paterson.
7.16 Morelli Tom-2 Torchère. Photo courtesy of Michel Morelli.
7.17 Laughton/Deacon Strala. Photo courtesy of Scot Laughton. Photo by Robert Burley.
7.18 Piccaluga Aztec wall lamp. Photo courtesy of Francesco Piccaluga.
7.19 Jacques Zenith. Photo courtesy of Jean-François Jacques.
7.20 Mosna Dragonfly spot. Photo courtesy of Kirk Mosna.

8 Textiles

8.1 Cosgrove Tree with Leaves. Photo courtesy of Musée des Beaux-Arts (D87.250.1) Liliane and David M. Stewart Collection. Photo by Giles Rivest.
of Edward Steeves. Photo by Ray Weber.
8.2 Kinghorn/Crawford Dead Guy. Photo courtesy of Lucia Kinghorn.
8.3 Bulow blinds. Photo courtesy
8.4 Bulow fabric samples. Design Exchange Collection (999.6.2-999.6.3). Gift of Edward Steeves. Photo by Pete Paterson.
8.5 Vilsons fabric. Photo courtesy of Velta Vilsons.
8.6 Hansen Jack-in Pulpit. Design Exchange. Gift of 20th Century Gallery. Photo by Pete Paterson.
8.7 Hansen Sunridge. Design Exchange. Gift of 20th Century Gallery. Photo by Pete Paterson.
8.8 Gallop Meadow. Design Exchange Collection (001.7.1). Gift of Joanne Brook. Photo by Pete Paterson.
8.9 J & J Brook Tobacco Leaf. Design Exchange Collection (001.7). Gift of Joanne Brook. Photo by Pete Paterson.
8.10 Dallegret Kiik. Photo courtesy of the Musée du Québec (95.481). Photo by Patrick Altman.
8.11 Beauregard Fleurs. Photo courtesy of SÉRI +.
8.12 Lamarre Oxide. Photo courtesy of SÉRI +.

9 Consumer Electronics

9.1 Duern 701. Photo courtesy of Electrohome.
9.2 Winter Rogers. Radio courtesy of Popular Culture. Photo by Pete Paterson.
9.3 Clairtone G3. Clairtone Sound Corporation archive, Design Exchange. Gift of Peter Munk.
9.4 Hugh Spencer portrait. Hugh Spencer file, Design Exchange. Gift of Anna Chester.
9.5 Clairtone GTV. Clairtone Sound Corporation archive, Design Exchange. Gift of Peter Munk.
9.6 Addison Model 2, R5. Design Exchange Collection (999.3.1). Gift of George Hartman. Photo by Pete Paterson.
9.7 Addison Model 5. Radio courtesy of Andre Nolf. Photo by Andre Nolf.
9.8 Northern Electric Baby Champ. Design Exchange. Gift of George Hartman. Photo by Pete Paterson.
9.9 Northern Electric Midge. Design Exchange Collection (999.3). Gift of George Hartman. Photo by Pete Paterson.
9.10 Ducharme Bean. Design Exchange. Photo by Pete Paterson.
9.11 Westinghouse Personalities. Radios courtesy of George Found. Photo by Pete Paterson.
9.12 Morin pink Forma. Photo courtesy of André Morin.
9.13 Morin green Forma. Photo courtesy of André Morin.
9.14 Morin white Forma. Photo courtesy of André Morin.
9.15 Spencer Project G. Collection of Peter Munk. Photo courtesy of Canada Post Corporation/Guy Lavigueur, Productions Punch Inc.
9.16 Spencer Faux G2. Design Exchange Collection (991.1). Gift of Frank Davies. Photo courtesy of Canada Post Corporation/Guy Lavigueur, Productions Punch Inc.
9.17 Duern Circa 75 stereo and sound chair. Photo courtesy of Electrohome.
9.18 Duern 703/704. Photo courtesy of Electrohome.
9.19 Apollo tabletop. Electrohome file, Design Exchange. Gift of Electrohome.
9.20 Duern/McQuarrie Circa 711. Photo courtesy of Electrohome.
9.21 Electrohome Perceptions. Photo courtesy of *Canadian Architect.*

10 Ceramics

10.1 Garnier teapot. Musée du Québec (96.67.01-.02). Gift of Daniel Cogné. Photo by Patrick Altman.
10.2 Tableware group. Medicine Hat Potteries Hatina ware courtesy of Before My Time; Medalta blue hotelware, Design Exchange.

Gift of Roger and Cora Golden. Photo by Pete Paterson.
10.3 Hycroft factory. Photo courtesy of Friends of Medalta Society.
10.4 Cartier casserole. Photo courtesy of the Museé du Québec (96.71) Photo by Patrick Altman.
10.5 Gerz vase. Design Exchange. Gift of Popular Culture. Photo by Pete Paterson.
10.6 Goyer-Bonneau teapot. Photo courtesy of Goyer-Bonneau. Photo by Jean Longpré.
10.7 Harlander vase. Vase courtesy of period (.) design. Photo by James Rae and Fraser Smith.
10.8 Ostermann platter. Photo courtesy of Mathias Ostermann.
10.9 Deichmann tureen. Photo courtesy of the Royal Ontario Museum ROM (987.23.1.1-2). Gift of Fern Weston.
10.10 Sovereign dinnerware. Cup and saucer courtesy of Before My Time; Chanticleer plate, Design Exchange; Jamboree plate and bowl, gift of Liz Crawford; Highlight bowl, gift of Rick McNulty. Photo by Pete Paterson.
10.11 Medalta Stardust. Photo courtesy of Friends of Medalta Society.
10.12 Hycroft Jack Straw. Photo courtesy of Friends of Medalta Society. Photo by Pete Paterson.
10.13 Beaudin mug. Design Exchange. Gift of 20th Century Gallery and Liz Crawford. Photo by Pete Paterson.
10.14 Laurentian mug. Gift of Liz Crawford. Photo by Pete Paterson.
10.15 Beaudin carafe and goblets. Musée du Québec (96.7901-96.7903). Gift of Daniel Cogné. Photo by Patrick Altman.
10.16 De Winter porcelain. Design Exchange Collection (994.7). Gift of High Tech. Photo courtesy of Canada Post Corporation/ Guy Lavigueur, Productions Punch Inc.
10.17 Goyer-Bonneau bowl. Photo courtesy of Goyer-Bonneau. Photo by Jean Longpré.

11 Glass and Miscellany

11.1 Dominion glass kitchenware. Glass courtesy of Walter T. Lemiski. Photo by Pete Paterson.
11.2 Dominion glass tableware. Glass courtesy of Walter T. Lemiski. Photo by Pete Paterson.
11.3 Corning glassware. Glass courtesy of Walter T. Lemiski. Photo by Pete Paterson.
11.4 Leser bottle. Design Exchange. Photo by Gill Alkin.
11.5 Goodman bowl. Design Exchange Collection (999.7.2). Gift of designer. Photo by Pete Paterson.
11.6 Copping bottles. Photo courtesy of Inniskillin Wines.
11.7 Corn Flower glasses. Glass courtesy of Walter T. Lemiski. Photo by Pete Paterson.
11.8 Alta Glass dish. Design Exchange. Gift of Linda Lewis. Photo by Pete Paterson.
11.9 Held goblets. Photo courtesy of Robert Held.
11.10 Melmac group. Tableware courtesy of Lloyd Gray, snack plate, Design Exchange. Photo by Pete Paterson.
11.11 Wong Sonoma tableware. Photo courtesy of Precidio.
11.12 Luck coffee pot. Design Exchange Collection (997.2). Gift of Ben Moogk. Photo by Pete Paterson.
11.13 Luck cookware. Design Exchange. Gift of Roger and Cora Golden. Photo by Pete Paterson.
11.14 Bersudsky Magnajector. Sid Bersudsky archives, Design Exchange. Gift of Joanne Bersudsky. Photo by Peake and Whittingham.
11.15 Tyson Contempra telephone. Design Exchange Collection (994.6). Photo courtesy of Canada Post Corporation/Guy Lavigueur, Productions Punch Inc..
11.16 Northern Telecom Imagination series. Telephones courtesy of Metro Retro. Photo by Pete Paterson.
11.17 Lamb bowls. Photo courtesy of Thomas Lamb archives.
11.18 Marcel Girard and Ian Bruce, Tukilik. Photo courtesy of Thomas Lamb archives.
11.19 Santella CD holder. Photo courtesy of Michael Santella.
11.20 Dallaire L' Attaché. Photo courtesy of Michel Dallaire.
11.21 Sturdy table/bowl. Photo courtesy of Martha Sturdy.
11.22 Rashid Garbo. Photo courtesy of Umbra.

12 Small Appliances

12.1 Moffatt dome kettle. Design Exchange Collection (999.3.10). Photo courtesy of Canada Post Corporation/Guy Lavigueur, Productions Punch Inc.
12.2 Fan/heater group. Design Exchange. Gift of Roger and Cora Golden. Photo by Pete Paterson.
12.3 McIntosh hair dryer. Design Exchange. Gift of Lawrie McIntosh.
12.4 CGE ad. Photo courtesy of Fred Moffatt.
12.5 DKR staff. KAN Industrial Design archive, Design Exchange. Gift of KAN Industrial Design.
12.6 Bersudsky kettle. Sid Bersudsky archives, Design Exchange. Gift of Joanne Bersudsky.
12.7 Penrose kettle. Design Exchange. Photo by Pete Paterson.
12.8 Rowan kettle. KAN Industrial Design archive, Design Exchange. Gift of KAN Industrial Design.
12.9 McIntosh Superior kettle. Photo courtesy of Superior Electrics.
12.10 Adamson Life Long. Photo courtesy of Jerry Adamson.
12.11 Fred Moffatt K 840. Photo courtesy of Fred Moffatt.
12.12 Glenn Moffatt kettle. Photo courtesy of Glenn Moffatt.
12.13 Toastess plastic kettle. Photo courtesy of Toastess.
12.14 Bersudsky iron. Design Exchange archives. Photo by the National Film Board of Canada.
12.15 Penrose iron. Design Exchange Collection (994.9). Photo courtesy of Canada Post Corporation/Guy Lavigueur, Productions Punch Inc.
12.16 McIntosh iron. Design Exchange. Gift of Lawrie McIntosh.

13 Metal Arts

13.1 Stacey candy dish. Photo courtesy of Callie Stacey.
13.2 Stacey at work. Photo courtesy of Callie Stacey.
13.3 Fussell shakers. Photo courtesy of Ritchies. Photo by Peter Ure.
13.4 Stacey coffee pot. Photo courtesy of Callie Stacey.
13.5 Boyd pitcher. Photo courtesy of the Metal Arts Guild, Toronto. Photo by Ed Gatner.
13.6 Fussell bowl. Photo courtesy of the Metal Arts Guild, Toronto. Photo by Ed Gatner.
13.7 Petersen tray. Photo courtesy of Sotheby's.
13.8 Petersen Dolphin flatware. Private collection. Photo by Pete Paterson.
13.9 Barros spoon. Photo courtesy of Anne Barros.
13.10 Kerr Ellipse. Photo courtesy of Kerr and Company. Photo by Rob Davidson.

Biographies and Corporate Histories

Spread: DKR offices. KAN Industrial Design archive, Design Exchange.Gift of KAN Industrial Design.

Authors photo: Gotlieb and Golden. Photo by Pete Paterson.

OBJECTS FROM THE DESIGN EXCHANGE COLLECTION

www.designexchange.com

THE DX HAS EXAMPLES OF THE FOLLOWING ITEMS IN ITS COLLECTION.

1.4 COURT NOXON, hat and coat rack, Metalsmiths, gift of 20th Century Gallery (996.2), p. 6

2.4 JULIAN ROWAN, thermos, Canadian Thermos Products, gift of manufacturer, p. 18

2.9 SCOT LAUGHTON, JAMES BRUER, Jim stool, gift of Pure Design, p.25

2.10 MILES KELLER, Os[5] chair, Allseating, gift of manufacturer, p. 27

2.11 TODD WOOD, Plus Four garden tools, Marketing and Design, gift of manufacturer (992.4), p. 27

2.13 THOMAS LAMB, Roo chair, Plydesigns, (001.3.1), p. 28

3.3 PATTY JOHNSON, stool/table (001.4), p. 35

4.3 WILLIAM WIGGINS, Ball-B-Q, Shepherd Products (996.1), p. 39

5.9 KARIM RASHID, Oh chair, Umbra, gift of manufacturer, p. 55

6.2 JACQUES GUILLON, Cord chair, Modern Art of Canada, gift of Yabu Pushelberg (997.11), p. 58

6.6 HUGH SPENCER, club chair U30 (swivel), Opus International, gift of Sam and Jack Markle, p. 65

6.7 SCOT LAUGHTON, Juxta storage, Umbra, gift of manufacturer, p. 67

6.9 WACLAW CZERWINSKI/HILARY STYKOLT, dining chair, Canadian Wooden Aircraft, gift of Marie DunSeith (998.4), p. 70

6.19 ARCHIBALD KING/BALFOUR SWIM, Coastline chair, Ven-Rez Products Company, gift of Joan King, p. 78

6.20 JAMES DONAHUE, lounge chair, gift of Yabu Pushelberg (997.7), p. 79

6.21 HUGH DODDS, The Dodds stacking chair, Aero Marine Industries (998.11) p. 80

6.22 LAWRIE McINTOSH, plywood chair, Aero Marine Industries, gift of designer (994.1), p. 80

6.25 JAN KUYPERS, Nipigon armchair, Imperial Manufacturing, DX acquisition fund (997.4), p. 83

6.26 JAN KUYPERS, Helsinki vanity/desk, Imperial Manufacturing, DX acquisition fund (997.3), p. 83

6.30 JOHN STENE, vanity/piano stool, Brunswick Manufacturing, gift of designer (001.1), p. 86

6.34 WALTER NUGENT, #22 side table, Walter Nugent Design, gift of 20th Cenury Gallery, p. 89

6.36 STEFAN SIWINSKI, 3-legged chair (994.8), p. 91

6.37 ROBIN BUSH, Lollipop seating, COSF, gift of Toronto City Centre Airport and the Toronto Harbour Commission (996.7), pp. 92–93

6.45 DOUGLAS BALL, System S, Sunar Industries, gift of Air Canada, p. 99

6.47 JERRY ADAMSON, Habitat chair and ottoman, IIL, gift of 20th Century Gallery (998.3), p. 101

6.48 MICHAEL STEWART/KEITH MULLER, MS stacking chair, Ambiant Systems, gift of Keith Muller (996.6), p. 102

6.49 MICHAEL STEWART/KEITH MULLER, Image armchair, Du Barry Furniture, gift of Michael Stewart (997.1), p. 103

6.53 PAUL BOULVA, Lotus chair, upholstered version, Artopex (001.5), p. 106

6.54 BOB FORREST, 2001 chair, L'image Design, gift of Honey and Alan Stark (001.2), p. 107

6.55 PHILIP SALMON/HUGH HAMILTON, stool, Kinetics Furniture, gift of Gerald Van Wyngaarden (997.8), p. 108

6.56 BILL LISHMAN, rocker, gift of designer (000.3), p. 109

6.57 THOMAS LAMB, Steamer occasional prototype, gift of Marianne Lamb (999.4.4), pp. 110–11

6.59 PAUL EPP, Nexus chair, Ambiant Systems, gift of designer (000.2), p. 113

6.60 VELLO HUBEL, Clover Leaf tables, Creative Space, gift of manufacturer (997.6), p. 114

6.61 JONATHAN CRINION, Gazelle chair, AREA, gift of manufacturer (992.14), p. 115

6.64 TOM DEACON, Tom chair, Keilhauer, gift of manufacturer, p. 117

7.1 WILLIAM TROTT, Sunspot floor lamp, gift of 20th Century Gallery (997.5), p. 118

7.2 FRANK REED, Ring Master pendant light, John C. Virden Lighting, p. 120

7.7 ADDISON LANIER, Peruse/Illumine lamp, gift of designer, p. 123

7.9 LOTTE BOSTLUND, table lamp 900, Bostlund Industries (000.1), p. 125

7.11 WILLIAM CAMPBELL, desk lamp, William J. Campbell Company, gift of Mary-Ann Metrick (996.3), p. 127

7.17 SCOT LAUGHTON/TOM DEACON, Strala lamp, Portico, gift of James Bruer and Elizabeth Eakins (997.12), p. 131

7.20 KIRK MOSNA, Dragonfly spot, Egoluce, gift of designer, p. 133

8.3 KAREN BULOW, fabric samples, Canadian Homespuns, gift of Edward Steeves (999.6.2–99.6.3), p. 137

8.6 THOR HANSEN, Jack-in-the-Pulpit, A.B. Caya, gift of 20th Century Gallery, p. 140

8.7 THOR HANSEN, Sunridge fabric sample, A.B. Caya, gift of 20th Century Gallery, p. 141

8.8 JOHN GALLOP, Meadow fabric sample, Contemporary Distribution, gift of Joanne Brook (001.7.1), p. 142

8.9 J & J BROOK, Tobacco Leaf fabric sample, Contemporary Distribution, gift of Joanne Brook (001.7), p. 143

9.7 ADDISON RADIO MODEL R5, Addison Industries, gift of George Hartman (999.3.1), p. 157

9.8 BABY CHAMP RADIO, Northern Electric Company, gift of George Hartman, p. 158

9.9 MIDGE RADIO, Northern Electric Company, gift of George Hartman (999.3), p. 159

9.10 MAX DUCHARME, Bean radio, Philips Electronics Industries, p. 160

9.15 HUGH SPENCER, Project G, Clairtone Sound Corporation, gift of Peter Munk (998.9), pp. 164–65

9.16 AL FAUX/HUGH SPENCER, G2, Clairtone Sound Corporation, gift of Frank Davies (991.1), p. 166

9.20 GORDON DUERN/KEITH McQUARRIE, Circa 711 stereo, Electrohome, gift of manufacturer (996.4), p. 168

10.2 CERAMIC TABLEWARE, Medalta Potteries, gift of Roger and Cora Golden, p. 173

10.5 HERTA GERZ, vase #2816, B.C. Ceramics, gift of Popular Culture, p. 174

10.10 MARY AND RUSSEL WRIGHT, ceramic tableware, Sovereign Potters, gift of Rick McNulty, p. 179

10.10 JAMBOREE TABLEWARE, Sovereign Potters, gift of Liz Crawford, p. 179

10.13 GAÉTAN BEAUDIN, mug, Décor Pottery, gift of 20th Century Gallery, p. 180

10.14 MUG, Laurentian Pottery, gift of Liz Crawford, p. 180

10.15 GAÉTAN BEAUDIN, Oval carafe and goblets, Sial II, gift of Daniel Cogné, p. 181

10.16 KOEN DE WINTER, Porcelaine de Chine, Danesco, gift of High Tech (994.7), p. 182

11.4 MAX LESER, perfume bottle, Elika, gift of designer (992.11), p. 188

11.5 JEFF GOODMAN, Compass bowl, gift of designer (999.7.2), p. 188

11.6 BRAD COPPING, Icewine bottle collection, gift of Inniskillin (000.5.1–000.5.8), p. 189

11.8 JOHN FURCH, dish, Altaglass, gift of Linda Lewis, p. 191

11.10 RPL, Cup and snack plate, p. 192

11.11 WILLA WONG, melamine dinnerware, Precidio, gift of manufacturer, p. 193

11.12 JACK LUCK, coffee pot, Aluminum Goods, gift of Ben Moogk (997.2), p. 194

11.13 JACK LUCK, Wear-Ever cookware, Supreme Aluminum Industries, gift of Roger and Cora Golden, p. 194

11.14 SID BERSUDSKY, Magnajector, Kelton Corporation, gift of Joanne Bersudsky (993.1), p. 195

11.15 JOHN TYSON, Contempra telephone, Northern Electric Company, (994.6), p. 197

11.17 THOMAS LAMB, Lummus bowls, gift of Marianne Lamb (994.5), p. 198

11.18 MARCEL GIRARD/IAN BRUCE, Tukilik salt and pepper shakers, Danesco, gift of Linda Lewis (996.8), p. 199

11.19 MICHAEL SANTELLA, Compakt CD holder, Dibis, gift of manufacturer (992.8), p. 199

11.20 MICHEL DALLAIRE, L'Attaché Collection, Resentel, gift of manufacturer (995.1), p. 200

11.22 KARIM RASHID, Garbo, Umbra, gift of manufacturer (995.1), p. 201

12.1 FRED MOFFATT, K42 electric kettle, CGE (993.10), p. 202

12.2 A.P. WHELAN, Fan no. 349, Superior Electrics, gift of Roger and Cora Golden, p. 205

12.2 D.K. STYLES, Canadian Beauty electric heater, Renfrew Electric and Refrigerator Co., gift of Roger and Cora Golden, p. 205

12.2 FRED MOFFATT, Teardrop space heater, CGE, gift of Roger and Cora Golden, p. 205

12.3 LAWRIE McINTOSH, Lady Torcan hair dryer, Rotor Electric (994.5), p. 205

12.6 SID BERSUDSKY, electric kettle, GSW, gift of Joanne Bersudsky (999.3), p. 208

12.7 THOMAS PENROSE, electric kettle, Canadian Westinghouse, p. 208

12.8 JULIAN ROWAN/DKR, electric kettle, Filtro Electric (993.11), p. 208

12.9 LAWRIE McINTOSH, electric kettle, Superior Electrics, gift of manufacturer (993.2), p. 208

12.10 JERRY ADAMSON/KAN, Life Long electric kettle, Proctor-Silex (993.8), p. 209

12.11 FRED MOFFATT, electric kettle, CGE (993.6), p. 209

12.12 GLEN MOFFATT, electric kettle, Superior Electrics, gift of designer, p. 209

12.13 ELECTRIC KETTLE, Toastess, gift of manufacturer, p. 209

12.15 THOMAS PENROSE, IB22 electric iron, Canadian Westinghouse (994.9), p. 210

12.16 LAWRIE McINTOSH, electric iron, Steam Electric Products, gift of designer (993.2), p. 211

13.3 ANDREW FUSSELL, salt and pepper shakers (000.4), p. 217

13.9 ANNE BARROS, baby spoons, gift of designer, p. 223

13.10 KERR AND COMPANY, Ellipse flatware, Gourmet Settings, gift of manufacturer, pp. 224–25

ACKNOWLEDGMENTS

A book of this scope and magnitude results from the dedication and support of hundreds of individuals, companies and institutions. First and foremost, we thank our financial partners: those who supported the project from its inception and later contributors who responded to an urgent need. They are The McLean Foundation, and, in particular, Michael Stewart, The Canada Council for the Arts, The Macdonald Stewart Foundation, The Peter Munk Charitable Foundation and Yabu Pushelberg.

We also appreciate the critical support of the Canadian designers whose work we honour. Although a complete list of interviews appears in Sources, a few require special recognition. They are Douglas Ball, Joanne Brook, Koen de Winter, Fred and Glenn Moffatt, Klaus Nienkämper, Court Noxon, Murray Oliver and Christen Sorensen. Others, who gracefully stood in for the designers, include Bill McGregor at Electrohome, Anna Chester for Hugh Spencer, Jamie Snyder for Snyder's and Callie Stacey for Harold Stacey.

We are grateful to a variety of institutions across the country that provided access to their collections and shared research. Of particular note are Paul Bourassa at the Museé du Québec in Quebec City, and in Montreal, Rosalyn Peppal and Diane Charbonneau at the Musée des Beaux-Arts. Isabel Jones at the Canadian Museum of Civilization in Hull, Bryan Dewalt at the National Museum of Science and Technology in Ottawa and Carol Baum and Alexandra Palmer at the Royal Ontario Museum in Toronto also deserve our thanks.

The following individuals generously shared their research and helped to locate artifacts and designers. We are grateful to Ian Anthony, Gary Borton, Gail Crawford, Robert Fones, Lloyd Gray, Leslie Hendy, Fred Holtz, Walter Lemiski, Gerald Levitch, Linda Lewis, Anthony Matthews, Skye Morrison, Greg Perras, Douglas Richardson, Craig Soper, Kim Stephenson and Ross Young, all in Toronto. Across the country, others lending support include George Found in Kitchener, Ronald Getty in Edmonton, Alan Elder in Ottawa, Daniel Cogné in Hull, and Adrian Bleasdale and Allan Collier in Vancouver.

We acknowledge all those who provided professional counsel and expertise, especially Dean Cooke, Dwayne Dobson, Jonathan Howells, Diane Martin, Deirdre Molina, Michael Mouland, Pete Paterson and Alison Reid, as well as Louise Dennys for believing in this project from its beginning. We are grateful for the persistent support of Luigi Ferrara and Lynda Friendly at the Design Exchange. Rachel would also like to thank Liz Crawford, Rebecca Duclos, Dov Goldstein, Marc Gotlieb, Allan and Sondra Gotlieb, and Elise Hodson.

And finally, this book would not have been possible without the unfailing support of our husbands, Rob Dickson and Roger Golden.

Design and typesetting by Dinnick & Howells
Linears by Brian Marchand (Dinnick & Howells)
Prepress by Quadratone
Printed and bound by Friesens Corporation

INDEX

PAGE CITATIONS IN **BOLD** REFER TO ILLUSTRATIONS IN THE TEXT.

A

Aalto, Alvar, 71, 78
ABS (plastic), **18**, 19–20
acrylic (plastic), 16, **18**, 41, **200**
Adamson, Jerry, 13, 20, 43, **101**, **209**, 241
Addison, Harry and Jack, **156-57**, 228–29
Addison Industries, 147–48, **156-57**, 228–29
Addison Lanier (co.), 122, **123**, **156-57**
Aero Marine Industries, 15, 26, **80**
Affleck, Desbarats, et al. (architects), 22
AKA Works, 13, 64
Alcan. *See* Aluminum Company of Canada
Alcoa. *See* Aluminum Co. of America
Allen Simpson Marketing and Design, 26, **27**
Allseating Corporation, 21, 24, **27**
aluminum, xi
  cookware, 22, 24, **194**, **198**
  design and, 21–26
  furniture, **25**, **27**
  lighting, **25**, **118**, **128**, **132–33**
  manufacturing techniques with, 24–26
  products, **27**, **94**, **198–99**
Aluminum Company of America, 22
Aluminum Company of Canada (Alcan), 5, 8, 10, 22, **23**, 96
Ambiant Systems, 29, 34, 45–46, **102–3**, 113, 229
American Cynamid, 16
Amisco, 68, **108**
appliances (small), **205**
  design of, 203–4
architecture, modernism and, 8, 10
AREA (furniture co.), 35, 45, **45**, 49, 235
Arens, Egmont, 4, 19
Arno, Paul, 13
Arp, Jan, 29
Arteluce, 122, **123**
Art Gallery of Ontario, 5, 15
Artistic Lighting, 187
Arts and Crafts movement, 31–34, 172
  silver and, 213–14
Ashley, William. *See* William Ashley
Association of Canadian Industrial Designers (ACID), 5
attache case(s), L'Attache Collection, **200**
audio equipment manufacturing, 151, 153
Authentics (German plastic furniture mfg.), 21
autoclave oven, **14**
Avanti Furniture Manufacturing, 40
Axis Lighting, 24, **25**, 122

B

Baird, George, 49
Bakelite (plastic), 16, **195**, 207
Bakery Group, 122
Baldwin, Michael, 40, **120**, 122
Ball, Douglas, ix, 11, 13, 40, 43, 52, **62**, 68, **99**, **116**, **129**, 228
Ball-B-Q (barbeque), 24, 38, **39**
Banff School of Fine Arts, 188
Banham, Reyner, 41
Barford, Ralph, 203
Barnim, Scott, 177
Baronet Corporation, 46, **69**
Barros, Anne, 216, **223**, 228–29
Bata Shoe Museum, 49
Bauhaus (school of design), 10
Bazz (lighting co.), **121**, 122, 229
B. C. Ceramics, 172, **174**, 229–30
B. C. Electric building, 10, 255
Beauce, Argile Vivante, 122, **170**, 233
Beaudin, Gaétan, 175, **180–81**, 230, 251
Beaugrand, Gilles, 213–14
Beaulac, Henri, 136
Beauregard, Monique, 40, 136, **145**, 250
bed, Baronet, **69**
Behrens, Peter, 204
Bélanger, Christian, 49
Bendtsen, Niels, 34.
  See also Bensen (furniture co.)
Ben Ghozi, Simon, **121**, 229
Bennett, Margit, 63
Bensen (furniture co.), 34, 54
Berezowsky, John, **129**
Bernecker, Nancy, 244
Bersudsky, Sid, 4, 16, **17–18**, 19, 22, **192**, **195**, **208**, **210**, 230
Betteridge, Lois E., 216
B. F. Harber (furniture mfg.), 22
Binning, B. C., 4
Black & Decker, 46, 204, 233
Black, Mary, 31–32
Blue Mountain Pottery, 172
Bonneau, Alain, 175, **183**, 238
Borduas, Paul-Émile, 136
Bostlund, Gunnar, 10
Bostlund, Gunnar and Lotte, 34, **125**, 231
Bostlund, Lotte, 10, **124–25**, 231
Bostlund Industries, 122, **124**, 231
bottle, Circular Scent, **188**
Boulva, Paul, 20, **106**
Bousquet, Louise, 175
bowl(s)
  bronze, **220**
  cast acrylic, **200**
  Compass, **188**
  CST, plastic, **192**
  Lummus, **198**
  "pie-crust," glass, **186**
  porcelain, **183**
  Sonoma, plastic, **193**
  Spike, glass, **186**
Boyd, Douglas, 8, 214, **219**, 231
Boym, Constantin, 35
Brault, Maurice, 214
Breuer, Marcel, 10–11
Brodovitch, Alexei, 58
Bronfman Family Foundation, 33
Brook, Joanne and John, **143**, 231–30.
  See also J & J Brook
Brown, Denise Scott, 45
Bruce, Ian, **199**
Bruer, James, 24, **25**, 242
Brûlé, Tyler, 54
Brunswick Contract Furniture, 69
Brunswick Manufacturing Company, **86–87**, 253
Buchanan, Donald, 3–4, 15, 32
Bulow, Karen, 10, 32, 135, **138**, 230–31
Bülow-Hübe, Sigrun, 8, 13, 64, **84–85**, 231–32
Bunting Furniture Co., 68
Burgee, John, 49
Burnham, Dorothy and Harold, 136
Burry, David, **48**, 49
Busby, Peter, 123
Bush, Robin, 7–8, 10–11, 13, 20, 33, 63–64, **72–73**, **88**, **92–93**, **104–5**, 232–33, 246–47, 251.
  See also Robin Bush Associates

C

cabinet, Juxta storage modules, **67**
Caiger-Smith, Alan, 177
Cain, Maryanne, 40
Calvert, Robert, 63
Campbell, William, 119, **127**
Canada Mortgage and Home Corp., 22
Canadart, **134**, 136
Canadian Appliance Manufacturing Co. (CAMCO), 204, 233
Canadian Clay and Glass Gallery (Waterloo), 33
Canadian Craft Museum (Vancouver), 33
Canadian General Electric Co. (CGE), 147–48, 150–51, **202**, 203–4, **205–6**, **209**, 233
Canadian Guild of Potters, 32, 175
Canadian Handicrafts Guild, 31–32, 175.
  See also Ontario Crafts Council
Canadian Homespuns, **138**, 230–31
Canadian Marconi Co., 147–48, 150–51
Canadian Museum of Civilization, 33, 113
Canadian Office and School Furniture (COSF), 63, **88**, **93**, **104-5**
Canadian Standard Tableware, **192**
Canadian Thermos Products, **18**, 19
Canadian Westinghouse Co., 147, 150–51, **161**, 203, 233–32
Canadian Wooden Aircraft, 26, **61**, 70–71, **70–71**, 232
Caplan, Ralph, 20
Carmichael, William Maurice, 214
Cartier, Jean, **174**, 233
Cathedral Place (Vancouver), 46
CBC Broadcast Centre, 49
CD rack
  Compakt, **199**
  Mantis, **35**
Central Technical School (Toronto), 175
Centrale d'Artisanant (Montreal), 31
Centre des Métiers du Verre du Québec, **190**
ceramic
  bowls, **178–79**
  carafe, **181**
  casserole, **174**
  mugs & cups, **180–81**
  plates, **179**
  teapots, **174**, **182**
  tureens, **178**
  vases, **174**, **176**
Ceramic Arts (pottery studio), 175
ceramics, 171–77
Céramique de Beauce, 46, 119, 122, 171, **174**, 232–33
Chabauty, Jean-Guy, 49
chair(s), **89**, **94–95**, **105**, **107**, **115**, **117**
  2001, **107**
  ABS, **18**
  acrylic, **18**, 19, **107**
  arm, **82–83**, **98**, **103**, **112**
  Barcelona, 90, 94
  Bowling, 49
  Carlisle club, 49
  Catalina dining, **75**
  children's, 29
  Club U30, **65**
  Coastline, **78**
  Contour, **81**
  dining, **70**, **77**, **86**, **91**
  The Dodds stacking, **80**

early designs of, 20
fibreglass, **78**
Fou du Roi, 49
Gazelle, 45, **115**
Habitat, 13, 20, **101**
Image series, **103**
Ladybug, 41, **41**
Lotus, 20, **106**
lounge, 34, **71–73**, **79**, **84–85**, **90**, **94**, **95**, **98**, **105**
Max, 49, **55**
moulded plywood, **28**, **70–71**, **80**, **102**, **113**
MS stacker, 29, **102**
Muffin task, 41, **41**, 43
Nexus, **113**
Number One, **112**
Oh, 54, **55**
Os[5], **27**
Parachute cord, 16, **17**, **58**, 63
plastic, **78**, **100–1**, **106**
Ribbon, 40
rocking, **109**
Ruspan Originals, **74**
Skogan side, **9**
sleigh-based, 66
sling, 5
Spring-back, 29, **77**
stacking, **28**, **80**, **102**
Steamer, 43, **110–11**
Stiletto Shoe, **48**, 51
swing-back, 29
Tom task, 49, **117**
tub, **97**
utility, 22
Wassily, 11
Wave beach, 20
webbed, 7
Chalet Artistic Glass, 187
Chalmers Fund, 33
Chatelaine (magazine), 13, 38
Chaudron, Bernard, 216
Chicago Furniture Mart, 5
Chicago Institute of Design, ix
Clairtone Sound Corp., ix, 11, **36**, 37, **38**, **149**, 151, **154**, **164–66**, 164–66, 233–34
Clapperton, 187
Clique (mfg. co.), 122
Club Monaco, 52
Coates, Wells, 147
cocktail pitcher, **219**
coffee pot, 194, 218
Collison, Gloria, 38
Colombo, Joe, 20
Community Arts Council (CAC; Vancouver), 3–4
Concertphone (audio mfg. co.), 151
Conran, Terence, 38
consumer demand, increase in, ix
Consumers Glass, 185, 237
Continental Glass, 187
Copeland, Stephan, 52, 122, **123**
Coper, Hans, 177
Copping, Brad, **189**, 190
Corning, 185
Co-Ro-Lite (resin plastic), 19
Corriveau, Jacques, 136
Cosgrove, Stanley, **134**, 136
Cotton, Peter, 29, 63, **77**, **126**, 234–35
C. P. Petersen & Sons, 214, **221–22**, 250
crafts, studio, 31–32
Cranbrook Academy of Art, 216
Crate & Barrel, 52
Crawford, Liz, **137**
Creative Appliance (mfg), 207
Creative Matters, 136
Crescent Silver Tone (audio mfg. co.), 151
Crichton, Daniel, 190
Crinion, Jonathan, 13, 21, 35, 45, 52, 68, **115**, 235
Cronenberg, David, 216
Crown Electrical Manufacturing Co.,119, **127**
Crown Zellerbach Building Materials, 26
Curvply Wood Products, 26
Czerwinski, Waclaw, 15, **70–71**, 232

D

Dair, Carl, 233
Dalí, Salvador, 51
Dallaire, Michel, ix, 43, **200**, 235–34
Dallegret, François, 37–38, **39**, **128**, **144**
Danesco, 40, 234
D'Angelo, Nunziata, 175
Davies, Frank, 38, **207**, 233
Deacon, Tom, x, 21, 35, **117**, **131**, 235, 242, 246
de Blois, Martin, 46, **69**
deconstructivism, 49
Décor Pottery, 175, **180**, 230
Deichmann, Erica and Kjeld, 32, 175, **178**, 235–36
Deilcraft, 11, 46, 60, 63, 119. *See also* Electrohome
Delrue, Georges, 214
Depping, Walter, 20
Desbarats, Jean, 7
design
ceramics and, 171–77
competitions, 5
computer aided (CAD), 51
emerging consciousness of, ix
environmentally sound, 54
free trade pressures and, 52
glass and, 185–87
household furniture and, 59
promotion of, 52
recessions and, 46–51
silver and, 213–17
spherical, 24, 38
stereos and, 151–53.
See also appliances (small), design of; lighting, design and; product extensions, design and; radio(s), design and; television(s), design and; textiles, design and
Design Canada, 43, 46
Design Centre, 5
Design Collaborative, 238
Design Cooperative, 136
Design Emphasis, **48**, 49
Design Exchange, x, 52
Design Index (catalogue), 4, 7
desk(s), 28, 83
Alumna, **96**
System S, **99**
wall-mounted, **28**
Desmarais & Robitaille, 213
de Winter, Koen, 24, **25**, 45, 122, **182**, 234–35
Dexter, Walter, 175
Diamond & Myers (architects), 38
di Castri, John, 7
Dickinson, Peter, 10
Didur, David, 216
Die-Plast Co., 16
DISMO International, 49, 236
Dixon Designs, 64
Dixon, Jack, 11, **28**, 64, **94**
Dodds, Hugh, 26, **80**, 236–37
Dominion Chair Company, 43
Dominion Glass Co., **184**, 185, **186**, 237
Donahue, James, ix, 4, 19, 34, **78–79**, 237–36
Downs, Barry, 10
dresser(s)
Baronet, **47**
Viator, 51, **53**
Du Barry Furniture, 41, 103
Dubé, Alfred, 172
Ducharme, Max, 160, 236
Ducharme, Michel, 136
Dudas, Frank, 241
Dudas Kuypers Rowan (DKR), 11, 13, **18**, 19, **207–8**, **227**, 241. *See also* KAN Industrial Design
Duern, Gordon, 11, **146**, **167–68**, 237
Dufresne, Marie-Josée, 123
Dunlea Plastics, 41
Dunne, Murray, 5

E

Eames, Charles, 3, 66, 74, 78, 105, 153, 232
Eatons. *See* T. Eaton Co.
Ecanada Art Pottery, 172, 236–37
École des Arts Domestiques, 31
École du Meuble, 31, 60, 135–36, 175
Egoluce, 122, **133**
Ehasalu, Helme, 136
Eitel, George, 153, 237
Electric Circus, 38
Electrohome, 7, 11, 37, 40, 46, 66, **120**, **146**, 147, 151, 153, **167–69**, 237–38. *See also* Deilcraft
Emery, George and George Jr., 236–37
Englesmith, George, 15, 80
Epp, Paul, 21, 33–34, **113**, 229
Erickson, Arthur, 10
ESA (Canada), 63
Etherington, Alfred, 251
Eureka, 123
Expo 67, 11–13, **12**, 101

F

Faucher, Sylvain, 236
Faux, Al, ix, 11, 29, 40, 64, 66, **166**, 233, 238–39, 242
Featherweight Aluminum Products (co.), 21–22
Fessenden, Reginald, 147
Finkel, Henry, 16, 239
Five Potters Studio, 175
Fleetwood Electrical Products, 151
Forma Collection (stereo system), 153, **162–63**
Forrest, Bob, **107**, 243–44
Fortune, Michael, 33–34, **112**
Foti, Michel, 123
Foulem, Leopold, 177
Free Trade Agreement, North American, 68
Freygood, Peter, 136
Furch, John, 187
furniture,
"Ability" system, **50**, 51
Clipper office, **116**
mass-production of, 60
Prismasteel office, **88**
RACE system, **62**
Radial office, **104**
System S, **99**
systems, x
Fussell, Andrew, 8, **30**, 214, **217**, **220**, 239–38

G

Galaxi Lighting, 122
Gallop, John, 40, **142**
garbage can, Garbo, **201**
Garnier, Jacques, 122, **170**, 233
Gauvreau, Jean-Marie, 31
Gehry, Frank, 49
Geiger International, 68. *See also* Interiors International Limited

General Agreement on Tariffs and Trade (GATT), 204
General Steel Wares (GSW), 203, **208**
Georg Jensen (co.), 8, 68
Gerz, Herta, 172, **174**
Gerz, Herta and Walter, 229
Gibson, Sidney, 41
Gilmour, David, 11, 37, 151, 233
Girard, Marcel, **199**
Girard Bruce Garebedian & Associates, 24
glass
  bottles, 185, **188–89**
  bowl(s), **188**, **191**
  cut, 185
  Depression, 185
  tableware, **184**, **186**, **191**. *See also* design, glass and
Glende, Ingo, 207
Glenn S. Woolley & Co., 19
Gluckstein, Brian, 136
Goodman, Jeff, 34, **188**, 190, 238
Gott, Earl, **127**
Gottlieb, Harry, 119
Goyer, Denise, 175, **183**, 238
Goyer-Bonneau, **174**, 238
Grabe, Klaus, 63
Graves, Michael, 45, 52
Grcic, Konstantin, 21
Greenaway, Suzann, 33
Griffin, D. S., 40
Gropius, Walter, 10
Guillon, Jacques, 10, 13, 16, **17**, 34, **58**, 63. *See also* Jacques Guillon & Associates

H
Habitat (Montreal), 4
Hamilton, Hugh, **108**
H&K Metal Products, 21
Hansen, Thor, 136, **140–41**
Harlander, Susan and Theo, **176**, 177
Harrison, Stephen, 41
Harter (furniture co.), **41**, 43
Hastings, Donna, 136
Hauser, John, 66
Hawker Siddeley Canada, 24
Hay, Norman, 8
Hébert, Julien, 5, **81**, 235, 239–40
Heinemann, Steven, 33
Held, Robert, 190, **191**, 240
Hellwig, John, **50**, 51
Henderson Furniture, 40
Henningsen, Poul, 122
Henry Birks & Sons, 213–14
Henry Morgan Co., 7
Herbst, Herman, 207
Herman Miller (furniture co.), 10, 66, 232
Hollingsworth, Fred, 10
Holtzman, William, 21
Homer Laughlin China Co., 171
Honderich, Elizabeth, 68
Hooser, Honey, 136
Hoselton Studio, 24, 216
Hothouse Design Studio, 51
Houdé, François, 190
Howarth, Z. "Bobs" Coghill, 175
H. Singer Furniture, 66
Hubel, Vello, **1**, 29, 46, **47**, **114**, 241
Huber Manufacturing, 68
Hughes, W. J. *See* W. J. Hughes Co.
Hughes, William, **191**
Hulme, Thomas, 172, 244
Huzan, Luba, 136
Hycroft China, 46, **173**, **179**, 244–45

I
IKEA, 52
Imperial Furniture Manufacturing Co., 7, 59–60, 241–40
Imperial School Desk Company, 63
Industrial Design Assistance Program (IDAP), 43
Institut de Design Montreal, 52
Interiors International Limited (IIL), 40. *See also* Geiger International
International Contemporary Furniture (ICF), 52
International Council of Societies of Industrial Design (ICSID), 54
IPL (L'Industrie Provinciale), 19, 40
iron(s), electric, **210–11**
Irvine, Herbert, 5
Isaacs Gallery, 20
Italinteriors, 122

J
Jacobson, Lee, 235
Jacques, Jean-François, 46, 49, **133**, 240
Jacques Guillon & Associates (designers), 10–11, 13, 24, **96**, 235, 238–39
J & J Brook (designers), 7, 10, 40, 136, **142–43**, 231–30
Jarvis, Alan, 3
J. A. Wilson Lighting & Display. *See* Wilson, J. A.
Jean Raymond Manufacturing, 22, 24
Jensen, Georg (co.). *See* Georg Jensen (co.)
John C. Virden Lighting. *See* Virden Lighting
John Hauser Iron Works, 63
Johnson, Patty, 34, **35**
Johnson, Philip, 49
Jones, Andrew, **28**, 29, 54, 122
Jones, Edward, 45, **47**
Juhl, Finn, 60

K
Kaiser, Robert, 8, **41**, 43, 64, **82**, **209**, 240–41
KAN Industrial Design, 241. *See also* Dudas Kuypers Rowan (DKR)
Karelia International, 68
Karen Bulow Ltd., 11, **137**, 230–31. *See also* Bulow, Karen
Kaufman Furniture, 40
Keilhauer (mfg. co.), x, 21, 34, 49, 52, 54, 68, 242
Keller, Miles, 21, 24, **27**, **44**, 51, 242–43
Kelly, Giles Talbot, 33
Kerr, Helen, x, 13, **44**, 46, 51, 217, **224–25**, 242–43
Kerr Keller Design, 21, **44**, 242–43
kettle(s)
  chrome dome, **viii**
  electric, 46, **202**, 204, **208–9**
  plastic, 207
Kidick, Tess, 175
Kinetics Furniture, 68, **108**
King, Archibald, 64, **78**, 255
Kinghorn, Lucia, **137**
Kirkland, Michael, 45, **47**
Klein, Toan, 187
Klusacek, Allan, 34
Knoll, Florence, 66, 153
Knoll International, 10–11, 64
Kohn, Martin, 54
Koller, Reinhart, 64
Kominik, Victor and John, **180**
Korina Designs, 34, 64, **90–91**, 251
Kramer, Burton, 233
Kravis, Janis, **12**
Krenov, James, 33–34
Krug Co., 59
Kuypers, Jan, ix, 7–8, **9**, 26, 41, **41**, 43, 60, **83**, **207**, 241, 243–42

L
Lamarre, Robert, 40, 136, **145**, 250
Lamb, Thomas, 11, 13, 20, **28**, 29, 33, 35, 43, 45–46, 68, **110–11**, 113, **198**, 229, 242, 246
Lambert, David, 175
Lambert Potteries, 175
lamp(s)
  Aztec wall, **132**
  designs of, 40
  desk, **123**, **127**
  Dragonfly, 122
  floor, **118**, **130–31**
  Glo-Up, 40, **129**
  halogen, 122
  KiiK, **128**
  Mushroom, **120**
  Peruse, **123**
  Ringmaster, 24, 119, **120**
  Strala, 46, **131**
  Sunspot floor, **118**
  table, **125**, **133**
  Tango desk, 122
  Tizio, 122
  Tom-2, **130**
  tripod, **126**
  Winglite, **25**, 122
  Zenith, **133**. *See also* lighting
Lanier, Addison, 122, **123**
Lapp, Donald, **98**
Laughton, Scot, 24, **25**, 34, 46, **67**, **131**, 242–43, 246
Laurentian Pottery, 172, **180**
Leach, Bernard, 175, 235
Le Corbusier (furniture), 45
Le Drug Discothèque & Café, **39**
Ledalite Architectural Products, 123
Lee, Elena, 190
Legault, Pierre, 230
Leif Jacobsen (co.), 10–11, 66, **94**, 243
Leithead, James, 63
Leser, Max, **188**
Leslie, Bailey, 175
Levine, Les, 20
Lewis, Marion, 175
Lewis, Raymond, 233
Libbey Glass, 237
Libeskind, Daniel, 49
lighting
  design and, 119–23
  halogen, 123
  pendant, **120–21**. *See also* lamp(s)
Lighting Materials (co.), **118**, 119, 255
Lilge, Geoffrey, 35, 248
L'image Design, 107, 243–44
Lindoe, Luke and Vivian, 175
Lishman, Bill, **109**, 244
Lorraine Glass Industries, 187
Louis Poulsen & Co., 122
Luck, Jack, 4, 22–23, **194**, 245–46

M
Mac Craft Industries, 26
MacMillan, Paula, 136, 248
Magder, Max, 41
Magistretti, Vico, 20
Magnajector, **195**
Manning, Edward, 151
manufacturing
  just-in-time, 51

nature of, 34–35
Marconi, Guglielmo, 147
marketing, branding and, 204
markets, ix–x
Markson, Jerome, 38
Markson, Mayta, 175
Martinez, Gustavo, 40, **130**, 248
mass production. *See* production, mass
Massey, Geoffrey, 10
Massey Foundation, 8, 33
Massey Medal Awards, 8
Mathieu, Paul, 177
Maur, Giovanni, **18**, 20
McAleese, David, 217
McClain, Marianne McCrea, 136
McCobb, Paul, 60, 74
McCormack, Donald, 119
McGraw-Edison of Canada, 19
McIntosh, Lawrie, ix, 8, 19, **80**, **205**, **208**, **211**, 245
McKinley, Donald, 33, 175
McKinley, Ruth Gowdy, 175, **198**
McLuhan, Marshall, 38
McNab, Duncan, 10
McQuarrie, Keith, 11, 37, **168**
Medalta Potteries, 32, 119, 122, 171–72, **173**, **179**, 244
Medicine Hat Potteries, 119, 122, 171–72, **173**. *See also* Hycroft China
Meggit, Stanley, **127**
Meitner, Ed, 153
melamine (plastic), Melmac, 16, 19, **192–93**
Memphis (design group), 45–46
Merrick, Paul, 46
Metal Arts Guild of Ontario, 32, 214
Metalsmiths (co.), 10, 40–41, 63–64, 245
Midanik, Dorothy, 175
Milan Triennale (design exposition), **2**, 5, 8, 64, 211, 246, 251
Millar, Blakeway, 38
Minden, George, 38
Mississauga City Hall, **47**
Miyake, Issey, 51
Modernart (mfg. co.), 34, **58**
modernism, 54
  Canadian growth of, 7–13
  expo and, 11
  nature of, ix, 3, 7.
  See also architecture, modernism and
Modulite (lighting mfg.), 119
Moffat (appliances mfg.), 204
Moffatt, Fred, **viii**, 46, **202**, 203, **205–6**, **209**, 233, 245–46
Moffatt, Glenn, **209**
Montgomery, Robert, 172
Morelli, Michel, 52, 122, **130**
Morin, André, 19, 148, 153, **162–63**, 246
Moriyama, Raymond, 49
Morrison, Earle, 7, **73**, 246–47
Mosna, Kirk, 122, **133**
moulding. *See* plastic moulding; plywood, moulded
Mourgue, Oliver, 66
MP Lighting, 123
Muller, Keith, 13, 29, 40, **102–3**, 229
Müller, Mark, 26, 49, **55**, **116**, 246
Munk, Peter, 37, 151, 233.
Murphy, Peter, 123
Murray, James, 66, **98**, 247
Murray, W. A. D., 7
Musée des Arts Décoratifs (Liliane and David M. Stewart Collection, Musée des Beaux-Arts de Montreal), x, 58, 134
Musée du Québec, x, 52
Museum of Modern Art, 40, 49
Myers, Earl, 187

N

Nambé, 51
National Film Board (of Canada), 3, 22
National Gallery of Canada, 3–5, 15
National Industrial Design Committee (NIDC), 4–5, 7–8, 15, 119, 135, 214. *See also* Design Canada
Nelson, George, 5, 66, 153, 232
Nielsen, Svend, **94**, 243
Nienkämper, 26, 34–35, 45–46, 49, 52, **55**, 66, 68, **116**, 246
Nienkämper, Klaus, 246
Norman Wade Co., 64, 238
Normandeau, Pierre-Aimé, 172
Nortel Networks. *See* Northern Electric Company
Northern Electric Company, 148, **158–59**, **197–98**, 246–47
Nova Scotia Design Institute, 43
Novascotian Crystal, 187
Noxon, Court, **6**, 7, 10–11, 41, **41**, 64, **84–85**, **97**, 245
Noxon, Kenneth, 245
Nugent, Doug, 248
Nugent, Walter, 11, 13, 64, 66, **89**, 247–48

O

Ontario Arts Council, 52
Ontario College of Art (now Ontario College of Art and Design), 32, 34, 60, 135–36, 190, 214
Ontario Craft Foundation, 33
Ontario Crafts Council, 32, 34
Ontario Silver Corp., 213
Oracle Audio Corp., 153
Origina Canada, 122, **130**, 248–49
Orme, Jack, 237
Orr Associates, 22
Ostermann, Matthias, 33, **176**, 177
Ostrom, Walter, 33, 177

P

Papanek, Victor, 43
Parkin, John C., 4–5, 10, 38, 40
Patkau Architects, 49
Paulin, Pierre, 66
Pego's (retail furniture, Montreal), 239
Pei, I. M., 10, 22
Pellan, Alfred, 136
Penrose, Thomas, 203, **208**, **210**, 232
pepper mills, Rocket, **44**
Perkins, Alan, 216
Peters, Alan, 33
Petersen, Carl Poul, ix, 32, 214, **221–22**
Petersen, C. P. & Sons, ix
Petersen, Knud, 234
Petri, Manfred, 13, 68
pewter, 216–17
Phelan, Jan, 177
Philco (appliances), 150–51
Philips (electronics mfg.), 151, **160**, 207
Phillips, George, 187
Phillipson, Ed, 171
Piccaluga, Aldo & Francesco, 122, **132**, 248–49
Piper, David, 63
Place Ville-Marie (Montreal), 4, 10, 22
Plastene Co., 20
plastic(s), ix
  design and, 16–21
  moulding, 19
  Pop and, 40–41
  problems with, 16, 20–21
  recycling of, 21
  tableware, **192–93**
  translucent, 21, **42**.
  See also chair(s); table(s)
Platner, Warren, 11
Plexiglas, 40
Plouk Design, 49
pluralism, 45–51
Plydesigns, 33, 35
plywood, furniture, **61**
plywood, moulded, ix, 15, 26, **28**, **61**, **70–71**, **80**, **102**, **113**
Pollock, Arthur, 237–38
Pollock, Carl, 151, 237–38
polypropylene (plastic), **18**, 19, 21, **55**, **106**, **200–1**
Pop (style), 37–43, 66, 122
Porter, John, 175
Portico (studio mfg.), 34, 46
Pospisil, Mirek, 123
postmodernism, 45–51
Poterie Vandesca, 172
pottery. *See also* ceramics, 31–32
Pottery Barn, 52
Pottery Guild. *See* Canadian Guild of Potters
Poulsen, Louis. *See* Louis Poulsen & Co.
Prangnell, Peter, 49
Precidio, **193**, 248
Prii, Uno, 40
Prime Gallery, 33
Proctor-Silex, 43, **209**
product extensions, design and, 150–51
production
  limited run, ix
  mass, ix, 15, 31
Project G, 37–38, 153, **164–65**
Project G2, **36**, **166**
Project G3, **149**, 233–34
Proud, Norm, 135
Prus, Victor, 20
Punch Designs, 68
Pure Design, 24, **35**, 35, 51, 248–49
Pyrex, 185, 187

Q

Quigg, John, 11

R

RACE (furniture system), 43, **62**, 228
radio(s), **149**, **158–61**
  Addison Model 2, **156**
  Addison Model 5, 148, **157**
  automotive design and, 150
  Bakelite plastic, 147, **158–59**
  Canadian Westinghouse, **161**
  design and, 147–53
  Northern Electric Baby Champ, **148**, **158**
  Northern Electric Midge, **159**
  Philips Bean, **160**
  plastics and, 147–48
Rainbow Plastics, **192**
Rashid, Karim, x, 13, 21, 29, 46, 51–52, **53**, **55**, **201**, 249
RCA Victor Co., 150, 153, **162–63**
Reed, Frank, 24, 119, **120**
Reff (furniture mfg.), 68
Reich, Dave, 153
Renzius, Rudolph, 214
Revell, Viljo, 10, 64
Reynolds Metal Company, 22
Riendeau, Marcel, 153
Ripple Design, 51
Risom, Jens, 60
Robert Simpson Co. *See* Simpson's
Robin Bush Associates, 11, 232. *See also* Bush, Robin
The Rockwell Group, 136